# Mathematics and Calculations for Agronomists and Soil Scientists

## About the Authors:

Dr. David Clay is Professor of Soil Science, South Dakota State University
Brookings, SD 57007

Dr. C. Gregg Carlson is Professor of Agronomy, South Dakota State University
Brookings, SD 57007

Dr. Sharon Clay is Professor of Weed Science, South Dakota State University
Brookings, SD 57007

Dr. T. Scott Murrell is Northcentral Director, International Plant Nutrition Institute
West Lafayette, IN 47906

**Fifth Edition**

First published by the International Plant Nutrition Institute in 2010, in cooperation with South Dakota State University

Funding support provided by USDA-CSREES Higher Education Grant 2005-38411-15864

# Mathematics and Calculations for Agronomists and Soil Scientists

Many natural resource managers have reduced the amount of time doing manual labor and increased the time devoted to problem solving. This change in resource allocation is typical of the information age. However, because natural resource management is a tradition-dominated industry, other managers have only slowly adopted information age technologies. Looking back in history, the transition from horse power to tractor power was truly an inevitable and monumental change. There were, however, a large number of farmers who perceived this change as being a mistake. The first tractors were built in the early 1900s, but did not substantially replace the horse until the 1940s and 1950s. The transition was a generational change as much as a change in thought processes. The transition of agriculture into the information age is happening and is every bit as dramatic as the transition from horses to tractors.

The language of the information age is that of mathematics and computers. Natural resource managers traditionally have been trained in the biological sciences with a focus on developing cognitive rather than mathematical skills. The lack of advanced mathematical skills hinders the ability to fully integrate information age technologies into decision processes. Being able to integrate mathematics and technological advances into decisions requires:

- understanding the scientific method,
- understanding how experiments are conducted and analyzed, and
- knowing how to develop and test conceptual and mathematical models.

Most courses are topic specific and compartmentalized. Even though solutions to many problems require the ability to integrate information using scientifically-based approach, few classes teach students how to accomplish this task. In many situations, natural resource managers are noticeably apprehensive about using mathematics. Managers need to learn how to use this important tool for solving everyday, practical problems.

The goals of this book are to teach current and future natural resource managers how to: 1) integrate information from different disciplines, and 2) run innovative management scenarios using the best available science. The manual is organized into three general sections. In the first section, students are introduced to a number of examples on unit conversions in many different formats. In the second section, general background information about the scientific method is covered. Included are principles of experimentation, sampling approaches, using models as tools to improve understanding of systems, and a review of concepts of economic analysis. In the third section, examples of integrating information from many disciplines are provided in order to show how concepts learned in the two sections can be used to solve real problems. The skills taught in this section have applications at local, regional, and national scales. These chapters increase in difficulty as the student proceeds through the manual. An overall goal of this manual is to teach students how to propose, test, and implement innovative management strategies that are better positioned to improve profiability, productivity, and environmental protection.

–D.E. Clay, C.G. Carlson, S.A. Clay, and T.S. Murrell

# TABLE OF CONTENTS

# REVIEW OF ADDITION, SUBTRACTION, MULTIPLICATION, AND DIVISION

**Key problems:** Knowledge of the properties of basic mathematical functions is key to solving more complex problems.

**Mathematical skills:** Addition, subtraction, multiplication, division (including fractions)

**Chapter concepts:** This chapter provides a background on numbers and the properties of the basic mathematical fuctions. Use of those functions with fractions is also reviewed.

## Positive and Negative Numbers

A **positive number** is a number that is greater than zero. Normally, it has no sign in front of it, but sometimes it is denoted with a "+" (**Box 1-1**). So +6 and 6 both mean that 6 is a positive number. A **negative number** is a number that is less than zero. It has a minus "-" sign in front of it. For instance, -6 is a negative number.

**Box 1-1.** Commonly used mathematical symbols.

| Function | Notation | Examples |
|---|---|---|
| Positive number | + or no sign | +6 or 6 |
| Negative number | - | -6 |
| Absolute value | \|\| | \|5\| = 5 and \|-5\| = 5 |
| Addition | + | 6 + 2 = 8 |
| Subtraction | - | 6 – 2 = 4 |
| Multiplication | × | 6 × 2 = 12 |
| | · | 6 · 2 = 12 |
| | * | 6 * 2 = 12 |
| | ( ) | (6)(2) = 6(2) = (6)2 = 12 |
| Division | ÷ | 6 ÷ 2 = 3 |
| | $\overline{)\phantom{0}}$ | $\begin{array}{r}3\\2\overline{)6}\end{array}$ |
| | — | $\frac{6}{2}=3$ |
| | / | 6/2 = 3 |

**Box 1-2** shows two ways of thinking about negative numbers. In this figure, two number lines are shown, one vertical and one horizontal. The vertical number line on the left has positive numbers at the top and negative numbers at the bottom. The horizontal line has positive numbers to the right of zero and negative numbers to the left. As we move up the number line, positive numbers that are greater indicate that we have more of something. An obvious example is money. Having $2 to spend is better than having only $1. On the other hand, what about the part of the number line below zero? What would having -$2 mean? This would be money that is owed. For instance, let's say you borrowed $6 from your best friend. Since this is money you currently owe, locate -6 on one of the number lines in **Box 1-2**. Let's say you pay back $4 to your friend. For each $1 paid, move up (vertically) or to the right (horizontally) one space. So starting at -6, move 4 units to the right and end at -2. After paying $4, $2 is still owed. It's not until the next $2 is paid that zero is reached, where nothing more is owed. So one way of understanding negative numbers is to think of them as indicating how large a deficit is.

**Box 1-2.** Two number lines, displayed either horizontally or vertically, showing the addition of 4 to an initial value of -6, resulting in -2.

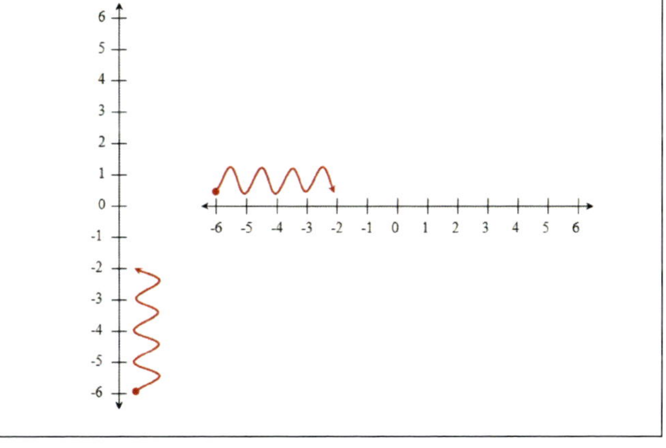

**Exercise 1-1. Indicate whether the numbers are positive or negative.**

| Examples: |
|---|
| -17 is a negative number |
| 23 is a positive number |

| | | Positive | Negative |
|---|---|---|---|
| 1-1. | 12 | | |
| 1-2. | -16 | | |
| 1-3. | -79 | | |
| 1-4. | 105 | | |

## Absolute Value

Sometimes we are concerned not so much with whether a number is positive or negative, but just how far it is from 0. As an example, consider both +5 and -5 in **Box 1-3**. We see that regardless of whether we move in a positive or negative direction from zero, the distance moved is 5. **Absolute value** is the distance a number is from zero, regardless of direction. It is denoted by vertical bars "| |" (**Box 1-1**). In the example in **Box 1-3**, the absolute value of -5 is denoted $|-5|$ and $|-5| = 5$. The absolute value of +5 is $|5| = 5$. So $|-5|$ and $|5|$ are both equal to 5. Absolute value is useful whenever we are concerned with the numerical value of the number, not whether it is positive or negative.

**Box 1-3.** Two number lines, displayed either horizontally or vertically, showing the concept of absolute value, using either |-5| or |5| as an example.

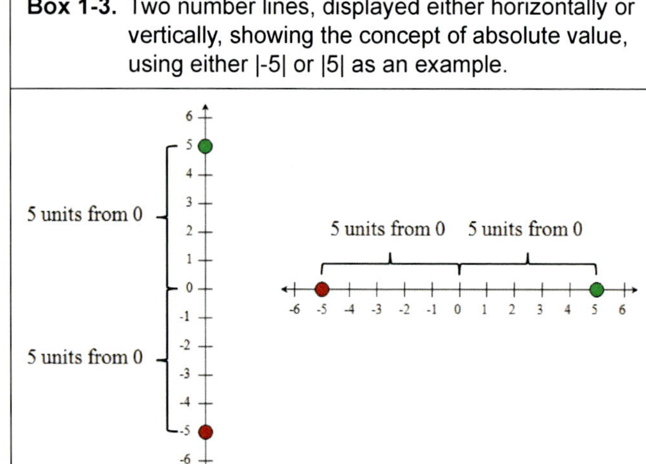


**Exercise 1-2. Evaluate the expressions.**

| Examples: $|-6| = 6$ |
|---|
| $|9| = 9$ |

1-5. $|-121| =$ ______

1-6. $|-80| =$ ______

1-7. $|76| =$ ______

1-8. $|54| =$ ______

## Addition

The **sum** is the result of adding two or more numbers. The symbol for addition is "+" (**Box 1-1**). The numbers being added are called **addends** (**Box 1-4**). For instance, 8 is the sum of 6 + 2. This may be written in the form of an **equation**, or an expression of equality (using an equal sign "=") of mathematical expressions: 6 + 2 = 8.

**Box 1-4.** Terminology of the four basic mathematical functions.

| Function | Terminology |
|---|---|
| Addition | addend + addend = sum |
| Subtraction | minuend – subtrahend = difference |
| Multiplication | factor × factor = product |
| Division | dividend ÷ divisor = quotient<br>numerator/denominator = quotient |

Adding zero to one or more addends does not change the sum. For instance, 0 + 11 = 11. Because the addition of zero does not change the sum, it is referred to as the **identity for addition** (**Box 1-5**).

**Box 1-5.** Properties of the four basic mathematical functions.

| Function | Property | Example |
|---|---|---|
| Addition | commutative | 11 + 10 = 10 + 11 |
| | associative | (10 + 11) + 12 = 10 + (11 + 12) |
| | identity | 0 + 11 = 11 |
| Subtraction | inverse for addition | -6 + 6 = 0 |
| Multiplication | commutative | 7 × 5 = 5 × 7 |
| | associative | (3 × 4) × 5 = 3 × (4 × 5) |
| | identity | 1 × 9 = 9 |
| | distributive over addition | 2(3 + 4) = 2(3) + 2(4) |
| Division | inverse for multiplication or reciprocal | $9\left(\frac{1}{9}\right)=1$ |

Commutative and associative properties are important for addition (**Box 1-5**). The **commutative property** for addition applies when two addends are considered and states that the order of two addends can be interchanged without affecting the sum. For instance 11 + 10 produces the same sum as 10 + 11, or in equation form, 11 + 10 = 10 + 11. The **associative property** for addition is similar to the commutative property, but holds for any three or more numbers added together. This property states that the order in which three or more addends are added together does not affect the sum. For example if three numbers are used, the middle number may be added initially to either the first or last number without changing the sum. To indicate association, or which addends are added initially, parentheses "( )" are used. Consider the statement, 10 + 11 + 12. The result will be the same if 10 is added to 11 first, and then 12 is added, written (10 + 11) + 12, or if 11 is added to 12 first, and then 10 added, written 10 + (11 + 12). In both cases, the sum is 33. In equation form, this example of the associative property is written (10 + 11) + 12 = 10 + (11 + 12).

#### Exercise 1-3. Use the commutative property of addition to write an equivalent expression.

Example: 6 + 3 = 3 + 6

| | | | |
|---|---|---|---|
| 1-9. | 14 + 5 | = | ________ |
| 1-10. | 18 + 20 | = | ________ |
| 1-11. | 3 + 8 | = | ________ |
| 1-12. | 9 + 15 | = | ________ |

#### Exercise 1-4. Re-write the expressions, using the associative property of addition.

Example: (7 + 8) + 3 = 7 + (8 + 3)

| | | | |
|---|---|---|---|
| 1-13. | (3 + 9) + 12 | = | ________ |
| 1-14. | 4 + (6 + 8) | = | ________ |
| 1-15. | 38 + (56 + 90) | = | ________ |
| 1-16. | (41 + 9) + 34 | = | ________ |

## Subtraction

A **difference** is the result of subtracting two numbers (**Box 1-4**). The symbol for subtraction is "–" (**Box 1-1**). The difference of 6 – 2 is 4, or expressed as an equation: 6 – 2 = 4. The larger number is termed the **minuend** (6) while the smaller number is called the **subtrahend** (2). Subtraction can also be thought of as adding a negative number. For instance, 6 – 2 can also be thought of as adding –2 to 6, or 6 + -2 = 4. One way to grasp this concept is to think again about negative numbers being debt. Two people pool their money together. One has $5,000 in savings, while the other has $2,000 of credit card debt. Pooling their money, they calculate their net worth. The net worth is the combination of $5,000 and a -$2,000 of debt, or $5,000 + -$2,000 = $5,000 - $2,000 = $3,000.

Unlike addition, subtraction does not have either the commutative or the associative property. Considering the commutative property for addition, remember that the order of the addends does not affect the sum. In subtraction, however, the order does matter. For instance, 11 – 10 is not the same as 10 – 11. In the first case, the answer is 1 and in the second case, the answer is -1. When we think of subtraction as adding a negative, we see why these two expressions are not equal. The expression 11 – 10 is equivalent to 11 + -10, while 10 – 11 is the same as 10 + -11. So the reason why subtraction is not commutative is because we end up trying to equate the sum of two different numbers added together. The associative property for addition indicates that when three or more numbers are added, the middle number can be added either to the first or the last number first without affecting the sum. This is not the case for subtraction. Consider 15 – 5 – 1. If we subtract 5 from 15 first, then subtract 1, we get (15 – 5) – 1 = 10 – 1 = 9. Now if we subtract 1 from 5, then subtract that difference from 15, we get 15 – (5 – 1) = 15 – 4 = 11. As explained above, when we try to change the associations in subtraction, we end up adding different numbers together than those we started with in the original expression.

In the example we went through in **Box 1-2**, our debt was expressed as -$6 and the amount of money that we paid to settle the debt was +$6. Expressed as an equation, -6 + 6 = 0. This actually turns out to be a property of addition. The **inverse for addition** is the negative number that, when added to a positive number of the same magnitude, results in a sum equal to zero (**Box 1-5**).

#### Exercise 1-5. Find the inverses for addition for the numbers.

Example: The inverse of 6 is -6: 6 - 6 = 0

| | | | |
|---|---|---|---|
| 1-17. | 8 | = | ________ |
| 1-18. | 92 | = | ________ |
| 1-19. | 20 | = | ________ |
| 1-20. | 63 | = | ________ |

## Multiplication

The numbers that are multiplied together are called **factors**, and the result is termed the **product** (**Box 1-4**). The symbol most people associate with multiplication is "×" (**Box 1-1**). In the equation 6 × 2 = 12, 6 and 2 are the factors, and 12 is the product. There are other symbols that are also used to denote multiplication. Many people use dot "·" notation: $6 \cdot 2 = 12$. In computer software applications such as spreadsheets or programming languages, the asterisk "*" is used and is often referred to as the "star:" 6 * 2 = 12. By far, however, the most frequently used notation is the parentheses. Separating two numbers by parentheses, without any other operators between them, indicates that they are being multiplied. For instance 6(2) is the same as $6 \times 2$. It doesn't matter which number is in parentheses, just as long as one or both are. The expressions 6(2), (6)2, and (6)(2) are all equivalent. The reason that parentheses are most commonly used is that they can be any size, allowing multiplication to be expressed more understandably in complex equations.

Like addition, multiplication can be done in any order without affecting the result. For instance, the product 35 can be attained by multiplying 7 by 5 or 5 by 7. The **commutative property of multiplication** states that the order of the two factors can be interchanged without affecting the product (**Box 1-5**).

**Exercise 1-6. Use the commutative property of multiplication to rewrite the expressions.**

Examples: $6 \times 8 = 8 \times 6$
$5(12) = 12(5)$

1-21. $7 \times 9$ = ______

1-22. $11 \times 5$ = ______

1-23. 6(4) = ______

1-24. 3(7) = ______

Another property that multiplication and addition share is the associative property. The **associative property of multiplication** states that when two or more factors are multiplied together, the order in which they are multiplied does not affect the product (**Box 1-5**). For instance, let's consider the numbers 3, 4, and 5 being multiplied together. The 3 and 4 can be multiplied first, then that product multiplied by 5: $3 \times 4 = 12$, then $12 \times 5 = 60$. Like addition, parentheses are used to indicate which operation is performed first. In the above example, the mathematical expression would be $(3 \times 4) \times 5$. However, 4 and 5 could be multiplied together first to get 20 and then this product multiplied by 3: $3 \times (4 \times 5) = 3 \times 20 = 60$. So the order in which two or more factors are multiplied together does not affect the product.

**Exercise 1-7. Use the associative property of multiplication to rearrange the expressions.**

Examples: $2 \times (4 \times 8) = (2 \times 4) \times 8$
$5(6 \times 4) = (5 \times 6)4$

1-25. $6 \times (2 \times 3)$ = ______

1-26. $10 \times (3 \times 4)$ = ______

1-27. $2(3 \times 4)$ = ______

1-28. $8(4 \times 5)$ = ______

Remember that in addition we learned that adding zero to any number did not change that number. This characteristic made zero the identity for addition. Multiplication has an identity with the same characteristic. If a number is multiplied by one, the number remains unchanged. For instance, multiplying $9 \times 1 = 9$. The same holds true for negative numbers: $-9 \times 1 = -9$. Therefore, the **identity for multiplication** is 1. Multiplying a factor by one does not change the product (**Box 1-5**).

Often, equations combine multiplication and addition. It is very common to see an expression like: $2(3 + 4)$. This expression indicates that the sum $(3 + 4)$ is being multiplied by 2. In such cases, we do what is in parentheses first. So we first add $3 + 4$ to obtain 7, then multiply the sum by 2 to get 14. There are often times, however, when we don't want to add 3 and 4 first, as we shall see later. In such cases, we can distribute the multiplication over the addition as follows: $2(3) + 2(4)$. This yields the same result: $6 + 8 = 14$. Thus, the **property of distribution over addition** states that when a sum is multiplied by a factor, the result is unchanged if each addend is first multiplied by the factor and then summed (**Box 1-5**).

The property of distribution over addition can also be extended to subtraction. To demonstrate this extension, consider $5(6 - 2)$. If we perform the subtraction in the parentheses first, we get $6 - 2 = 4$, which is then multiplied by 5 for a product of $5(4) = 20$. If we distribute the factor 5 over the difference in parentheses we can rewrite the expression as $5(6) - 5(2)$. This expression also results in a difference of 20: $(30 - 10)$. The reason distribution works for both addition and subtraction is that subtracting a number is the same as adding its negative, as discussed above.

**Exercise 1-8. Use the property of distribution over addition to re-write the expressions.**

Examples: 3(5 - 4) = 3(5) - 3(4)
10(8) + 10(3) = 10(8 + 3)

1-29. 5(6 + 3) = ____

1-30. 8(3 + 7) = ____

1-31. 10(9 – 7) = ____

1-32. 5(4 – 2) = ____

1-33. 5(2) + 5(8) = ____

1-34. 12(4) + 12(3) = ____

1-35. 9(6) – 9(2) = ____

1-36. 3(10) – 3(7) = ____

## Division

The symbol for division that most of us recognize immediately is "÷" (**Box 1-1**). In the equation 6 ÷ 2 = 3, the number being divided, in this case 6, is termed the **dividend** (**Box 1-4**). The number by which the dividend is divided, in this case 2, is called the **divisor**. The result of division is known as the **quotient**. Like multiplication, division has many symbols, too. One of the symbols is $\overline{)\ \ }$. Using this symbol, the expression 6 ÷ 2 = 3 looks like $2\overset{3}{\overline{)6}}$. Another commonly used symbol for division is the horizontal line "—". In this case, the dividend is on top of the line and is termed the **numerator**. The divisor is under the line and termed the **denominator**. So in the expression

$$\frac{6}{2}$$

6 is the numerator or dividend and 2 is the denominator or divisor. When this notation is used, the expression is often termed a **fraction**, which is simply the ratio of two numbers or expressions. In computer software, the "/" is often used for division instead of the straight line. The numerator is to the left of the "/" and the denominator is to the right: 6/2 = 3.

Like subtraction, the order of division does matter, so it does not have the commutative or associative properties. For instance, 10/5 = 2, but 5/10 = 0.5.

Like multiplication, any quotient divided by 1 remains unchanged. For instance, consider the equation $\frac{(10/5)}{1}=\frac{2}{1}=2$

The quotient of (10/5) is 2. Dividing 2 by 1 does not change the result. It is still 2. This can sometimes be useful. For instance, we can write 9 as 9/1 and both expressions are equivalent. An important property that we will use time and again is the reciprocal, or inverse of multiplication (**Box 1-5**). To understand the reciprocal, we need first to understand multiplying and dividing fractions.

## Multiplying Fractions

Multiplying fractions is straightforward. Numerators are multiplied by numerators, and denominators are multiplied by denominators. As an example, consider 2/3 multiplied by 1/5:

$$\frac{2}{3}\times\frac{1}{5}=\frac{2\times1}{3\times5}=\frac{2}{15}$$

To solve this multiplication problem, the numerators are multiplied: 2 × 1 = 2, and this product becomes the numerator of the new fraction. The denominators of the two fractions are then multiplied, 3 × 5 = 15, and the product becomes the denominator of the new fraction. When multiplying numerators and denominators, it is important to remember that all of the properties of multiplication still apply (**Box 1-5**).

**Exercise 1-9. Multiply the following fractions.**

Example: $\frac{2}{3}\times\frac{4}{7}=\frac{8}{21}$

1-37. $\frac{3}{4}\times\frac{5}{7}$ = ____

1-38. $\left(\frac{1}{2}\right)\frac{9}{10}$ = ____

1-39. $\frac{2}{3}\cdot\frac{5}{9}$ = ____

1-40. $\frac{5}{9}\left(\frac{4}{6}\right)$ = ____

## Dividing Fractions

Dividing fractions relies on multiplying by 1. We don't use the number 1 itself, rather an expression that is equivalent to 1. To demonstrate, consider the following division problem:

$$\frac{2}{3} \div \frac{1}{5}$$

The expression can also be written:

$$\frac{\frac{2}{3}}{\frac{1}{5}}$$

First, we want to find some way of getting the denominator (1/5) to equal 1, so we don't have to work with such a confusing expression. Multiplying 1/5 by 5/1 will accomplish this goal:

$$\frac{1}{5} \times \frac{5}{1} = \frac{1 \times 5}{5 \times 1} = \frac{5}{5} = 1$$

However, we can't just multiply the denominator by 5/1 or else we end up changing the expression. But if we multiply both the denominator and the numerator by 5/1, we end up multiplying by a factor equivalent to 1, leaving the original expression unchanged, shown below:

$$\frac{\frac{2}{3}}{\frac{1}{5}} \times \frac{\frac{5}{1}}{\frac{5}{1}} = \frac{\frac{2}{3}}{\frac{1}{5}} \times 1 = \frac{\frac{2}{3}}{\frac{1}{5}}$$

Now let's perform the multiplication:

$$\frac{\frac{2}{3}}{\frac{1}{5}} \times \frac{\frac{5}{1}}{\frac{5}{1}} = \frac{\frac{2}{3} \times \frac{5}{1}}{1}$$

Because any number divided by 1 remains unchanged, the expression simplifies to:

$$\frac{\frac{2}{3} \times \frac{5}{1}}{1} = \frac{2}{3} \times \frac{5}{1}$$

Solving the problem now requires simply multiplying the numerators and denominators together, as discussed in the previous section:

$$\frac{2}{3} \times \frac{5}{1} = \frac{10}{3}$$

The fraction 5/1 that we used in our expression equivalent to 1 is the reciprocal of the fraction 1/5 that was in denominator of the original division problem. For any number, the **reciprocal**, also known as the **inverse for multiplication**, is one divided by that number (**Box 1-5**). As a simple example, the reciprocal of 6 is 1/6. Remembering that 6 can also be expressed as 6/1, we see that:

$$\frac{6}{1} \times \frac{1}{6} = \frac{6}{6} = 1$$

Thus, when a number is multiplied by its reciprocal, the result is 1.

To see how 5/1 is the reciprocal of 1/5, we use the same procedure we just reviewed. Remembering that the reciprocal of a number is one divided by that number, we find that the reciprocal of 1/5 is:

$$\frac{1}{\frac{1}{5}}$$

Using the same approach of multiplying both the denominator and numerator by an expression that makes the denominator equal to 1, we use the fraction 5/1:

$$\frac{1}{\frac{1}{5}} \times \frac{\frac{5}{1}}{\frac{5}{1}} = \frac{1 \times \frac{5}{1}}{\frac{1}{5} \times \frac{5}{1}} = \frac{\frac{5}{1}}{1} = \frac{5}{1}$$

So the reciprocal of 1/5 is 5/1.

You may have remembered the reciprocal of a fraction as being a "flipped" version of the original fraction. As examples, consider the following reciprocals:

a. the reciprocal of 2/3 is 3/2

b. the reciprocal of 4/5 is 5/4

c. the reciprocal of 9/10 is 10/9

Although it appears that we just flip a fraction over to find its reciprocal, we actually take one and divide it by the fraction, as discussed above. The result, however, is the same for both approaches.

**Exercise 1-10. Find the reciprocals of the following fractions.**

Example: The reciprocal of $\frac{2}{3}$ is $\frac{3}{2}$

1-41. $\frac{12}{13}$ is ________

1-42. $\frac{5}{8}$ is ________

1-43. $\frac{7}{16}$ is ________

1-44. $\frac{23}{50}$ is ________

You may have been taught that when you divide two fractions, you "flip" the bottom fraction and multiply it by the top one. Consider $\frac{\frac{2}{7}}{\frac{1}{3}}$

Using this commonly taught approach, we first "flip" the bottom fraction, resulting in 3/1, then multiply it by the top fraction:

$$\frac{2}{7} \times \frac{3}{1} = \frac{6}{7}$$

This process is actually a shortcut for first finding the reciprocal of the bottom fraction (taking 1 and dividing it by 1/3), creating an expression equivalent to one that uses this reciprocal (3/1 divided by 3/1), and then multiplying this new expression by the original one. While a convenient shortcut, students have often used it without fully understanding the underlying processes involved. In this chapter, we went through the more complete procedure to demonstrate the power of multiplying by expressions equal to one – an approach we will use time and again in the rest of this book.

**Exercise 1-11. Perform the following divisions.**

Example: $\frac{1}{3} \div \frac{2}{5} = \frac{1}{3} \times \frac{5}{2} = \frac{5}{6}$

1-45. $\frac{2}{5} \div \frac{1}{2}$ __________

1-46. $\frac{3}{10} \div \frac{2}{3}$ __________

1-47. $\dfrac{\frac{1}{6}}{\frac{2}{5}}$ __________

1-48. $\dfrac{\frac{5}{12}}{\frac{2}{3}}$ __________

A handy shortcut when multiplying fractions is to look for numbers or expressions that are present in both the numerator and the denominator. Those can be "crossed off" or eliminated from the fraction. As an example, consider the multiplication of 1/5 and 5/7. Since 5 is in the denominator of the first fraction and in the numerator of the second, we can cross it off and remove it from the multiplication: $\frac{1}{\cancel{5}} \times \frac{\cancel{5}}{7} = \frac{1}{7}$

The reason we can eliminate like terms when they are found in both the numerator and denominator is because their quotient is equal to one. To show this with our example, we first show the multiplication in the numerator and denominator: $\frac{1 \times 5}{5 \times 7}$

Next we use the commutative property of multiplication to rearrange the order of the expression in the denominator: $\frac{1 \times 5}{7 \times 5}$

Now we can "break apart" this expression into two fractions multiplied together: $\frac{1}{7} \times \frac{5}{5}$

Since any number divided by itself is equal to 1, we see that 5/5 = 1. So the expression becomes: $\frac{1}{7} \times 1$

Using the identity for multiplication, the expression above can be re-written simply as 1/7. So this exercise shows that "crossing off" items is actually founded on the properties of multiplication.

We can also simplify fractions when 1) the numerator is a factor of an expression in the denominator or 2) vice versa. For instance, consider the fraction

$$\frac{5 \times 6}{3 \times 7}$$

Note that 3 in the denominator is a factor of 6 in the numerator. To show this, we rewrite 6 as 2 × 3:

$$\frac{5 \times 2 \times 3}{3 \times 7}$$

Because 3 occurs in both the numerator and the denominator, we can eliminate it both places, since, 3/3 is equivalent to 1: $\frac{5 \times 2 \times \cancel{3}}{\cancel{3} \times 7}$

Eliminating factors present in both the denominator and numerator allows us to solve a problem more rapidly.

**Exercise 1-12. Find the common factors in the numerator and denominator of the fractions and solve.**

Example: $\frac{3}{4} \times \frac{2}{9} = \frac{\cancel{3}}{\cancel{2} \times 2} \times \frac{\cancel{2}}{\cancel{3} \times 3} = \frac{1}{2} \times \frac{1}{3} = \frac{1}{6}$

Note: when we cross out factors, they can be replaced by a 1, since, for example 3/3 = 1 = 1/1.

1-49. $\frac{4}{5} \times \frac{3}{16}$ __________

1-50. $\frac{5}{6}\left(\frac{1}{25}\right)$ __________

1-51. $\frac{2}{3} \times \frac{12}{15}$ __________

1-52. $\frac{5}{7}\left(\frac{14}{15}\right)$ __________

## Adding and Subtracting Fractions

Adding and subtracting fractions is not as easy as multiplying or dividing them. To add or subtract fractions, their denominators must be the same. When this is the case, addition or subtraction occurs only with the numerators. For instance, let's subtract 1/5 from 3/5: $\frac{3}{5}-\frac{1}{5}=\frac{2}{5}$

Notice that the denominator is unchanged. The same holds true for addition.

The difficulty comes when we want to add or subtract fractions with denominators that are unequal. In such cases, we must find a common denominator. We illustrate the procedure using an example. Let's say we want to add $\frac{1}{3}+\frac{1}{4}$.

To determine the common denominator, we want to find a product for which both denominators are factors. For instance, both 3 and 4 are factors of 12. A quick way to find a common denominator is to multiply together the denominators of the two fractions you want to add. This ensures that both denominators are factors of the product; however this approach may not always produce the lowest product. For instance, if we had 3 and 9 in the denominators, we would come up with 27 as the common denominator if we just multiplied the two factors together. In fact, 9 is a lower common denominator, since it is divisible by both 9 and 3. The lowest product for which two or more denominators are factors is called the **least common denominator**.

Returning to the example, we found that a common denominator of 1/4 and 1/3 is 12. To convert these fractions into new ones with 12 as the denominator, we will multiply each one by another fraction equivalent to 1. For instance, to convert 1/3 to a fraction with 12 in the denominator, we multiply it by 4/4, which is equivalent to 1. To convert 1/4 to a fraction with 12 in the denominator, we multiply by 3/3, another expression equivalent to 1. These multiplications by one, or expressions equivalent to one, are illustrated below:

$$\left(\frac{4}{4}\right)\frac{1}{3}+\frac{1}{4}\left(\frac{3}{3}\right)$$

Now we perform the multiplications to come up with our newly transformed fractions:

$$\frac{4\times 1}{4\times 3}+\frac{1\times 3}{4\times 3}=\frac{4}{12}+\frac{3}{12}$$

Now that we have common denominators, we can proceed with the addition:

$$\frac{4}{12}+\frac{3}{12}=\frac{7}{12}$$

**Exercise 1-13. Add or subtract the fractions, finding common denominators when necessary.**

Examples: $\frac{3}{7}+\frac{2}{7}=\frac{5}{7}$

$\frac{2}{5}+\frac{1}{3}=\left(\frac{3}{3}\right)\frac{2}{5}+\frac{1}{3}\left(\frac{5}{5}\right)=\frac{6}{15}+\frac{5}{15}=\frac{11}{15}$

1-53. $\frac{4}{15}+\frac{8}{15}=$ ________

1-54. $\frac{9}{13}+\frac{2}{13}=$ ________

1-55. $\frac{8}{9}-\frac{4}{9}=$ ________

1-56. $\frac{12}{17}-\frac{9}{17}=$ ________

1-57. $\frac{3}{7}-\frac{1}{21}=$ ________

1-58. $\frac{5}{6}+\frac{1}{12}=$ ________

1-59. $\frac{1}{3}+\frac{1}{4}=$ ________

1-60. $\frac{3}{4}-\frac{2}{5}=$ ________

# SI UNITS AND RULES FOR UNIT CONVERSION

**Key problems:** SI unit conversions

**Mathematical skills:** Addition, subtraction, multiplication, division and fractions, rules of conversions

**Chapter purpose:** This chapter provides explanations of SI units, equalities for conversions, and demonstrates how unit conversions are used.

## The SI Unit System

One of the pillars of science is the ability to measure and quantify amounts of something of importance. The ability to quantify properties is a prerequisite of science and communication. For example, rather than saying, "this thing is heavy," we can say, "this thing weighs 20 kilograms." A complicating aspect of measurement is that different areas of the world use different units. For instance, in the USA, a farmer will measure the distance between rows of crops in inches (2.54 cm). In Germany, that same distance may be measured in centimeters, whereas in China, the unit may be a chi (33.3 cm) and in Egypt, the traditional measurement has been a cubit (20.6 cm). The use of different unit systems hinders worldwide communication. To make it easier to convey measurements across political boundaries, an international system of units was established in 1960 at the General Conference on Weights and Measures. The system was termed, in French, Système International d'Unités and abbreviated SI (Taylor and Thompson, 2008). The SI unit system has been the standard adhered to by scientists and is also used worldwide in commerce.

Although scientists can communicate across political boundaries, the SI unit system often does not allow easy communication directly with practitioners in their own countries. People who are directly involved in production within a country are often not conversant in SI units, but may use a unit system with a longer history. Consequently, for results of scientific investigations to be meaningful to people making management decisions, SI units must usually be converted back to the more traditional unit system of the country of interest.

The SI unit system is divided into two basic classes: base units and derived units. The base units are the foundation of the system and are independent, meaning that they are not derived from one another. The base quantities are length, mass, time, electric current, thermodynamic temperature, amount of substance, and luminous intensity (**Box 2-1**). The SI units used to describe these base quantities are meter, kilogram, second, ampere, kelvin, mole, and candela (**Box 2-2**).

**Box 2-1.** Base quantities and units in the SI system (Taylor and Thompson, 2008).

| Base quantity name | Base quantity symbol | Base unit | Base unit symbol |
|---|---|---|---|
| Length | l, x, r, etc. | meter | m |
| Mass | m | kilogram | kg |
| Time | t | second | s |
| Electric current | I, i | ampere | A |
| Thermodynamic temperature | T | kelvin | K |
| Amount of a substance | n | mole | mol |
| Luminous intensity | $I_v$ | candela | cd |

Derived units are obtained by multiplying or dividing base units. For instance, area is expressed in square meters ($m^2$), volume in cubic meters ($m^3$), rate in kilograms per square meter (kg $m^{-2}$), and speed in meters per second (m $s^{-1}$). There is virtually no limit on how base measurements can be manipulated to come up with derived units. The generalized formula for derived units is:

$$\text{Derived unit} = m^a kg^b s^c A^d K^e mol^f cd^g$$

Where m, kg, s, A, K, mol, and cd are the symbols for the base units and *a-g* are exponents, either positive or negative. Therefore, any derived unit is the product of base units raised to various powers.

**Box 2-2.** Base unit definitions (Taylor and Thompson, 2008).

| Base unit | Definition |
|---|---|
| Meter | The meter is the length of the path travelled by light in a vacuum during a time interval of 1/299 792 458 of a second. |
| Kilogram | The kilogram is the unit of mass; it is equal to the mass of the international prototype of the kilogram. |
| Second | The second is the duration of 9 192 631 770 periods of the radiation corresponding to the transition between the two hyperfine levels of the ground state of the cesium 133 atom. |
| Ampere | The ampere is that constant current which, if maintained in two straight parallel conductors of infinite length, of negligible circular cross-section, and placed 1 meter apart in vacuum, would produce between these conductors a force equal to $2 \times 10^{-7}$ newton per meter of length. |
| Kelvin | The kelvin, unit of thermodynamic temperature, is the fraction 1/273.16 of the thermodynamic temperature of the triple point of water. |
| Mole | The mole is the amount of substance of a system which contains as many elementary entities as there are atoms in 0.012 kilogram of carbon 12; its symbol is "mol." When the mole is used, the elementary entities must be specified and may be atoms, molecules, ions, electrons, other particles, or specified groups of such particles. |
| Candela | The candela is the luminous intensity, in a given direction, of a source that emits monochromatic radiation of frequency $540 \times 10^{12}$ hertz and that has a radiant intensity in that direction of 1/683 watt per steradian. |

Examples of derived units and their respective base unit derivations are provided in **Box 2-3**. Some derived units have their own unit names and symbols. For instance, the newton, a unit of force and abbreviated as "N," is derived from the base unit formula of m kg $s^{-2}$.

To represent multiples and submultiples of SI units, SI prefixes were developed. The list of acceptable prefixes and their abbreviations are listed in **Box 2-4**. For instance, the prefix "kilo" is used in front of the unit "meter" to create "kilometer," which expresses that 1,000, or $10^3$, meters are being reported. Kilometer is abbreviated "km". Similarly, the prefix "milli" placed in front of "meter," creates "millimeter" and means one thousandth of a meter (0.001 or $10^{-3}$), abbreviated "mm". The kilogram is the only base unit whose name and symbol include a prefix. This is done for historical reasons.

## Conventions for Writing SI Units

When writing units, there are some conventions that are followed (Taylor and Thompson, 2008). Either a space or a dot is used to represent multiplication of units. For instance, the base units for a coulomb are the multiplication of a second with an ampere. This product can be written symbolically as

**Box 2-3.** Commonly used SI derived units (Taylor and Thompson, 2008).

| Derived quantity name | Derived unit name | Derived unit abbreviation | Expressed in other SI units | Expressed in SI base units |
|---|---|---|---|---|
| Absorbed dose, specific energy (imparted), kerma | Gray | Gy | J $kg^{-1}$ | $m^2$ $s^{-2}$ |
| Absorbed dose rate | Gray per second | — | Gy $s^{-1}$ | $m^2$ $s^{-3}$ |
| Acceleration | Meter per second squared | — | -- | m $s^{-2}$ |
| Activity referred to a radionuclide | Becquerel | Bq | -- | $s^{-1}$ |
| Amount concentration, concentration | Mole per cubic meter | — | -- | mol $m^{-3}$ |
| Angular velocity | Radian per second | — | rad $s^{-1}$ | m $m^{-1}$ $s^{-1}$ |
| Angular acceleration | Radian per second squared | — | rad $s^{-2}$ | m $m^{-1}$ $s^{-2}$ |
| Area | Square meter | — | -- | $m^2$ |
| Capacitance | Farad | F | C $V^{-1}$ | $m^{-2}$ $kg^{-1}$ $s^4$ $A^2$ |
| Catalytic activity | Katal | kat | -- | $s^{-1}$ mol |
| Catalytic activity concentration | Katal per cubic meter | — | kat $m^{-3}$ | $m^{-3}$ $s^{-1}$ mol |
| Celsius temperature | Degree Celsius | °C | -- | K |
| Current density | Ampere per square meter | — | -- | A $m^{-2}$ |
| Density, mass density | Kilogram per cubic meter | — | -- | kg $m^{-3}$ |
| Dose equivalent, ambient dose equivalent, directional dose equivalent, personal dose equivalent | Sievert | Sv | J $kg^{-1}$ | $m^2$ $s^{-2}$ |

(...continued on next page.)

**Box 2-3.** Continued.

| Derived quantity name | Derived unit name | Derived unit abbreviation | Expressed in other SI units | Expressed in SI base units |
|---|---|---|---|---|
| Dynamic viscosity | Pascal second | — | Pa s | $m^{-1}$ kg $s^{-1}$ |
| Electric charge, amount of electricity | Coulomb | C | — | s A |
| Electric charge density | Coulomb per cubic meter | — | C $m^{-3}$ | $m^{-3}$ s A |
| Electric conductance | Siemens | S | A $V^{-1}$ | $m^{-2}$ $kg^{-1}$ $s^3$ $A^2$ |
| Electric field strength | Volt per meter | — | V $m^{-1}$ | m kg $s^{-3}$ $A^{-1}$ |
| Electric flux density, electric displacement | Coulomb per square meter | — | C $m^{-2}$ | $m^{-2}$ s A |
| Electric potential difference, electromotive force | Volt | V | W $A^{-1}$ | $m^2$ kg $s^{-3}$ $A^{-1}$ |
| Electric resistance | Ohm | Ω | V $A^{-1}$ | $m^2$ kg $s^{-3}$ $A^{-2}$ |
| Energy, work, amount of heat | Joule | J | N m | $m^2$ kg $s^{-2}$ |
| Energy density | Joule per cubic meter | — | J $m^{-3}$ | $m^{-1}$ kg $s^{-2}$ |
| Exposure (x and γ rays) | Coulomb per kilogram | — | C $kg^{-1}$ | $kg^{-1}$ s A |
| Force | Newton | N | — | m kg $s^{-2}$ |
| Frequency | Hertz | Hz | — | $s^{-1}$ |
| Heat capacity, entropy | Joule per kelvin | — | J $K^{-1}$ | $m^2$ kg $s^{-2}$ $K^{-1}$ |
| Heat flux density, irradiance | Watt per square meter | — | W $m^{-2}$ | kg $s^{-3}$ |
| Illuminance | Lux | lx | lm $m^{-2}$ | $m^{-2}$ cd |
| Inductance | Henry | H | Wb $A^{-1}$ | $m^2$ kg $s^{-2}$ $A^{-2}$ |
| Luminance | Candela per cubic meter | — | — | cd $m^{-3}$ |
| Luminous flux | Lumen | lm | cd sr | cd |
| Magnetic field strength | Ampere per meter | — | — | A $m^{-1}$ |
| Magnetic flux | Weber | Wb | V s | $m^2$ kg $s^{-2}$ $A^{-1}$ |
| Magnetic flux density | Tesla | T | Wb $m^{-2}$ | kg $s^{-2}$ $A^{-1}$ |
| Moment of force | Newton meter | — | N m | m2 kg $s^{-2}$ |
| Mass concentration | Kilogram per cubic meter | — | — | kg $m^{-3}$ |
| Molar energy | Joule per mole | — | J $mol^{-1}$ | $m^2$ kg $s^{-2}$ $mol^{-1}$ |
| Molar entropy, molar heat capacity | Joule per mole kelvin | — | J $mol^{-1}$ $K^{-1}$ | $m^2$ kg $s^{-2}$ $K^{-1}$ $mol^{-1}$ |
| Permeability | Henry per meter | — | H $m^{-1}$ | m kg $s^{-2}$ $A^{-2}$ |
| Permittivity | Farad per meter | — | F $m^{-1}$ | $m^{-3}$ $kg^{-1}$ $s^4$ $A^2$ |
| Plane angle | Radian | rad | 1 | m $m^{-1}$ |
| Power, radiant flux | Watt | W | J $s^{-1}$ | $m^2$ kg $s^{-3}$ |
| Pressure, stress | Pascal | Pa | N $m^{-2}$ | $m^{-1}$ kg $s^{-2}$ |
| Radiance | Watt per square meter steradian | — | W $sr^{-1}$$m^{-2}$ $nm^{-1}$ | W $m^{-2}$ $sr^{-1}$ |
| Radiant intensity | Watt per steradian | — | W $sr^{-1}$ | W $sr^{-1}$ |
| Refractive index | One | — | — | 1 |
| Relative permeability | One | — | — | 1 |
| Solid angle | Steradian | sr | 1 | $m^2$ $m^{-2}$ |
| Specific energy | Joule per kilogram | — | J $kg^{-1}$ | $m^2$ $s^{-2}$ |
| Specific heat capacity, specific entropy | Joule per kilogram kelvin | — | J $kg^{-1}$ $K^{-1}$ | $m^2$ $s^{-2}$ $K^{-1}$ |
| Specific volume | Cubic meter per kilogram | — | — | $m^3$ $kg^{-1}$ |
| Speed, velocity | Meter per second | — | — | m $s^{-1}$ |
| Surface charge density | Coulomb per square meter | — | C $m^{-2}$ | $m^{-2}$ s A |
| Surface density | Kilogram per square meter | — | — | kg $m^{-2}$ |
| Surface tension | Newton per meter | — | N $m^{-1}$ | kg $s^{-2}$ |
| Thermal conductivity | Watt per meter kelvin | — | W $m^{-1}$ $K^{-1}$ | m kg $s^{-3}$ $K^{-1}$ |
| Volume | Cubic meter | — | — | $m^3$ |
| Wavenumber | Reciprocal meter | — | — | $m^{-1}$ |

either "s A," or "s · A." When units are divided, the quotient can be represented as a horizontal line, a solidus (forward slash "/"), or a negative exponent. For instance, the following three expressions are all acceptable ways to write the base unit symbols for a volt: $\frac{m^2 kg}{s^3 A}$ or $m^2 kg/(s^3 A)$ or $m^2 kg s^{-3} A^{-1}$

The use of more than one solidus is not recommended. For instance, you should not write the base unit symbols for a volt as $m^2$ kg/$s^3$/A.

**Box 2-4.** SI prefix names and symbols (Taylor and Thompson, 2008).

| Factor | Name | Symbol | Factor | Name | Symbol |
|---|---|---|---|---|---|
| $10^1$ | deka | da | $10^{-1}$ | deci | d |
| $10^2$ | hecto | h | $10^{-2}$ | centi | c |
| $10^3$ | kilo | k | $10^{-3}$ | milli | m |
| $10^6$ | mega | M | $10^{-6}$ | micro | μ |
| $10^9$ | giga | G | $10^{-9}$ | nano | n |
| $10^{12}$ | tera | T | $10^{-12}$ | pico | p |
| $10^{15}$ | peta | P | $10^{-15}$ | femto | f |
| $10^{18}$ | exa | E | $10^{-18}$ | atto | a |
| $10^{21}$ | zetta | Z | $10^{-21}$ | zepto | z |
| $10^{24}$ | yotta | Y | $10^{-24}$ | yocto | y |

**Box 2-5.** Equalities for power, energy, and pressure.

| |
|---|
| 1 kilowatt = 3,412.76 BTU (IST)/hour |
| 1 pound/in$^2$ = 27.6807 inch water |
| 1 atmosphere = 1.01325 bars |
| 1 megapascal = 9.9 atmospheres |
| 1 megapascal = 10 bars |
| 1 pascal = 1.45 • $10^{-4}$ lb/in$^2$ |
| 1 horsepower = 0.7457 kilowatts |

**Box 2-6.** Equalities for dry volume.

| Unit Converting From (UF) \ Unit Converting To (UT) | Bushel | Peck | Dry quart | Cubic yard | Cubic foot | Cubic meter | Cubic centimeter | Liter | Milliliter |
|---|---|---|---|---|---|---|---|---|---|
| | bu | pk | qt-d | yd$^3$ | ft$^3$ | m$^3$ | cc | L | ml |
| Bushel | — | 4 | 32 | 0.0461 | 1.244 | 00352 | 35,239 | 35.24 | 35,239 |
| Peck | 0.25 | — | 8 | 0.0115 | 0.311 | 0.0088 | 8,810 | 8.81 | 8,810 |
| Dry Quart | 0.03125 | 0.125 | — | * | 0.039 | * | 1,101 | 1.101 | * |
| Cubic Yard | 21.7 | 86.785 | * | — | 27.0 | 0.7646 | 764,555 | 764.555 | 764,555 |
| Cubic Feet | 0.804 | 3.214 | 25.71 | 0.037 | — | 0.0283 | 28,317 | 28.317 | 28,317 |
| Cubic Meter | 28.378 | 113.51 | 908 | 1.308 | 35.315 | — | 1,000,000 | 1,000 | 1,000,000 |
| Cubic Centimeter | * | * | * | * | * | * | — | 0.001 | 1.0 |
| Liter | 0.0284 | 0.1135 | 0.908 | 0.00131 | 0.0353 | 0.001 | 1,000 | — | 1,000 |
| Milliliter | * | * | * | * | * | 0.000001 | 1.0 | 0.001 | — |

* Values are either too large or small to be useful for conversion. If it is necessary to convert to these units, convert to a larger or smaller unit and then convert that result to the desired unit. Conversion multiplier values have been rounded but are appropriate in most situations. To change units to other equalities in this table, multiply the number of units to be converted by the appropriate equality. For example: 1 bu = 4 pecks or 1 = 1 bu/4 pecks.

**Box 2-7.** Equalities for distance.

| Unit Converting From (UF) \ Unit Converting To (UT) | Mile | Rod | Yard | Foot | Inch | Kilometer | Meter | Centimeter | Millimeter |
|---|---|---|---|---|---|---|---|---|---|
| | mi | rd | yd | ft | in. | km | m | cm | mm |
| Mile | - | 320 | 1760 | 5280 | 63,360 | 1.609 | 1609 | 160,934 | * |
| Rod | 0.003125 | - | 5.5 | 16.5 | 198 | 0.00503 | 5.03 | 502.9 | * |
| Yard | 0.000568 | 0.1818 | - | 3 | 36 | 0.00091 | 0.914 | 91.44 | 914.4 |
| Foot | 0.000189 | 0.0606 | 0.333 | - | 12 | 0.00031 | 0.3048 | 30.48 | 304.8 |
| Inch | * | 0.0051 | 0.028 | 0.083 | - | * | 0.0254 | 2.54 | 25.4 |
| Kilometer | 0.6214 | 198.84 | 1,093.6 | 3,280.8 | 39,370 | - | 1,000 | 100,000 | * |
| Meter | 0.00062 | 0.1988 | 1.0936 | 3.28 | 39.370 | 0.001 | - | 100 | 1,000 |
| Centimeter | * | * | 0.0109 | 0.0328 | 0.3937 | * | 0.01 | - | 10 |
| Millimeter | * | * | * | 0.00328 | 0.0394 | * | 0.001 | 0.1 | - |

* Values are either too large or small to be useful for conversion. If it is necessary to convert to these units, convert to a larger or smaller unit and then convert that result to the desired unit. Conversion multiplier values have been rounded but are appropriate in most situations. Example: 1 mile = 320 rods; 1 = 1 mile/320 rods

**Box 2-8.** Equalities for area.

| Unit Converting From (UF) | Unit Converting To (UT): Square mile | Acre | Square yard | Square foot | Square inch | Square kilometer | Hectare | Square meter | Square centimeter |
|---|---|---|---|---|---|---|---|---|---|
| | mi² | Ac | yd² | ft² | in² | km² | ha | m² | cm² |
| Square Mile | - | 640 | * | * | * | 2.59 | 259 | * | * |
| Acre | 0.0015625 | - | 4,840 | 43,560 | * | 0.00405 | 0.4047 | 4,047 | * |
| Square Yard | * | * | - | 9.0 | 1,296 | * | * | 0.8361 | 8,361 |
| Square Foot | * | * | 0.1111 | - | 144 | * | * | 0.0929 | 929 |
| Square Inch | * | * | * | 0.0069 | - | * | * | 0.0006 | 6.4516 |
| Square Kilometer | 0.3861 | 247 | * | * | * | - | 100 | 1,000,000 | * |
| Hectare | 0.003861 | 2.47 | * | * | * | 0.01 | - | 10,000 | * |
| Square Meter | * | * | 1.196 | 10.764 | 1,550 | * | 0.0001 | - | 10,000 |
| Square Centimeter | * | * | * | 0.00108 | 0.155 | * | * | 0.0001 | - |

* Values are either too large or small to be useful for conversion. If it is necessary to convert to these units, convert to a larger or smaller unit and then convert that result to the desired unit. Conversion multiplier values have been rounded, but are appropriate in most situations. Example: 1 sq mile = 640 A; 1 = 1 sq mile/640 A

**Box 2-9.** Equalities for liquid volume.

| Unit Converting From (UF) | Unit Converting To (UT): Acre inch | Cubic foot | Gallon | Quart | Pint | Cup | Fluid ounces | Tablespoon | Teaspoon | Cubic meter | Liter |
|---|---|---|---|---|---|---|---|---|---|---|---|
| | Ac*In | ft³ | gal | qt | pt | c | fl. oz | tbsp | tsp | m³ | L |
| Acre*Inch | - | 3,630 | 27,154 | * | * | * | * | * | * | 102.79 | * |
| Cubic Foot | 0.000275 | - | 7.48 | 29.92 | 59.84 | * | * | * | * | 0.0283 | 28.32 |
| Gallon | * | 0.1337 | - | 4 | 8 | 16 | 128 | 256 | 768 | 0.003785 | 3.785 |
| Quart | * | 0.0334 | 0.25 | - | 2 | 4 | 32 | 64 | 192 | 0.000946 | 0.946 |
| Pint | * | 0.0167 | 0.125 | 0.5 | - | 2 | 16 | 32 | 96 | 0.000473 | 0.473 |
| Cup | * | 0.0084 | 0.0625 | 0.25 | 0.5 | - | 8 | 16 | 48 | 0.000237 | 0.236 |
| Fluid Ounces | * | 0.0011 | 0.0078 | 0.03125 | 0.0625 | 0.125 | - | 2 | 6 | 0.000029 | 0.029 |
| Tablespoon | * | * | 0.0039 | 0.01562 | 0.03125 | 0.0625 | 0.5 | - | 3 | * | * |
| Teaspoon | * | * | 0.0013 | 0.00521 | 0.01042 | 0.0208 | 0.167 | 0.334 | - | * | * |
| Cubic Meter | * | | | | | | | | | | |
| Liter | * | 0.0353 | 0.2642 | 1.05669 | 2.11337 | 4.2268 | 33.814 | 67.628 | 202.884 | 0.001 | - |

* Values are either too large or small to be useful for conversion. If it is necessary to convert to these units, convert to a larger or smaller unit and then convert that result to the desired unit. Conversion multiplier values have been rounded, but are appropriate in most situations. Example: 1 sq mile = 640 A; 1 = 1 sq mile/640 A

**Box 2-10.** Equalities for mass.

| To convert column 1 into column 2, multiply by | Column 1 SI unit | Column 2 non-SI unit | To convert column 2 into column 1, multiply by |
|---|---|---|---|
| 0.0352 | gram | ounce | 28.4 |
| 2.205 | kilogram | pound | 0.454 |
| 1.102 | tonne (called "metric ton" in US) | ton, short (2,000 lb) | 0.907 |
| 0.1575 | kilogram | stone | 6.350 |
| 0.01543 | milligram | grain | 64.799 |

## Rules for Unit Conversions

1. Multiplying a value by 1 does not change the value of the number. One can be written in many different ways, $\frac{12\,\text{in.}}{1\,\text{ft}}=1, \frac{1\,\text{ha}}{2.47\,\text{acres}}=1, \frac{1\,\text{bu}}{35.231}=1, \frac{1\,\text{acre}}{43,560\,\text{ft}^2}=1,$ and $\frac{1\,\text{kg}}{2.205\,\text{lb}}=1.$ The data from Boxes 2-5 – 2-10 can be used to create a multitude of ratios that are equal to 1.
2. Follow the units carefully; common units in the numerator and denominator must cancel each other out.

3. Check the units in the final answer to determine if the units are correct. If the units are not correct, the answer is not correct.

4. Identical units in the numerator and denominator cancel each other out. For example, in this equation, $10\ \text{sec}\left(\frac{10\ \text{ft}}{\text{sec}}\right)$, the seconds cancel each other out, resulting in an answer that is in feet.

## Unit Conversions

In many situations, English units must be converted to metric units or visa versa. Examples of unit conversions are below.

**Exercise 2-1. Convert 152 g to kg.**

Solution:

$$152\ \text{g} \cdot \frac{1\ \text{kg}}{1{,}000\ \text{g}} = 0.152\ \text{kg}$$

In this problem the ratio $\frac{1\ \text{kg}}{1{,}000\ \text{g}}$ is one because $\frac{1\ \text{kg}}{1{,}000\ \text{g}} = 1.$

Multiplying a value times one does not change the value of the number.

**Exercise 2-2. Convert these units.**

a. 3.2 mg to μg

b. 1 ft 4 in. to feet

**Exercise 2-3. Convert 2.76 lb to kg.**

**Exercise 2-4. Convert $\frac{100\ \text{kg}}{\text{ha}}$ to $\frac{\text{lb}}{\text{acre.}}$**

**Exercise 2-5. Convert 112 kg/ha to lb/acre.**

**Exercise 2-6. The tires on the bike are from England and are rated to hold 300 kPa (kilopascals). When you go to a pump in the USA, the output is in lb/in.$^2$ (psi). What is the maximum tire pressure for inflation?**

Solution:

1 pascal = $1.4504 \cdot 10^{-4}$ lb/in.$^2$
1 kilopascal = 1,000 pascal = $10^3$ pascal
1 kilopascal = $1.4504 \cdot 10^{-1}$ lb/in.$^2$
therefore 1 kilopascal (kPa) = 0.145 lb/in.$^2$
300 kPa • 0.14504 lb/in.$^2$/kPa = 43.5 lb/in.$^2$
The maximum tire pressure for safe inflation is 43.5 lb/in.$^2$ or 43.5 psi. The equation is shown below.

$$300\ \text{kPa} \times \frac{1{,}000\ \text{Pa}}{1\,\text{kPa}} \times \frac{1.45 \times 10^{-4}\ \text{psi}}{1\,\text{Pa}} = 43.5\ \text{psi}$$

**Exercise 2-7. A sprayer in an apple orchard uses 75 psi. What is the spray pressure in kilopascals (kPa)?**

## References

Taylor, B.N. and A. Thompson. 2008. The international system of units (SI). NIST Spec. Publ. 330. 2008 ed. Available online at http://physics.nist.gov/Pubs/SP330/sp330.pdf (accessed 22 Apr. 2008; verified 22 Apr. 2008). United States Dep. Commerce Natl. Inst. Standards Tech., Gaithersburg, MD.

# UNIT CONVERSIONS INVOLVED IN TEMPERATURES, GROWING DEGREE DAYS, AND THERMAL CONDUCTIVITY

**Key problems:** Temperature conversions and solving temperature equations that have biological meaning.

**Mathematical skills:** Solving equations.

**Chapter concept:** For many problems, temperature must be converted from one unit to another. This chapter shows how to convert °F to °C and determine heat units.

Equations must be used when converting temperatures from the Fahrenheit (F) scale to the Celsius (C) scale or when calculating growing degree days (GDD). An **equation** is a statement of equality, i.e. one object equals another object. A **variable** is any object that can take on a range of values within an equation.

## Rules for Solving Equations:

**1. Common units should be used. To change units, multiply the value by the conversion equality equal to one.**

Example: 1 gallon = 4 quarts OR 1 = 4 qt/1 gal calculate how many quarts are equal to 15 gal?

Solution: 15 gal • (4 qt/1 gal) = 60 qt

**2. To add values, the units must be identical.**

Example: 15 lb + 8 oz ≠ 23 lb, because units are not equal

For addition, quantities must have the same units. To change units, use equalities that are equal to one. In the problem, 15 lb + 8 oz, ounces must be converted to pounds.

15 lb + [8 oz • (1 lb/16 oz)] =

15 lb + 0.5 lb = 15.5 lb

Alternatively, the answer could be expressed in ounces rather than pounds. Then,

1 = 16 oz/1 lb would be used and the problem solved as:

[15 lb • (16 oz/1 lb)] + 8 oz = 240 oz + 8 oz = 248 oz

**3. To add (or subtract) a value to one side of an equation requires that the same value must be added (or subtracted) from the other side.**

Example: $a = b$ then

$a + c = b + c$ or $a - c = b - c$

**4. Multiplying (or dividing) by a number must be done on both sides of the equation.**

Example: $a = b$ then

$a \cdot c = b \cdot c$ or $a/c = b/c$

**Box 3-1.** Temperature equalities.

- °Celsius (°C) = (°F-32) • 5/9
- °Fahrenheit (°F) = (9/5 • °C) + 32
- Kelvin (K) = °C + 273

**Box 3-2.** Relationship between °C and °F.

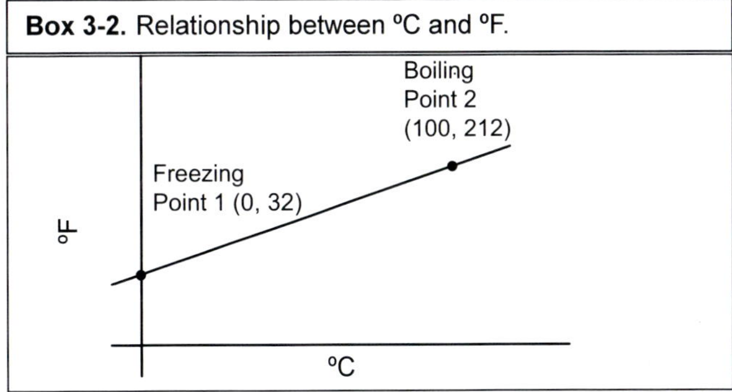


## Developing Equations

There are many approaches to developing equations. One approach is to combine several statements. For example, the linear equation

$$y = mx + b$$

expresses the change in $y$ as $x$ varies. There are two

unknowns in that equation, $m$ and $b$. The $m$ is called the slope of the line, and indicates the rate of change in $y$ as $x$ changes. The $b$ is called the "y-intercept", and it corresponds to the value of $y$ when $x = 0$.

The equation that converts degrees Celsius (°C) to degrees Fahrenheit (°F) is derived by substituting known values of °C into the linear equation

$$°F = m\ (°C) + b$$

At the freezing point of water, the temperature in degrees Celsius is 0 °C and in degrees Fahrenheit is 32 °F

$$32\ °F = m\ (0\ °C) + b, \text{ which is equivalent to } 32\ °F = b$$

At the boiling point of water, the temperature in degrees Celsius is 100 °C, and in degrees Fahrenheit is 212 °F

$$212\ °F = m\ (100\ °C) + 32\ °F$$

Solving for $m$ results in $m = 9/5$, which indicates that for each change of 5 °C, the temperature in °F changes by 9 degrees. Thus, the equation that converts degrees Celsius (°C) to degrees Fahrenheit (°F) is

$$°F = (9\ °F / 5\ °C)\ (°C) + 32\ °F$$

## Temperature Conversions

Temperature data are routinely converted from one scale to another. Two commonly used transformations are the conversion of Fahrenheit to Celsius temperatures and converting degrees to heat units.

### Exercise 3-1. Convert 85 °F to °C.

Solution: $°C = \left[(85\ °F - 32\ °F) \cdot \frac{5°C}{9°F}\right] = 29.4\ °C.$

In this equation, the 5/9 is equal to the ratio between the units from boiling to freezing in the Celsius scale relative to the Fahrenheit scale (100/180 = 5/9). The 32 in the equation represents the freezing point in Fahrenheit. Temperature conversions are done by inserting the appropriate numbers into the equation.

### Exercise 3-2. Convert 12 °C to °F.

### Exercise 3-3. Convert 95 °F to °C.

### Exercise 3-4. Convert 20 °C to °F.

## Converting Temperature Data to Growing Degree Days

Heat strongly influences the growth and development of plants and insects. The combination of temperature and time is used to calculate accumulated heat units. **Growing degree days (GDD)** are the accumulated amount of heat units during some period of time. GDD estimates are used for many diverse reasons, including hybrid selection, estimating water use, and forecasting weed and insect emergence. Different problems use different GDD equations. These data can then be used to determine scouting periods, and to schedule diverse operations such as irrigation, tillage, and pesticide applications. For example, 50% of the western bean cutworm emergence occurs when 1,422 GDD (base 50 °F) is reached, whereas 50% of the second generation European corn borer hatch is expected when GDD reach 1,550 (base 54 °F) .

GDD for corn growth is calculated using the equation,

$$GDD = \frac{T_{max} + T_{min}}{2} - T_{base}$$

where $T_{max}$ is the maximum daily temperature. $T_{max}$ is sometimes capped at a high temperature because plant or insect growth is slowed if temperatures are above that maximum value. If the maximum daily temperature is greater than the maximum value, $T_{max}$ is replaced with the maximum value. $T_{base}$ is the minimum temperature at which development occurs. $T_{min}$ is the lowest temperature of the day. If

$T_{min}$ is less than $T_{base}$, then $T_{min}$ is replaced with $T_{base}$. Different base values are used for different organisms. For example, alfalfa weevil has a $T_{base}$ value of 47 °F, whereas European corn borer has a $T_{base}$ of 54 °F. **Box 3-3** provides commonly used $T_{base}$ values for crops, weeds, and insect pests.

**Exercise 3-5. Determine the number of corn growing degree days for these temperatures.**

| | Maximum temperature ($T_{max}$) | Minimum temperature ($T_{min}$) |
|---|---|---|
| | °F | °F |
| Day 1 | 84 | 45 |
| Day 2 | 100 | 81 |

Solution:
Using $T_{base}$ and $T_{min}$ data for corn given in **Box 3-3**, the number of GDD for these two days were calculated.

Day 1: $GDD = \frac{84+50}{2} - 50$ and Day 2: $GDD = \frac{86+81}{2} - 50$

Total GDD is the sum of both days, 17 + 33.5 = 50.5 GDD. In these calculations, 50 is used rather than 45 in Day 1 because $T_{min}$ was 50 °F, and 86 is used rather than 100 in Day 2 because $T_{max}$ was 86 °F.

**Box 3-3.** Commonly used base values ($T_{base}$) when calculating heat units.

| | $T_{base}$ | | $T_{base}$ |
|---|---|---|---|
| Crop | °F | Weeds | °F |
| Corn | 50[1] | Barnyardgrass | 50 |
| Wheat | range from 32 to 40 | Yellow foxtail | 55 |
| Soybean | 50 | Common cocklebur | 40 |
| Sunflower | 44 | Sowthistle | 47 |
| Insects | | Wild buckwheat | 40 |
| European corn borer | 54 | | |
| Alfalfa weevil | 48 | | |
| Western corn rootworm | 53 | | |
| Gypsy moth | 50 | | |
| Stalk borer | 41 | | |

[1] $T_{max}$ value for corn is 86 °F.

**Exercise 3-6. Determine the number of stalk borer GDD using the data set in Exercise 3-5.**

**Exercise 3-7. Determine the number of wheat GDD if the base value is 40 °F.**

| | Maximum temperature | Minimum temperature |
|---|---|---|
| | °F | °F |
| Day 1 | 70 | 35 |
| Day 2 | 60 | 43 |

**Exercise 3-8. Calculate the number of wheat GDD for the data in Exercise 3-7 if the base temperature is 32 °F.**

**Exercise 3-9. Calculate the number of Corn GDD (base °C) using the same information given in Exercise 3-5 (base °F).**

## Thermal Conductivity

Heat movement through soil can be calculated by the Fourier heat flow equation, $Q = A \cdot K \cdot \frac{dt}{dx}$ where

$Q$ is the thermal flux,

$A$ is the cross sectional area

$K$ is the thermal conductivity, which represents the ability of a material to transmit heat, and

$dt/dx$ is the temperature gradient ($dt$) over the soil length of the soil column ($dx$)

The heat movement in soil is related to the rate at which soil cools down and heats up.

**Box 3-4.** Bulk density, specific heat capacity, and thermal conductivity of selected soil components (Coyne and Thompson, 2006).

| Selected soil component | Bulk density | Specific heat capacity | Thermal conductivity |
|---|---|---|---|
| | g/cm³ | cal/(g °C) | cal/(cm s K) |
| Sand | 1.5 | 0.3 | 0.0045-0.0055 |
| Organic matter | 1.3 | 0.60 | 0.0006 |
| Water | 1 | 1.00 | 0.00137 |
| Air | 0.00125 | 0.003 | 0.00006 |

**Exercise 3-10. What is the thermal flux if the surface temperature is 50 °C, the temperature at 30 cm is 25 °C, and the thermal conductivity is 0.0045 cal (cm sec °C)$^{-1}$. The surface area is 30 cm$^2$.**

Solution:

$$Q = A \cdot K \cdot \frac{dt}{dx}$$

$$Q = 30\ \text{cm}^2 \cdot \frac{0.0045\ \text{cal}}{\text{cm sec °C}} \cdot \frac{(50\ °\text{C} - 25)°\text{C}}{30\text{cm}} = \frac{0.1125\ \text{cal}}{\text{sec}}$$

## Specific Heat Capacity

Specific heat capacity is the amount of heat required to raise one gram of a substance 1 °C. The specific heat capacity influences how quickly a soil will heat up in the spring. The specific heat capacity of soil is defined by the equation,

$$\text{Specific heat capacity} = f_{\text{soil}} C_{\text{soil}} + f_{\text{water}} C_{\text{water}} + f_{\text{air}} C_{\text{air}}$$

where, $f$ is the fraction of soil, water, and air and $C$ is the specific heat capacities of the soil, water, and air.

**Exercise 3-11. Determine the specific heat capacity for a soil that contains 75% sand and 25% water. The specific heat capacity for sand and water is 0.3 cal/(g °C) and 1.0 cal/(g °C), respectively (Box 3-4).**

Solution:

Specific heat capacity = 0.75 • 0.3 cal/(g °C) + 0.25 • 1.00(cal)/(g °C) = 0.475 cal/(g °C)

**Exercise 3-12. Determine the specific heat capacity for a soil containing 15 g of sand and 6 g of water if the specific heat capacities of the soil components are 0.3 cal/(g °C) and 1.0 cal/(g °C), respectively.**

**Exercise 3-13. If a soil has a specific heat capacity of 0.5 cal/(g °C), how much heat is needed to increase 5 g of soil 10 °C?**

**Exercise 3-14. Find the linear equation** ($y = mx + b$) **for the following data.**

| x | y |
|---|---|
| 100 | 25 |
| 200 | 0 |

## References and Additional Information


Adams, N. Using growing degree days for insect management. 2008. At: http://extension.unh.edu/agric/GDDays/Docs/growch.pdf (verified 7/2008).

Coyne, M.S. and J.A. Thompson. 2006. Math for soil scientists. Thompson Demar Learning. Clifton Park, NY.

Steinmaus, S.J., T.S. Prather, and J.S. Holt. 2000. Estimation of base temperature for nine weed species. J. Experimental Bot. 51:275-286.

Using growing degree days for herbicide timing. 2008. At: http://www.crystalsugar.com/agronomy/gold/fact/growing_degree_days_lr.pdf (verified 7/2008).

Weedcast. 2008. At: http://www.weedcast.net/ (verified 7/2008).

# LATITUDE/LONGITUDE, CALCULATING LENGTH, AREA, AND RATES

**Key problems:** Determine end coordinates using global positioning system (GPS), calculating lengths, areas, volume, and calculating length of time to complete a job.

**Mathematical skills:** Base conversions, trigonometry equalities, distances between points, and areas.

**Chapter concepts:** The ability to calculate length, area, and conduct multiple conversions are needed for efficient resource management.

## Latitude, Longitude, and Universal Transverse Mercator System (UTM)

The lines of latitude are imaginary lines that circle the Earth. The largest circle is at the equator, where the latitude is zero. The radius of the circle surrounding the earth at 0° latitude is approximately 3,963 miles. The radius of the latitude at the poles (90° latitude) is zero miles.

There are imaginary lines that extend like an arc from pole to pole. These lines are called meridians. The distance between adjacent longitude degrees is dependent on its latitude. The distances between adjacent longitude degrees decreases as one moves north or south from the equator, because the radius of the imaginary circles decreases. A discussion for converting latitude and longitude values to coordinate system with the units measured in feet or meters is available in Carlson and Clay (1999).

Latitude and longitude values can be measured with the global positioning system (GPS). It is composed of a network of more than 24 satellites orbiting the Earth, GPS receivers, and control and monitoring stations on the Earth. Each satellite, in orbit 12,600 miles above the Earth, consists of a computer, clock, and radio. Each satellite continuously broadcasts its position and time (Pfrost et al., 1999). Satellite information is converted into latitude, longitude, and elevation information by a receiver. If information is available from one satellite, then the position of the receiver will be somewhere on the sphere surrounding the satellite. If information from two satellites is available, then the position is located on the intersection of the two spheres surrounding each satellite which narrows the position to two possible locations. If information is available from three satellites, the location is at the intersection of the three spheres, which results in one unique position.

**Box 4.1.** Diagrams showing latitude, longitude, and a real-time differential correction global positioning system (DGPS).

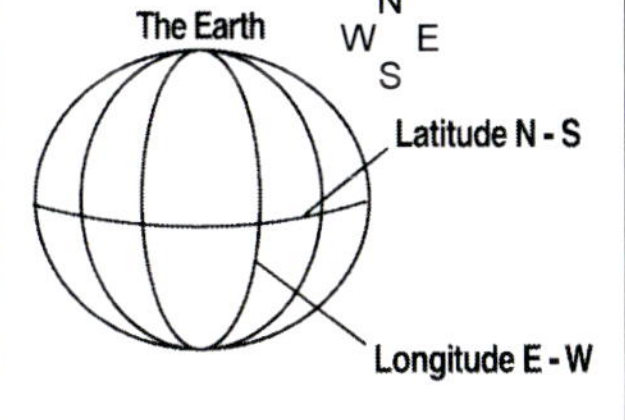


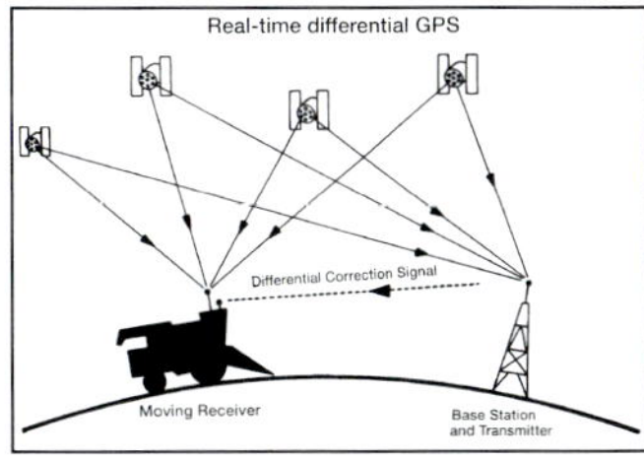


In natural resource management, GPS provides latitude, longitude, and altitude information for locations of interest. This information can be used in yield monitoring, grid soil sampling, determining the distance between two points, and determining areas in production fields. In many situations, the latitude/longitude values (degrees and records) must be converted to a different scale with units of meters or feet. One commonly used scale is the Universal Transverse Mercator (UTM) projection and grid system. The UTM system divides the Earth into 60 zones (numbered 1 to 60). Each zone is 6° of longitude wide. The first UTM zone (180° W to 174° W) is located at the international date line, longitude 180°. Each zone is divided into horizontal bands spanning 8° of latitude. These bands are lettered, south to north, beginning at 80° S with the letter C and ending with the letter X at 84° N. The letters I and O are skipped to avoid confusion with the numbers one and zero. A square grid is superimposed on each area. It's aligned so that vertical grid lines are parallel to the center of the zone, called the central meridian. UTM grid coordinates are expressed as a distance in meters to

**Box 4.2.** UTM zone number and designations.

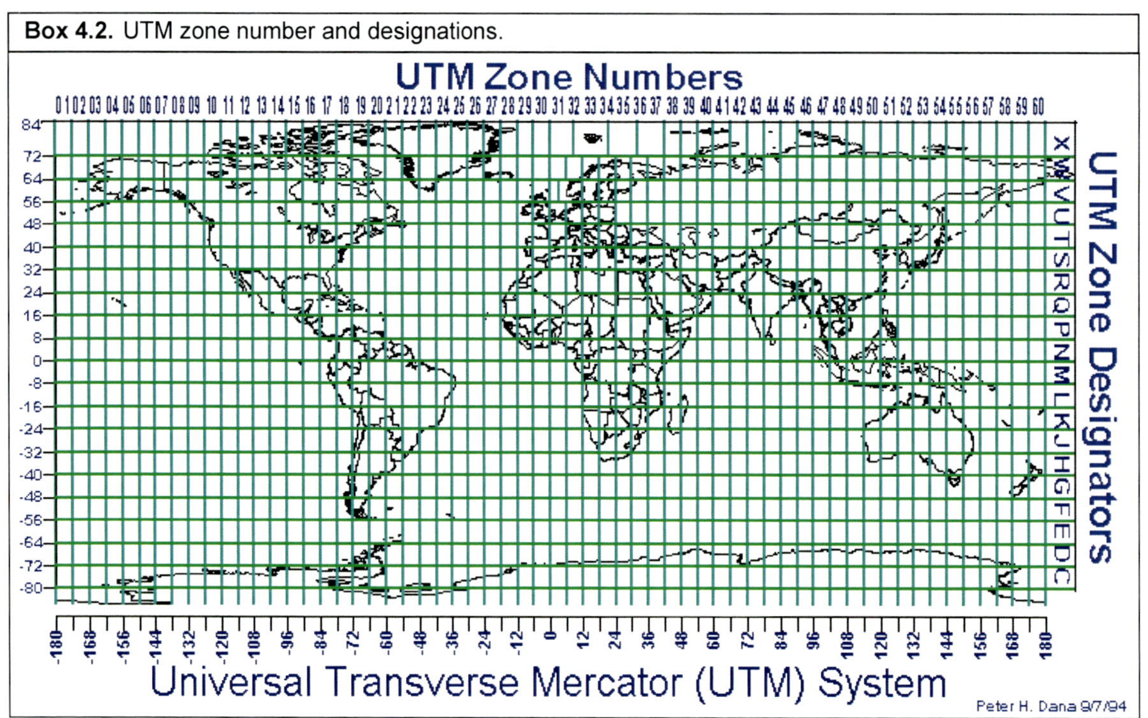


the east, referred to as the "easting", and a distance in meters to the north, referred to as the "northing". The advantage of converting latitude and longitude value to UTM is that it provides a constant distance relationship anywhere on the map.

**Exercise 4-1. Locate H26.**

## Importance of Digits and Converting Latitude/Longitude Values

When using GPS, it is important to understand the relationship between the number of digits in the location and precision. At 44° N latitude, the distance between adjacent 1 sec longitude values (96° 34' 22" and 96° 34' 23") is approximately 72.8 ft, whereas at the equator the distance between adjacent 1 sec latitude values is 101.3 ft. Note the symbols used to abbreviate minutes ( ' ) and seconds ( " ).

**Exercise 4-2. What is the distance between adjacent longitude values that differ by 0.1 sec at 44° N latitude?**

Solution:
The distance between adjacent latitude and longitude values depends on where you are located. At 44° N latitude, the distance between adjacent longitude values that differ by 1 sec is 72.8 ft. Therefore the difference between adjacent longitude values at 44° N latitude that differ by 0.1 sec is 7.28 ft (72.8 • 0.1 = 7.28).

**Exercise 4-3. What is the difference associated with latitude values that differ by 0.1 sec at the equator?**

Solution:
As discussed above, the difference between adjacent latitude values at the equator that differ by 1 sec is 101.3 ft. Therefore the difference between adjacent latitude values that differ by 0.1 sec is 10.13 ft. When using GPS the precision is directly related to the number of units in the latitude and longitude values.

To convert minutes and seconds to decimal degrees requires that numbers in one base system be converted to a different base system. Decimal degrees are base 100, while minutes and seconds use 60 and 3,600 as their bases, respectively. To convert minutes (base 60) to decimal degrees (base 100), minutes must be divided by 60. To convert seconds (base 3,600) to decimal degrees, seconds must be divided by 3,600.

For example,

10 min/60 = 0.1666 decimal degrees

10 sec/3,600 = 0.00002778 decimal degrees

Minutes can be written as decimal minutes or separated into minutes and seconds

57.557 min is equal to

$$57 \text{ min} + \left(0.557 \text{ min}\frac{60 \text{ sec}}{1 \text{ min}}\right) = 33.42 \text{ sec}$$

**Exercise 4-4. Convert 96° 47.46732' into decimal degrees.**

Solution:

This is accomplished by dividing the minutes by 60 and adding the degrees and minutes together.

$$\text{Decimal degrees} = 96° + 47.46732 \text{ min } \frac{1°}{60 \text{ min}} = 96.791122°$$

**Exercise 4-5. Convert 96.791122° to degrees, minutes, seconds, and decimal seconds.**

**Exercise 4-6. Convert 96° 50' 58" into decimal degrees.**

**Exercise 4-7. Convert 34.448729° to degrees, minutes, seconds.**

## Calculating Distance

Latitude and longitude values can be converted to meters or feet by collecting data in a UTM projection or by converting latitude and longitude values to distance values (Carlson and Clay, 1999). Once the latitude and longitude values are converted to feet or meters, the distance between two points $[(x_1, y_1)$ and $(x_2, y_2)]$ can be calculated using the equation,

$$\text{Distance}^2 = (x_1-x_2)^2 + (y_1-y_2)^2 \text{ or}$$

$$\text{Distance} = [(x_1-x_2)^2 + (y_1-y_2)^2]^{0.5}$$

**Exercise 4.8. Determine the distance between two points if the x and y coordinates are** (1, 5) and (-5, -5)

Solution:

$$\text{Distance} = [(1-(-5))^2 + (5-(-5))^2]^{0.5}$$
$$= [6^2+10^2]^{0.5}$$
$$= [136]^{0.5} = 11.7$$

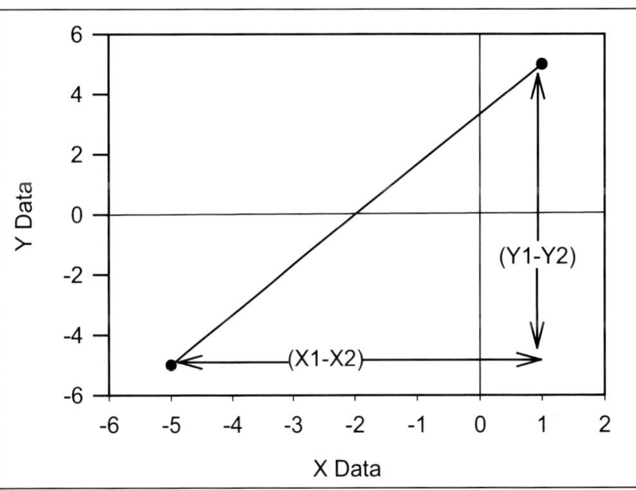

## Calculating Areas

The area within three points can be determined using standard trigonometric functions. Techniques for determining areas are shown below.

**Box 4.3.** Techniques used to determine areas.

**Coordinates are known, but areas are not known**

Areas can be determined for triangles based on the X, Y coordinates of the three corner points using the equation,

$$\text{Area} = 0.5\ |X_1Y_3 - X_1Y_2 + X_2Y_1 - X_2Y_3 + X_3Y_2 - X_3Y_1|$$

where, the X and Y values represent the coordinates of the three points and | | represent the absolute values (always positive).

**Length of sides are known**

If the lengths of the sides are known, the area can be calculated using Heron's formula.

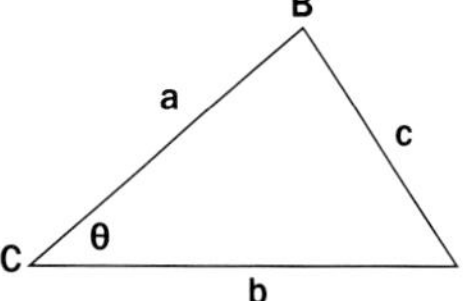


$$\text{Area} = [p\cdot(p-a)\cdot(p-b)\cdot(p-c)]$$

where p is half the perimeter of the triangle [(a+b+c)/2], and a, b, and c are the lengths of the three sides.

**One angle and two sides**

Triangle areas can also be determined if an angle and lengths of two sides adjacent to a known angle using the equation.

$$\text{Area} = \tfrac{1}{2} \cdot a \cdot b \sin\theta$$

Because the sin of 90° is 1, the area for a right triangle (90° angle) is calculated with the equation,

$$\text{The area of a triangle} = \left[\frac{\text{base}\cdot\text{height}}{2}\right] = \frac{ab}{2}$$

**Area for parallelograms**

For parallelograms, rectangles, and circles areas are calculated with the equations,

Area of a rectangle = width • height = ab

Area of parallelogram = a • b • sin θ or ab

Area of circle = $\pi \cdot r^2$, where $\pi$ is approximately equal 3.14159 and r is the radius (½ the diameter)

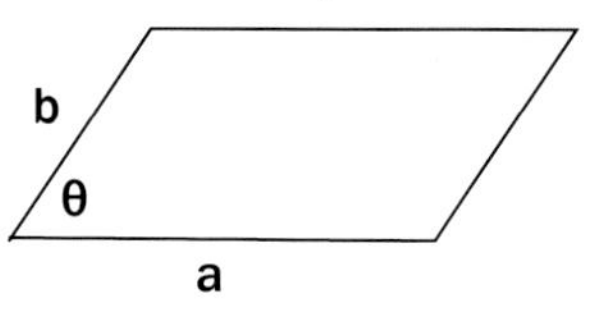


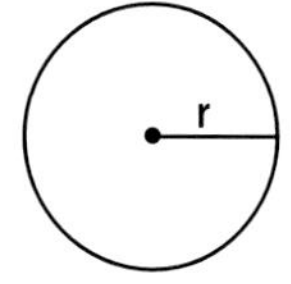


To calculate areas using GPS data, a program that accounts for the curvature of the earth is available at http://plantsci.sdstate.edu/precisionfarm/paper/publicationSoftware.aspx.

**Box 4-4.** Trigonometric equalities for sine (sin), cosine (cos), and tangent (tan).

sin (θ) = length of opposite side/length of hypotenuse

cos (θ) = length of adjacent side/length of hypotenuse

tan (θ) = length of opposite side/length of adjacent side

$\pi$ = pi = 3.14159

To convert degrees to radians, multiply degrees by $\pi/180°$. If problems are solved in Excel, degrees must be converted to radians.

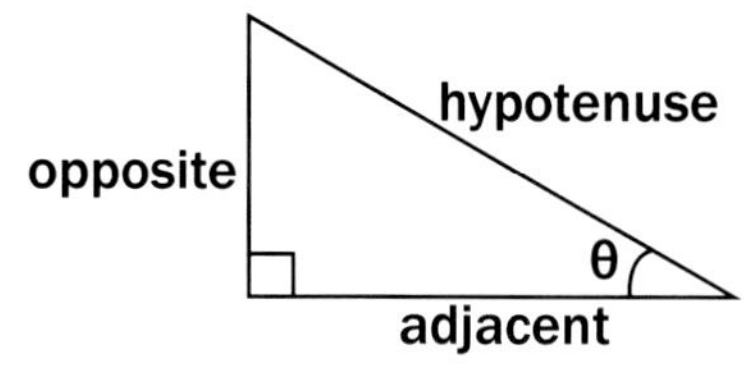


### Exercise 4-9. Determine the area in ft² and acres of a rectangle that has the dimensions of 600 by 900 ft.

Solution:

area = [600 ft • 900 ft] = 540,000 ft²

$$\text{acres} = \left[540{,}000\ \text{ft}^2 \cdot \frac{1\ \text{acre}}{43{,}560\ \text{ft}^2}\right] = 12.4\ \text{acres}$$

Degrees can also be reported as radians. A circle contains $2\pi$ radians. Radians and degrees can be converted back and forth with the equation

$$\text{radians}\left[\frac{180°}{\pi}\right] = \text{degrees, or}$$

$$\text{radians}\left[\frac{\pi}{180°}\right] = \text{degrees.}$$

### Exercise 4-10. Convert 150° to radians.

Solution:

degrees ($\pi/180°$) = radians

150°($\pi/180°$) = 2.618 radians

**Exercise 4-11. Convert 1 radian to degrees.**

**Exercise 4-12. Determine the area in hectares for a triangularly shaped field with where θ is 48° and the lengths of the adjacent sides are 500 and 100 m.**

Solution:

$$\text{Area} = \tfrac{1}{2} \cdot a \cdot b \cdot \sin\theta$$

$$\text{Area} = \tfrac{1}{2} \cdot 500\text{ m} \cdot 100\text{ m} \cdot 0.743$$

$$= 18{,}575\text{ m}^2$$

$$= 18{,}575\text{ m}^2 \cdot = \frac{1\text{ha}}{10{,}000\text{m}^2} = 1.86\text{ ha}$$

**Exercise 4-13. Determine the area of an irrigation circle (center pivot) in ft² and acres that has a diameter of 2,640 ft.**

**Exercise 4-14. Determine the area in acres if the coordinates of a triangle are 0, 2,600 ft (corner 1), 2,600, 0 ft (corner 2), and 0, 0 ft (corner 3).**

**Exercise 4-15. Determine the area in acres if the coordinates of a triangle are 100, 200 ft (corner 1), 50, 400 ft (corner 2), and 200, 300 ft (corner 3).**

**Exercise 4-16. Determine the area in acres and hectares for a right triangle that has a base of 200 ft and height of 900 ft.**

**Exercise 4-17. Convert 1.1 acre to ft$^2$.**

**Exercise 4-18. Convert 53,400 ft$^2$ to acres.**

## Converting Rate Units

To solve many problems, rates must be converted from one unit to another. In these problems both the numerator and denominator must be converted.

**Exercise 4-19. Convert $\frac{100\text{ kg}}{\text{ha}}$ to $\frac{\text{lb}}{\text{acre}}$**

Solution:

$$\frac{100\text{ kg}}{\text{ha}} \cdot \frac{\text{ha}}{2.47\text{ acre}} \cdot \frac{2.20\text{ b}}{\text{kg}} = \frac{89.1\text{b}}{\text{acre}}$$

**Exercise 4-20. Convert 112 kg/ha to lb/acre.**

Solution:

$$\frac{112\,kg}{ha} \cdot \frac{ha}{2.47\,acre} \cdot \frac{2.20\,lb}{kg} = \frac{99.8\,lb}{acre}$$

**Exercise 4-21. It takes 42 sec for a car to travel 500 ft. What is the speed in miles per hour (mph)?**

Solution:

Note that since the ratios $\frac{1\,mi}{5,280\,ft}, \frac{60\,sec}{1\,min}, \frac{60\,min}{1hr}$ are all equal to 1, ft/sec can be converted to miles/hour by multiplying the input value (ft/sec) by a series of ones. If the problem is set up correctly, the units will cancel.

$$\left[\frac{500\,ft}{42\,sec} \cdot \frac{1\,mi}{5280\,ft} \cdot \frac{60\,sec}{1\,min} \cdot \frac{60\,min}{1\,hr}\right] = \frac{8.12\,mi}{hr}$$

**Exercise 4-22. A tractor completes a 900 ft course in 98 sec. The speedometer reads 7 mph. Is the speedometer accurate?**

## Converting Area into Labor Requirements

For planning purposes, it is necessary to estimate the length of time to complete a job. This problem is solved by dividing the total amount of work by the rate that the work is conducted.

**Exercise 4-23. How much lawn in acres is mowed in an 8 hour day if the mower travels at 3.5 mi/hr and has a swath width of 20 ft?**

Solution:

This problem is solved by determining the distance traveled in 8 hours and then converting this to area.

$$\text{Distance traveled} = \frac{8\,hr \cdot 3.5\,mi}{hr} = 28\,mi$$

The total area is now determined using the equation for a rectangle (area = length x width). In this equation 28 miles is the length and 20 ft is the width. To solve this problem miles must be converted to ft and $ft^2$ converted to acres.

$$28mi \cdot 20ft \cdot \frac{5,280\,ft}{1\,mi} \cdot \frac{1\,acre}{43,560\,ft^2} = 67.9\,acres$$

So if a golf course is 180 acres, then t would take 3 days to mow.

**Exercise 4-24. How many acres are planted in a 10 hour day if the planter is 20 ft wide and it travels at 4 mi/hr?**

**Exercise 4-25. How many acres are planted in a 12 hour day if the planter is 90 ft wide and it travels at 6 mi/hr?**

**Exercise 4-26. How many acres are covered in a 6 hour day for an applicator traveling at 6 mi/hr if each pass is 120 ft wide?**

**Exercise 4-27. What is the time requirement to plant, cultivate, and combine a quarter section (160 acres) of land? The field will be planted at 3 mph with a 12 row planter with a row spacing of 30 in. Weeds will be controlled with a cultivator with a 40 ft swath traveling at 3.5 mph. The field will be harvested with a 6 row combine (30 in. between rows) traveling at 2.4 mph. (Assume that no time is needed for maintenance, refill, equipment, unload grain, etc.)**

Solution:

Time to plant

$$160\text{ acres}\cdot\frac{1\text{ hr}}{3\text{ mi}}\cdot\frac{1\text{ mi}}{5,280\text{ ft}}\cdot\frac{1}{12\text{ rows}}\cdot\frac{1\text{ row}}{30\text{ in}}\cdot\frac{12\text{ in.}}{1\text{ ft}}\cdot\frac{43,560\text{ ft}^2}{1\text{ acre}}$$

$$=14.7\text{ hr}$$

Time to cultivate

$$160\text{ acres}\cdot\frac{1\text{hr}}{3.5\text{ mi}}\cdot\frac{1\text{mi}}{5,280\text{ ft}}\cdot\frac{1}{40\text{ft}}\cdot\frac{43,560\text{ ft}^2}{1\text{acre}}$$

$$=9.4\text{ hr}$$

Time to combine

$$160\text{ acres}\cdot\frac{1\text{ hr}}{2.4\text{ mi}}\cdot\frac{1\text{ mi}}{5,280\text{ ft}}\cdot\frac{1}{6\text{ rows}}\cdot\frac{1\text{ row}}{30\text{ in.}}\cdot\frac{12\text{ in.}}{1\text{ ft}}\cdot\frac{43,560\text{ ft}^2}{1\text{ acre}}$$

$$=36.7\text{ hr}$$

Total time = 14.7 hr + 9.4 hr + 36.7 hr = 60.8 hr

## Additional Information


Carlson, C.G. and D.E. Clay. 1999. The earth model–calculating field size and distance between points using GPS coordinates SSMG #11. *In* Clay et al. (Ed) Site-Specific Management Guidelines. PPI. Norcross, GA. Available at http://www.ipni.net/ssmg.

Johansen, D.P., D.E. Clay, C.G. Carlson. K.W. Stange, S.A. Clay, and K. Dalsted. 1999. Selecting a DGPS for making topography maps SSMG #14. *In* Clay et al. (Ed) Site-Specific Management Guidelines. PPI. Norcross, GA. Available at http://www.ipni.net/ssmg.

Pfrost, D., W. Casady, and K. Shannon. 1999. Global positioning satellite receivers. SSMG #6. *In* Clay et al. (Ed) Site-Specific Management Guidelines. PPI. Norcross, GA. Available at http://www.ipni.net/ssmg.

# UNIT CONVERSIONS INVOLVING FERTILIZERS

**Key problems:** Using percentages to determine how much of an impure substance to apply to meet application requirements; using density to interchange mass and volume measures.

**Mathematical skills:** Percent, sequential conversions.

**Chapter concepts:** To determine application rates of fertilizer sources with varying percentages of nutrients and varying forms (dry or liquid).

## Percent and Fertilizer Grade

A key mathematical concept, percent, is used in this chapter. Percent comes from the Latin phrase "per centum," meaning "out of one hundred." A **percent** is a fraction with a denominator of 100 and with equal units in the numerator and denominator. The symbol for percent is "%." So 43% is equivalent to 43/100, and specifies how much of one substance (43 in this case) is in 100 units of another. In the case of fertilizers, we are interested in how much of a particular nutrient (lb) is in 100 lb of fertilizer material. For example, monoammonium phosphate (MAP) may contain 10% N. This is interpreted as 10 lb N/100 lb MAP. Notice that the unit lb is used in both the numerator and denominator.

By law, commercial fertilizer products must display on their label the minimum percentages of nutrients that the manufacturer can guarantee are present. Each percentage represents the pounds of nutrient present in 100 lb of fertilizer material. The grade is composed of at least three numbers separated by dashes. For instance, the grade of diammonium phosphate (DAP) is 18-46-0. The first number (18) indicates the percent N. The second number (46) is the percent $P_2O_5$, and the third number (0) is the percent $K_2O$. Some ranges in grades of various fertilizers are reported in **Box 5.1**. The oxide forms of P and K are used by convention. It is often necessary to convert between $P_2O_5$ and P as well as between $K_2O$ and K. To make these conversions requires knowing the atomic weights of P, K, and O as well as the formula weights of $P_2O_5$ and $K_2O$. These weights are listed in **Box 5.2**. The formula weights were calculated by adding the atomic weights of the various elements making up the oxide compounds. For instance, $K_2O$ has two atoms of

**Box 5.1.** Fertilizer grades of selected fertilizer materials.

| | N, % | $P_2O_5$, % | $K_2O$, % | Density*, lb/gal |
|---|---|---|---|---|
| **Solid fertilizers** | | | | |
| Ammonium sulfate | 21 | 0 | 0 | |
| Potassium sulfate | 0 | 0 | 50-52 | |
| Potassium magnesium sulfate | 0 | 0 | 22 | |
| Ammonium nitrate | 33-34 | 0 | 0 | |
| Diammonium phosphate (DAP) | 18-21 | 46-54 | 0 | |
| Monoammonium phosphate (MAP) | 10-11 | 48-55 | 0 | |
| Potassium chloride | 0 | 0 | 60-62 | |
| Potassium nitrate | 13 | 0 | 44 | |
| Urea | 45-46 | 0 | 0 | |
| **Liquid fertilizers** | | | | |
| 7-21-7 multigrade | 7 | 21 | 7 | 11.3 |
| 9-18-9 multigrade | 9 | 18 | 9 | 11.1 |
| Urea-ammonium-nitrate (UAN) | 28-32 | 0 | 0 | 10.6-11.1 |
| Ammonium polyphosphate (APP) | 10-12 | 34-41 | 0 | 12 |
| Phosphoric acid | 0 | 48-53 | 0 | 14.1 |
| **Gas fertilizers** | | | | |
| Anhydrous ammonia | 82 | 0 | 0 | |

* Densities are reported at the standard temperature of 20 °C (68 °F). Higher temperatures have lower densities than those reported, while lower temperatures have higher densities.

**Box 5.2.** Approximate atomic weights of oxide compounds and their constituent elements.

| Element/Molecule | Approximate atomic formula wt., g/mol |
|---|---|
| K | 39 |
| O | 16 |
| P | 31 |
| $K_2O$ | 94 |
| $P_2O_5$ | 142 |

K (2 x 39 g/mol = 78 g/mol) and one atom of O for a total formula weight of 78 g/mol + 16 g/mol = 94 g/mol.

Oxide/elemental conversions are based on the ratio of the mass of the element contained in the oxide compound. For instance, the conversion from 1 lb P to lb $P_2O_5$ is done as follows:

$$1\,\text{lb}\,\cancel{P} \times \frac{\cancel{\text{mol}\,P_2O_5}}{62\,\cancel{g}\,\cancel{P}} \times \frac{142\,\cancel{g}\,P_2O_5}{\cancel{\text{mol}\,P_2O_5}} = 2.29\,\text{lb}\,P_2O_5$$

Converting from $K_2O$ to K is done as follows:

$$1\,\text{lb}\,\cancel{K_2O} \times \frac{\cancel{\text{mol}\,K_2O}}{94\,\cancel{g\,K_2O}} \times \frac{78\,\text{g}\,K}{\cancel{\text{mol}\,K_2O}} = 0.83\,\text{lb}\,K$$

**Exercise 5-1. Convert 46 lb $P_2O_5$/100 lb DAP to lb P/100 lb DAP.**

Solution: $\frac{46\text{ lb }P_2O_5}{100\text{ lb DAP}} \times \frac{62\text{ lb P}}{142\text{ lb }P_2O_5} = \frac{20\text{ lb P}}{100\text{ lb DAP}}$

**Exercise 5-2. Convert 60 lb $K_2O$/100 lb potash to lb K/100 lb potash.**

## Density and Specific Gravity

Like dry fertilizers, the grade of liquid fertilizers denotes pounds of nutrient per 100 lb of fertilizer material. However, when the material is applied, application equipment uses volumetric rates (gal/acre), rather than mass rates (lb/acre). To convert from mass to volumetric rates, the density of the fertilizer material is used. **Density**, as we use it here, is the ratio of the mass of a liquid to a unit volume of that liquid. Density is used to convert liquid fertilizer values (lb/gal) to lb/acre. Density is defined by the equation,

$$density = \frac{mass}{volume}$$

Density is temperature dependent. At warmer temperatures, the density of liquid fertilizers decreases. At colder temperatures, densities increase. **Box 5.3** is an example of how water density changes with temperature. Because density is temperature dependent, the temperature at which the density was determined is usually included in reference materials.

**Box 5.3.** The influence of temperature on density of water (Wikipedia, 2009).

| Temp, °C | Density, g/cm³ |
|---|---|
| 100 | 0.9584 |
| 80 | 0.9718 |
| 60 | 0.9832 |
| 40 | 0.9922 |
| 30 | 0.9956502 |
| 25 | 0.9970479 |
| 22 | 0.9977735 |
| 20 | 0.9982071 |
| 15 | 0.9991026 |
| 10 | 0.9997026 |
| 4 | 0.999972 |
| 0 | 0.9998395 |
| −10 | 0.998117 |
| −20 | 0.993547 |
| −30 | 0.983854 |

Before applying a liquid fertilizer, its density should be measured or obtained from its reference material. A device commonly used for this purpose is a hydrometer. The hydrometer records specific gravity. **Specific gravity** is the ratio of the density of the fertilizer liquid to the density of water at a given temperature. Specific gravity and density have different definitions. The relationship between the two is defined by the equation

$$\text{Specific gravity fertilizer} = \frac{\text{Density of fertilizer}}{\text{Density of water}}$$

Because specific gravity is the ratio of two densities, both with the same units, all units cancel and therefore specific gravity is a unitless quantity. Because density and specific gravity change with temperature, the temperature of the liquid fertilizer must also be recorded when using the hydrometer. Once the temperature and specific gravity are known, the density of the solution can be determined.

**Exercise 5-3. Determine the lb/gal of UAN at 0 °C if the specific gravity of the fertilizer as measured with a hydrometer is 1.27.**

Solution: 0 °C

Specific gravity fertilizer • Density of water = Density of fertilizer

$$1.27 \times \frac{0.9998395\ g}{cm^3} = \frac{1.27g}{cm^3} = \text{Density of fertilizer}$$

$$\frac{lb\ UAN}{gal} = \frac{1.27\ g}{cm^3} \cdot \frac{1{,}000\ cm^3}{L} \cdot \frac{3.785\ L}{gal} \cdot \frac{lb}{454\ g} = 10.59\ lb/gal$$

**Exercise 5-4. Determine the lb/gal of UAN at 40 °C if the specific gravity of the fertilizer as measured with a hydrometer is 1.27.**

Solution: 40 °C

Specific gravity fertilize • Density of water = Density of fertilizer

$$1.27 \times \frac{0.9922\ g}{cm^3} = \frac{1.26\ g}{cm^3}$$

$$\frac{lb\ UAN}{gal} = \frac{1.26\ g}{cm^3} \cdot \frac{1000\ cm^3}{L} \cdot \frac{3.785\ L}{gal} \cdot \frac{lb}{454\ g} = 10.50\ lb/gal$$

This example shows that the density decreased from 1.27 g/cm$^3$ to 1.26 g/cm$^3$. This decrease is the result of the higher temperature. By comparing results from exercises 5-3 and 5-4, it shows that temperature corrections are very small for normal temperatures.

## Calculating Quantities of Nutrients in Fertilizer Materials

Density can also be measured by weighing a known volume.

**Exercise 5-5. What is the density in lb/gal and g/cm$^3$ of 5 gal of fertilizer if it weighs 65 lb?**

Solution:

$$\frac{65\ lb}{5\ gal} \cdot \frac{1\ gal}{3.785\ L} \cdot \frac{1\ L}{1{,}000\ cm^3} \cdot \frac{454\ g}{lb} = 1.559\ g/cm^3$$

**Exercise 5-6. What is the density of 5 gal of fertilizer in lb/gal and g/cm$^3$ if it weighs 70 lb?**

A common need is to determine the amount of a particular nutrient contained within a fertilizer material. To perform this calculation, the weight of the material is multiplied by the percent of the nutrient present, expressed as a fraction.

**Exercise 5-7. How much $K_2O$ is contained in a 50 lb bag of 10-7-7?**

Solution:

$$50\ lb\ fertilizer \cdot \frac{7\ lb\ K_2O}{100\ lb\ fertilizer} = 3.5\ lb\ K_2O$$

**Exercise 5-8. How much $P_2O_5$ is contained in a 50 lb bag of 6-6-8?**

Solution:

$$50 \text{ lb fertilizer} \cdot \frac{6 \text{ lb } P_2O_5}{100 \text{ lb fertilizer}} = 3 \text{ lb } P_2O_5$$

**Exercise 5-9. How much N is contained in 200 lb of monoammonium phosphate (MAP) if the fertilizer grade is 11-52-0?**

Solution:

$$200 \text{ lb MAP} \cdot \frac{11 \text{ lb N}}{100 \text{ lb MAP}} = 22 \text{ lb N}$$

**Exercise 5-10. How much N is contained in 300 lb of anhydrous ammonia ($NH_3$) if the fertilizer grade is 82-0-0?**

**Exercise 5-11. How much N is contained in 200 lb of diammonium phosphate (DAP) if the fertilizer grade is 18-46-0?**

**Exercise 5-12. How much $P_2O_5$ is contained in 200 lb of DAP if the fertilizer grade is 18-46-0?**

## Determining Weights of Fertilizers to Apply

To calculate the rate of fertilizer to apply, the recommended rate of the nutrient must be divided by the percent of that nutrient in the material. For instance, if DAP (18-46-0) is applied at a rate to supply 38 lb of $P_2O_5$, one would need to apply

$$38 \text{ lb } P_2O_5 \cdot \frac{100 \text{ lb fertilizer}}{46 \text{ lb } P_2O_5} = 82.6 \text{ lb DAP}$$

If the 82.6 lb DAP is applied, how much N is applied? This calculation is done as follows:

$$82.6 \text{ lb DAP} \times \frac{18 \text{ lb N}}{100 \text{ lb DAP}} = 14.9 \text{ lb N}$$

**Exercise 5-13. How much anhydrous ammonia should be applied to provide 150 lb N/acre?**

Solution:

$$\frac{150 \text{ lb N}}{\text{acre}} \times \frac{100 \text{ lb anhydrous ammonia}}{82 \text{ lb N}} = \frac{183 \text{ lb anhydrous ammonia}}{\text{acre}}$$

**Exercise 5-14. How much MAP (11-52-0) needs to be applied to provide 45 lb $P_2O_5$/acre?**

**Exercise 5-15. If 90 lb (18-46-0) DAP is applied per acre, how much N is applied?**

Solution:

$$\frac{90 \text{ lb DAP}}{\text{acre}} \cdot \frac{18 \text{ lb N}}{100 \text{ lb DAP}} = 16.2 \text{ lb N/acre}$$

## Determining Volumes of Fertilizers to Apply

Calculating the amount of a liquid fertilizer needed to supply a certain quantity of nutrient requires an additional step. This step involves including converting gallons to pounds or pounds to gallons.

**Exercise 5-16. Convert 100 lb N/acre to the number of gal/acre of UAN (28-0-0), assuming a density of 10.6 lb UAN/gal.**

Solution:
Urea ammonium nitrate (UAN) is a liquid fertilizer that typically has a fertilizer grade of 28-0-0. This fertilizer is purchased in gallons and applied as gal/acre. To apply the appropriate amount of gallons per acre, the density of the solution is needed to convert lb/acre to gal/acre (**Box 5.1**).

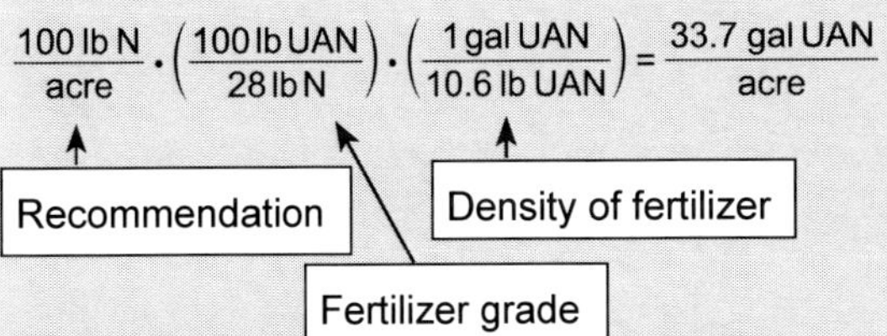


Based on these calculations, a quick estimate shows that each gallon of 28% contains approximately 3 lb of N.

**Exercise 5-17. What volume of ammonium (11-37-0) polyphosphate should be applied to provide 100 lb N/acre, assuming a density of 11.6 lb/gal?**

**Exercise 5-18. What volume of ammonium polyphosphate (11-37-0) needs to be applied to provide 40 lb $P_2O_5$/acre, assuming a density of 11.6 lb/gal?**

**Exercise 5-19. If APP (11-37-0) is at 0 °C, how many pounds of $P_2O_5$ are contained in 1,000 gal if the specific gravity is 1.412 at that temperature? (Box 5.3 for water density by temperature)**

Solution:

Specific gravity fertilizer · Density of water = Density of fertilizer

$$1.412 \cdot 0.9998395 \text{ g/cm}^3 = 1.412 \text{ g/cm}^3$$

$$\frac{1.412 \text{ g}}{\text{cm}^3} \cdot \frac{1{,}000 \text{ cm}^3}{\text{L}} \cdot \frac{3.785 \text{ L}}{\text{gal}} \cdot 1{,}000 \text{ gal} \cdot \frac{1 \text{ lb APP}}{454 \text{ g}} \cdot \frac{37 \text{ lb } P_2O_5}{100 \text{ lb APP}} =$$

$4{,}355$ lb $P_2O_5$

**Exercise 5-20. If APP (11-37-0) is at 40 °C, how many pounds of $P_2O_5$ are contained in 1,000 gal if the specific gravity is 1.412 at that temperature? (Box 5.3, water density)**

## Determining Costs of Nutrients in Single Nutrient Products

Fertilizer costs are provided per short ton (2,000 lb) of material, whether it is liquid or dry. The following exercise is used to illustrate the procedure for converting price per ton to price per pound of nutrient.

**Exercise 5-21. What is the cost per lb of N if urea (46-0-0) sells for $400/ton?**

Solution:

$$\frac{\$400}{\text{ton urea}} \cdot \frac{1\text{ ton urea}}{2{,}000\text{ lb urea}} \cdot \frac{100\text{ lb urea}}{46\text{ lb N}} = \frac{\$0.435}{\text{lb N}}$$

**Exercise 5-22. The liquid fertilizer 28-0-0 sells for $280/ton. What is the cost per lb of N?**

Solution:

$$\frac{\$280}{\text{ton UAN}} \times \frac{1\text{ ton UAN}}{2{,}000\text{ lb UAN}} \times \frac{100\text{ lb UAN}}{28\text{ lb N}} = \frac{\$0.50}{\text{lb N}}$$

**Exercise 5-23. How much will it cost to apply 100 lb N/acre to 100 acres if urea (46-0-0) costs $350/ton?**

## Determining Costs of Nutrients in Multiple Nutrient Products

Many fertilizers contain multiple nutrients. From the consumer's perspective, the cost associated with each nutrient is not known. However, such costs can be estimated if costs of single nutient sources are known. For example, say we know that the price of N in urea, a single nutrient product, is $0.50/lb N. If DAP sells for $1,000/ton, then the cost of the P can be determined following these steps.

First, estimate the cost of N in DAP:

$$\frac{2{,}000\text{ lb DAP}}{\text{ton DAP}} \cdot \frac{18\text{ lb N}}{100\text{ lb DAP}} \cdot \frac{\$0.50}{1\text{ lb N}} = \frac{\$180}{\text{ton DAP}}$$

Second, subtract the cost of N from the DAP cost:

$1,000-$180 = $820

Third, calculate the $P_2O_5$ cost.

$$\frac{1\text{ ton DAP}}{2{,}000\text{ lb DAP}} \cdot \frac{100\text{ lb DAP}}{46\text{ lb }P_2O_5} \cdot \frac{\$820}{\text{ton DAP}} = \frac{\$0.891}{\text{lb }P_2O_5}$$

This method provides one way of allocating costs to various nutrients in multi-nutrient sources.

## Calculating Application Rates of Multiple Fertilizer Sources

**Exercise 5-24. How much MAP (11-52-0) and urea (46-0-0) are needed on an acre basis if the recommendation is 85 lb $P_2O_5$/acre and 115 lb N/acre.**

Solution:

The following steps should be followed to solve this problem:

1. Calculate how much MAP is needed.
2. Calculate the amount of N contained in the MAP (11-52-0)(100 lb MAP contains 11 lb of N, 52 lb $P_2O_5$, and no $K_2O$).
3. Subtract the N contained in the DAP from the N recommendation.
4. Determine how much urea is needed.

1. $\frac{85\text{ lb }P_2O_5}{\text{acre}} \cdot \frac{100\text{ lb MAP}}{52\text{ lb }P_2O_5} = \frac{163\text{ lb MAP}}{\text{acre}}$

is needed to meet the $P_2O_5$ recommendation.

2. $\frac{163\text{ lb MAP}}{\text{acre}} \cdot \frac{11\text{ lb N}}{100\text{ lb MAP}} = \frac{18\text{ lb N}}{\text{acre}}$

This is the amount of N applied by the MAP.

3. 115 lb N - 18 lb N = 97 lb N

This is the remaining N requirement.

4. $\frac{97\text{ lb N}}{\text{acre}} \cdot \frac{100\text{ lb urea}}{46\text{ lb N}} = \frac{211\text{ lb urea}}{\text{acre}}$

This is the amount of urea that must be applied to fill that part of the N recommendation not met with the MAP application.

**Exercise 5-25. How much MAP (11-52-0) and anhydrous ammonia (82-0-0) should be applied if the recommendation is 70 lb $P_2O_5$ and 100 lb N/acre?**

**Exercise 5-26. How much DAP (18-46-0) and anhydrous ammonia should you apply if the desire is to add 70 lb $P_2O_5$ and 100 lb N/acre?**

**Exercise 5-27. How much N and $P_2O_5$ are contained in 120 gal of ammonium polyphosphate (10-34-0), assuming a density of 11.6 lb/gal?**

Determining the least expensive fertilizer mixture from multiple sources can become very complicated. Programs are available on-line to assist in these calculations. See http://ipni.net/toolbox

## References:

Wikipedia. 2009. Water (properties). Available at http://en.wikipedia.org/wiki/Water_(molecule)#Density_of_water_and_ice (accessed 5 Jun. 2009).

# NUTRIENT REMOVAL AND NUTRIENT BUDGETS

**Key problems:** Estimate or measure the quantity of various nutrients removed by a crop and compare it to the quantity of nutrients applied.

**Mathematical skills:** Review of basic functions.

**Chapter concepts:** The purpose of this chapter is to demonstrate first how nutrient removal is either estimated or measured. Secondly, this chapter will demonstrate how to compare nutrient removal to nutrient applications to determine a budget.

## Estimating Nutrient Removal from Published Data

To prevent the mining of soil nutrients, the amount of nutrients removed by a crop should equal additions. **Box 6-1** provides averages of nutrient removal rates published by many different information sources (Murrell, 2005). To use this table, simply multiply the crop yield by the value in the table for a particular nutrient. Doing so will provide the quantity of nutrient removed per acre. To calculate nutrient removal from a field, multiply this result by the area of the field.

**Exercise 6-1. Harvest removes 11 tons of corn silage per acre from a 100 acre field. How much N, $P_2O_5$, and $K_2O$ were removed?**

Solution:

$$\frac{11\text{ ton}}{1\text{ acre}} \cdot \frac{9.7\text{ lb N}}{1\text{ ton}} \cdot 100\text{ acres} = 10,670\text{ lb N}$$

$$\frac{11\text{ ton}}{1\text{ acre}} \cdot \frac{3.1\text{ lb } P_2O_5}{1\text{ ton}} \cdot 100\text{ acres} = 3,410\text{ lb } P_2O_5$$

$$\frac{11\text{ ton}}{1\text{ acre}} \cdot \frac{7.3\text{ lb } K_2O}{1\text{ ton}} \cdot 100\text{ acres} = 8,030\text{ lb } K_2O$$

**Exercise 6-2. Harvest removes 4 tons of fescue per acre from a 50 acre field. How much total N, $P_2O_5$, and $K_2O$ were removed using the values in Box 6-1?**

**Exercise 6-3. Calculate the amount of $P_2O_5$ removed per acre by a 60 bu grain/acre wheat crop, using the values in Box 6-1.**

Solution:

$$\left(\frac{60\text{ bu wheat grain}}{1\text{ acre}}\right)\left(\frac{0.60\text{ lb } P_2O_5}{1\text{ bu wheat grain}}\right) = \frac{36\text{ lb } P_2O_5}{1\text{ acre}}$$

**Exercise 6-4. Calculate the amount of $K_2O$ removed by 4 tons/acre of corn stover, using the values in Box 6-1.**

**Box 6-1.** Average rates of nutrient removal by several crops. (Murrell, 2005).

| | | Removal, lb/unit | | | | |
|---|---|---|---|---|---|---|
| **Crop** | **Unit of yield** | **N** | **$P_2O_5$** | **$K_2O$** | **Mg** | **S** |
| Alfalfa | ton | 51 | 12 | 49 | 5.4 | 5.4 |
| Alsike clover | ton | 41 | 11 | 54 | 7 | 3 |
| Barley grain | bu | 0.99 | 0.4 | 0.32 | 0.06 | 0.09 |
| Barley straw | bu | 0.4 | 0.16 | 1.2 | 0.08 | 0.1 |
| Barley straw | ton | 13 | 5.1 | 39 | 3 | 3 |
| Beans, dry | bu | 3 | 0.79 | 0.92 | 0.06 | 0.52 |
| Birdsfoot trefoil | ton | 45 | 11 | 42 | -- | -- |
| Bluegrass | ton | 30 | 12 | 46 | 3.5 | 5.0 |
| Bromegrass | ton | 32 | 10 | 46 | -- | 5.0 |
| Buckwheat | bu | 0.83 | 0.25 | 0.22 | -- | -- |
| Canola | bu | 1.9 | 1.2 | 2.0 | -- | 0.34 |
| Corn grain | bu | 0.90 | 0.38 | 0.27 | 0.09 | 0.08 |
| Corn stover | bu | 0.45 | 0.16 | 1.1 | 0.14 | 0.07 |
| Corn stover | ton | 16 | 5.8 | 40 | 5.0 | 3 |
| Corn silage | bu | 1.6 | 0.51 | 1.2 | 0.33 | 0.18 |
| Corn silage | ton | 9.7 | 3.1 | 7.3 | 2.0 | 1.1 |
| Fescue | ton | 37 | 12 | 54 | 3.7 | 5.7 |
| Flax grain | bu | 2.5 | 0.7 | 0.6 | -- | 0.19 |
| Flax straw | bu | 0.7 | 0.16 | 2.2 | -- | 0.15 |
| Millet | bu | 1.4 | 0.4 | 0.4 | 0.08 | 0.08 |
| Mint | lb oil | 1.9 | 1.1 | 4.5 | -- | -- |
| Oat grain | bu | 0.77 | 0.28 | 0.19 | 0.04 | 0.07 |
| Oat straw | bu | 0.31 | 0.16 | 0.94 | 0.1 | 0.11 |
| Oat straw | ton | 12 | 6.3 | 37 | 4 | 4.5 |
| Oat silage | ton | 9.0 | 11 | 45 | -- | -- |

*Continued on next page.*

**Exercise 6-5. A farmer usually grows corn one season and soybeans the next. His corn yield is usually 180 bu/acre and his soybean yield is about 55 bu/acre. How much $P_2O_5$ and $K_2O$ are removed per acre in two seasons (one corn crop and one soybean crop)? Use values in Box 6-1.**

Corn:

$$\left(\frac{180 \text{ bu corn grain}}{1 \text{ acre}}\right)\left(\frac{0.38 \text{ lb } P_2O_5}{1 \text{ bu corn grain}}\right) = \frac{68 \text{ lb } P_2O_5}{1 \text{ acre}}$$

$$\left(\frac{180 \text{ bu corn grain}}{1 \text{ acre}}\right)\left(\frac{0.27 \text{ lb } K_2O}{1 \text{ bu corn grain}}\right) = \frac{49 \text{ lb } K_2O}{1 \text{ acre}}$$

Soybean:

$$\left(\frac{55 \text{ bu soybean grain}}{1 \text{ acre}}\right)\left(\frac{0.84 \text{ lb } P_2O_5}{1 \text{ bu soybean grain}}\right) = \frac{46 \text{ lb } P_2O_5}{1 \text{ acre}}$$

$$\left(\frac{55 \text{ bu soybean grain}}{1 \text{ acre}}\right)\left(\frac{1.3 \text{ lb } K_2O}{1 \text{ bu soybean grain}}\right) = \frac{72 \text{ lb } K_2O}{1 \text{ acre}}$$

Total $P_2O_5$ removed with one season of corn and one season of soybean = 68 lb $P_2O_5$/acre + 46 lb $P_2O_5$/acre = 114 lb $P_2O_5$/acre.

Total $K_2O$ removed with one season of corn and one season of soybean = 49 lb $K_2O$/acre + 72 lb $K_2O$/acre = 121 lb $K_2O$/acre

| *Continued from previous page.* | | Removal, lb/unit | | | | |
|---|---|---|---|---|---|---|
| **Crop** | **Unit** | **N** | **$P_2O_5$** | **$K_2O$** | **Mg** | **S** |
| Orchardgrass | ton | 36 | 13 | 54 | 4.2 | 5.8 |
| Potato tuber | cwt | 0.32 | 0.12 | 0.55 | 0.03 | 0.03 |
| Potato vine | cwt | 0.2 | 0.05 | 0.3 | 0.04 | 0.02 |
| Red clover | ton | 45 | 12 | 42 | 7 | 3 |
| Reed canarygrass | ton | 28 | 9.7 | 44 | -- | -- |
| Rye grain | bu | 1.4 | 0.46 | 0.31 | 0.1 | 0.1 |
| Rye straw | bu | 0.8 | 0.21 | 1.5 | 0.07 | 0.14 |
| Rye straw | ton | 12 | 3.0 | 22 | 1 | 2.0 |
| Ryegrass | ton | 43 | 12 | 43 | 8 | -- |
| Sorghum grain | bu | 0.66 | 0.39 | 0.27 | 0.06 | 0.06 |
| Sorghum stover | bu | 0.56 | 0.16 | 0.83 | 0.12 | 0.12 |
| Sorghum stover | ton | 28 | 8.3 | 42 | 6.1 | 5.9 |
| Sorghum-sudan | ton | 30 | 9.5 | 34 | 6.8 | 5.8 |
| Soybean grain | bu | 3.8 | 0.84 | 1.3 | 0.21 | 0.18 |
| Soybean stover | bu | 1.1 | 0.24 | 1.0 | 0.22 | 0.17 |
| Soybean stover | ton | 40 | 8.8 | 37 | 8.1 | 6.2 |
| Soybean hay | ton | 45 | 11 | 25 | 9 | 5 |
| Sugarbeet root | ton | 3.7 | 2.2 | 7.3 | 0.95 | 0.45 |
| Sugarbeet top | ton | 7.4 | 4.0 | 20 | 1.1 | 0.40 |
| Sunflower grain | cwt | 2.7 | 0.97 | 0.90 | 0.25 | 0.25 |
| Sunflower stover | cwt | 2.8 | 0.24 | 4.1 | 1.6 | 0.6 |
| Sunflower stover | ton | 23 | 2.0 | 34 | 13 | 5 |
| Switchgrass | ton | 22 | 12 | 58 | -- | -- |
| Timothy | ton | 25 | 11 | 42 | 2 | 2 |
| Tobacco (leaves) | cwt | 3.6 | 0.90 | 5.7 | 0.45 | 0.6 |
| Vetch | ton | 57 | 15 | 49 | -- | -- |
| Wheat grain | bu | 1.5 | 0.60 | 0.34 | 0.15 | 0.1 |
| Wheat straw | bu | 0.7 | 0.16 | 1.2 | 0.1 | 0.14 |
| Wheat straw | ton | 14 | 3.3 | 24 | 2 | 2.8 |

**Exercise 6-6. A farmer grows soybean, corn, and sugarbeet crops in rotation. Soybeans typically yield 45 bu/acre, corn yields 165 bu/acre, and sugarbeet yields 18 ton/acre (root yield). How much $P_2O_5$ and $K_2O$ are removed per acre by one rotation of all three crops? Use values in Box 6-1.**

Exercise 6-7. A farmer is thinking of switching from a corn/soybean rotation to a corn/corn rotation. Typical soybean yield is 60 bu/acre and average corn yield is 190 bu/acre. Compare the amounts of $P_2O_5$ and $K_2O$ removed in two seasons of the corn/soybean rotation and the corn/corn rotation. How much more or less $P_2O_5$ and $K_2O$ are removed per acre by the corn/corn rotation? Use values in Box 6-1.

Corn:

$$\left(\frac{190\text{ bu corn grain}}{1\text{ acre}}\right)\left(\frac{0.38\text{ lb }P_2O_5}{1\text{ bu corn grain}}\right)=\frac{72\text{ lb }P_2O_5}{1\text{ acre}}$$

$$\left(\frac{190\text{ bu corn grain}}{1\text{ acre}}\right)\left(\frac{0.27\text{ lb }K_2O}{1\text{ bu corn grain}}\right)=\frac{51\text{ lb }K_2O}{1\text{ acre}}$$

Soybean:

$$\left(\frac{60\text{ bu soybean grain}}{1\text{ acre}}\right)\left(\frac{0.84\text{ lb }P_2O_5}{1\text{ bu soybean grain}}\right)=\frac{50\text{ lb }P_2O_5}{1\text{ acre}}$$

$$\left(\frac{60\text{ bu soybean grain}}{1\text{ acre}}\right)\left(\frac{1.3\text{ lb }K_2O}{1\text{ bu soybean grain}}\right)=\frac{78\text{ lb }K_2O}{1\text{ acre}}$$

Total $P_2O_5$ and $K_2O$ removed with two seasons of corn (corn/corn rotation) = 2(72 lb $P_2O_5$/acre) = 144 lb $P_2O_5$/acre.
= 2(51 lb $K_2O$/acre) = 102 lb $K_2O$/acre.

Total $P_2O_5$ and $K_2O$ removed with one season of corn and one season of soybean (corn/soybean rotation)
= 72 lb $P_2O_5$/acre + 50 lb $P_2O_5$/acre = 122 lb $P_2O_5$/acre
and 51 lb $K_2O$/acre + 78 lb $K_2O$/acre = 129 lb $K_2O$/acre

So comparing the corn/corn rotation to the corn/soybean rotation:

144 lb $P_2O_5$/acre - 122 lb $P_2O_5$/acre = 22 lb $P_2O_5$/acre.

102 lb $K_2O$/acre - 129 lb $K_2O$/acre = -27 lb $K_2O$/acre.

The corn/corn rotation removes more $P_2O_5$ and less $K_2O$ per acre than a corn/soybean rotation.

Exercise 6-8. A farmer is thinking of changing from his corn/soybean/wheat rotation to a corn/soybean/alfalfa rotation. Alfalfa will be grown for 2 years. Examine a 4-year period. For both rotations, start with corn and consider which crops will be grown in the 4 years. How much more or less $P_2O_5$ and $K_2O$ will be removed by the corn/soybean/alfalfa rotation than by the corn/soybean/wheat rotation? Use a corn yield of 150 bu/acre, a soybean yield of 45 bu/acre, a wheat yield of 50 bu/acre, and an alfalfa yield of 5 ton/acre/yr. Use values in Box 6-1.

Corn:

$$\left(\frac{150\text{ bu corn grain}}{1\text{ acre}}\right)\left(\frac{0.38\text{ lb }P_2O_5}{1\text{ bu corn grain}}\right)=\frac{57\text{ lb }P_2O_5}{1\text{ acre}}$$

$$\left(\frac{150\text{ bu corn grain}}{1\text{ acre}}\right)\left(\frac{0.27\text{ lb }K_2O}{1\text{ bu corn grain}}\right)=\frac{41\text{ lb }K_2O}{1\text{ acre}}$$

Soybean:

$$\left(\frac{45\text{ bu soybean grain}}{1\text{ acre}}\right)\left(\frac{0.84\text{ lb }P_2O_5}{1\text{ bu soybean grain}}\right)=\frac{38\text{ lb }P_2O_5}{1\text{ acre}}$$

$$\left(\frac{45\text{ bu soybean grain}}{1\text{ acre}}\right)\left(\frac{1.3\text{ lb }K_2O}{1\text{ bu soybean grain}}\right)=\frac{59\text{ lb }K_2O}{1\text{ acre}}$$

Wheat:

$$\left(\frac{50\text{ bu wheat grain}}{1\text{ acre}}\right)\left(\frac{0.60\text{ lb }P_2O_5}{1\text{ bu wheat grain}}\right)=\frac{30\text{ lb }P_2O_5}{1\text{ acre}}$$

$$\left(\frac{50\text{ bu wheat grain}}{1\text{ acre}}\right)\left(\frac{0.34\text{ lb }K_2O}{1\text{ bu wheat grain}}\right)=\frac{17\text{ lb }K_2O}{1\text{ acre}}$$

Alfalfa:

$$\left(\frac{5\text{ ton alfalfa}}{1\text{ acre}}\right)\left(\frac{12\text{ lb }P_2O_5}{\text{ton alfalfa}}\right)=\frac{60\text{ lb }P_2O_5}{1\text{ acre}}$$

$$\left(\frac{5\text{ ton alfalfa}}{1\text{ acre}}\right)\left(\frac{49\text{ lb }K_2O}{\text{ton alfalfa}}\right)=\frac{245\text{ lb }K_2O}{1\text{ acre}}$$

Four years of a corn/soybean/wheat rotation, starting with corn:

year 1 = corn, year 2 = soybean, year 3 = wheat, year 4 = corn. So there are 2 years of corn and one each of soybean and wheat.

Total lb $P_2O_5$/acre removed by the corn/soybean/wheat rotation in 4 years = 2(57 lb $P_2O_5$/acre) + 38 lb $P_2O_5$/acre + 30 lb $P_2O_5$/acre = 182 lb $P_2O_5$/acre.

Four years of a corn/soybean/alfalfa rotation with alfalfa grown for 2 years:

year 1 = corn, year 2 = soybean, year 3 = alfalfa, year 4 = alfalfa. So there are 2 years of alfalfa and one each of corn and soybean.

Total lb $P_2O_5$/acre removed by the corn/soybean/alfalfa rotation in 4 years = 57 lb $P_2O_5$/acre + 38 lb $P_2O_5$/acre + 2(60 lb $P_2O_5$/acre) = 215 lb $P_2O_5$/acre.

Comparing the corn/soybean/alfalfa rotation to the corn/soybean/wheat rotation:

215 lb $P_2O_5$/acre – 182 lb $P_2O_5$/acre = 33 lb $P_2O_5$/acre. So in 4 years, the corn/soybean/alfalfa rotation will remove 33 lb $P_2O_5$/acre more.

Total lb $K_2O$/acre removed by the corn/soybean/wheat rotation in 4 years = 2(41 lb $K_2O$/acre) + 59 lb $K_2O$/acre + 17 lb $K_2O$/acre = 158 lb $K_2O$/acre.

Total lb $K_2O$/acre removed by the corn/soybean/alfalfa rotation in 4 years = 41 lb $K_2O$/acre + 59 lb $K_2O$/acre + 2(245 lb $K_2O$/acre) = 590 lb $K_2O$/acre.

Comparing the corn/soybean/alfalfa rotation to the corn/soybean/wheat rotation:

590 lb $K_2O$/acre – 158 lb $K_2O$/acre = 432 lb $K_2O$/acre. So in 4 years, the corn/soybean/alfalfa rotation removes 432 lb $K_2O$/acre more than the corn/soybean/wheat rotation.

## Measuring Nutrient Removal from Laboratory Results

The estimates in **Box 6-1** are averages and may not accurately reflect the quantity of nutrients removed in a particular field in any given year. To measure the quantity of nutrients removed, a sample must be obtained from the harvested plant material and analyzed in a laboratory for nutrient content. When results are returned from the laboratory, they are either in percent or parts per million (ppm). The major nutrients (N, P, and K) and secondary nutrients (Ca, Mg, and S) are typically reported in percent, while micronutrients (Fe, Zn, Mo, Cu, B, Cl, and Mn) are reported in ppm. Therefore, micronutrient removal by crops is much less than macronutrient removal.

All concentrations reported by a laboratory are on a dry matter (DM) basis, meaning 0% moisture. Consequently, we need to know the amount of DM in a unit of harvested plant material. Dry matter weights for various grain crops at commonly accepted test weights and moisture percentages are in Box 6.2. For instance, corn grain has a test weight of 56 lb/bu but contains only 47.32 lb DM/bu. The lower values for DM reflect the reduction in weight after water has been removed. For forages, a ton of harvested plant material contains less than a ton of DM for the same reason. The fraction of DM in various forage crops at commonly estimated moisture percentages are presented in **Box 6-3**.

**Box 6-2.** U.S. Standard test weights and DM weights for various grains (Hirning et al., 1987).

| Crop | U.S. Standard test weight | U.S. Standard moisture | Dry matter contained in 1 bushel at 0% moisture |
|---|---|---|---|
| | (lb/bu) | (%) | (lb/bu) |
| Barley | 48.00 | 14.50 | 41.04 |
| Corn | 56.00 | 15.50 | 47.32 |
| Flax | 56.00 | 9.00 | 50.96 |
| Oats | 32.00 | 14.00 | 27.52 |
| Rye | 56.00 | 14.00 | 48.16 |
| Sorghum | 55.00 | 14.00 | 47.30 |
| Soybean | 60.00 | 13.00 | 52.20 |
| Sunflower | 100.00 | 10.00 | 90.00 |
| Wheat | 60.00 | 13.50 | 51.90 |

**Box 6-3.** Commonly estimated DM of forage crops (Koelsch et al., 2004). Units in the numerator and denominator are the same in the fractions listed below (e.g. the corn silage DM fraction is interpreted as 0.35 tons DM/ton of silage).

| Crop | Dry matter fraction |
|---|---|
| All hay | 0.85 |
| Alfalfa silage, mid-bloom | 0.40 |
| Barley straw | 0.90 |
| Corn silage | 0.35 |
| Corn stover | 0.85 |
| Oat straw | 0.90 |
| Rye straw | 0.90 |
| Small grain silage, dough stage | 0.35 |
| Sorghum silage | 0.30 |
| Sorghum-sudan silage | 0.30 |
| Sorghum stover | 0.80 |
| Wheat straw | 0.90 |

**Exercise 6-9. How much N is removed by corn yielding 150 bu/acre and containing 1.4% N? From Box 6-2, 1 bu of corn contains 47.32 lb DM/bu.**

Solution:

$$\frac{150\text{ bu corn}}{\text{acre}} \cdot \frac{47.32\text{ lb DM}}{\text{bu corn}} \cdot \frac{1.4\text{ lb N}}{100\text{ lb DM}} = \frac{99\text{ lb N}}{\text{acre}}$$

**Exercise 6-10. How much P and $P_2O_5$ is removed by 15 tons of corn silage containing 0.19% P? From Box 6-3, the fraction of DM in corn silage is 0.35 t DM/t of silage.**

## Nutrient Budgets

A basic concept in nutrient management is a budget that compares how much of a nutrient has been applied to how much has been removed during a crop harvest. This is similar to a bank account that keeps track of income and expenses. In the case of nutrients, applications are the income and crop removal is the expense. The basic equation for a budget is shown in **Box 6-4**.

| **Box 6-4.** Basic form of a nutrient budget. |
|---|
| Nutrient budget = sum of nutrient applications – sum of nutrient removals |

The equation requires that all nutrient applications and removals are accounted for during the time period of interest. The amount of a nutrient removed is totaled and then subtracted from the total amount of nutrient applied.

**Exercise 6-11. Potassium is applied at a rate of 180 lb $K_2O$/acre once every two years Corn removes 47 lb $K_2O$/acre and soybean removes 65 lb $K_2O$/acre. What is the difference between the application and the removal rate? The answer will be the $K_2O$ budget.**

Solution:
$K_2O$ budget = Total $K_2O$ applied – total $K_2O$ removed
$K_2O$ budget = 180 lb $K_2O$/A – 112 lb $K_2O$ = 68 lb $K_2O$/acre.

The answer is positive and indicates that 68 lb $K_2O$/acre more was applied than the crops removed between applications.

**Exercise 6-12. Potassium is applied at a rate of 180 lb $K_2O$/acre once. Alfalfa is cut 3 times during the season and the removal rates of $K_2O$ are 98 lb/A for the first cutting, 147 lb/A for the second cutting, and 110 lb/A for the third cutting.**

Solution:
$K_2O$ budget = 180 lb $K_2O$ applied – 355 total $K_2O$ removed = -175 lb $K_2O$/A

The answer is negative and indicates that more $K_2O$ was removed than applied.

**Exercise 6-13. A wheat crop was fertilized with 26 lb $P_2O_5$/acre. Harvest of the crop removed 30 lb $P_2O_5$/acre. What was the nutrient budget for the season?**

Solution:
$P_2O_5$ budget = 26 lb $P_2O_5$/acre – 30 lb $P_2O_5$/acre = -4 lb $P_2O_5$/acre

**Exercise 6-14. A soybean crop was fertilized with 80 lb $P_2O_5$/acre. Harvest removed 42 lb $P_2O_5$/acre. What was the $P_2O_5$ budget for the season?**

**Exercise 6-15. A farmer broadcast 52 lb $P_2O_5$/acre in the fall before growing corn in the first season. When corn was planted, the farmer applied an additional 21 lb $P_2O_5$/acre with the corn seed at planting. The corn crop removed 68 lb $P_2O_5$/acre. No additional fertilizer was applied before the next soybean crop, which removed 46 lb $P_2O_5$/acre. What was the budget for the two seasons?**

**Exercise 6-16. A farmer applied 120 lb $P_2O_5$/acre as swine manure in the spring before growing corn. Corn removed 57 lb $P_2O_5$/acre. The next season, wheat was grown and the farmer applied 15 lb $P_2O_5$/acre with the seed at planting. Wheat removed 21 lb $P_2O_5$/acre. Last, soybean was grown with no additional fertilizer applications and removed 29 lb $P_2O_5$/acre. What was the budget for the three seasons?**

Solution:

Total additions:
120 lb $P_2O_5$/acre + 15 lb $P_2O_5$/acre = 135 lb $P_2O_5$/acre

Total removals:
57 lb $P_2O_5$/acre + 21 lb $P_2O_5$/acre + 29 lb $P_2O_5$/acre = 107 lb $P_2O_5$/acre

$P_2O_5$ budget = 135 lb $P_2O_5$/acre – 107 lb $P_2O_5$/acre = 28 lb $P_2O_5$/acre

## References


Hirning, H., K. Hellevang, and J. Helm. 1987. Equivalent weights of grain and oilseeds. AE-945. North Dakota Coop. Ext., North Dakota State University, Fargo. (Available online with updates at http://www.ext.nodak.edu/extpubs/ageng/machine/ae945w.htm.) (accessed 18 Aug. 2004; verified 18 Aug. 2004).

Koelsch, R., C. Shapiro, and R. DeLoughery. 2004. Nebraska's comprehensive nutrient management plan (CNMP) manure application workbook. p. 76. Circ. EC02-720. Nebraska State Coop. Ext., Univ. of Nebraska, Lincoln. Available online at http://cnmp.unl.edu/ManureApplicationWorkbook04.pdf. (accessed 25 Jul. 2007; verified 25 Jul. 2007).

Murrell, T.S. 2005. Average nutrient removal rates for crops in the Northcentral Region [Online]. Available at http://www.ipni.net/northcentral/nutrientremoval (accessed 4 Sep. 2007; verified 4 Sep. 2007).

# SOIL PHYSICAL PROPERTIES AND SOIL WATER

**Key problems:** Calculating bulk density and amount of available water.

**Mathematical skills:** Algebra.

**Chapter concepts:** The chapter demonstrates how to convert gravimetric values to volumetric values.

## Bulk Density, Particle Density, and Porosity

Soil bulk density, an important physical characteristic in soil is generally given on a dry weight basis. Measuring soil bulk density is accomplished by taking a soil sample that occupies a known volume. This is often accomplished using a core sampler that is constructed from a tube of known dimensions. Once the sample is brought into the laboratory, it is generally oven dried at 105 °C.

The total volume of the sample is made up of the volume occupied by the soil solids ($V_s$), the soil water ($V_w$), and the soil air ($V_a$). So the equation for soil bulk density can be written as: (Logsdon et al., 2008)

$$\rho_b = \frac{M_s}{V_s + V_w + V_a}$$

The percent pore space f (% volume not filled with solids) is calculated with the equation:

$$f = 100\left(1 - \frac{\rho_b}{\rho_s}\right)$$

where $\rho_s$ is particle density (mass solids/volume solids)

In most highly productive agricultural soils, porosity ranges from 45 to 55%. A porosity value of 45% means that 45% of the soil volume is filled with either water or air, with the remainder filled with solids. Dry bulk density and porosity are important because they provide an indication of soil compaction and are required to convert ppm to lb/acre. In coarse and fine textured soils, roots will have a difficult time penetrating if $\rho_b$ values are greater than 1.6 and 1.4 g/cm$^3$, respectively.

**Exercise 7-1. A core sampler with an inner diameter of 5.7 cm is used to extract a soil sample 6 cm long. When this sample is taken back to the laboratory, it is dried at 105 °C and weighed. The dry weight is 230 g. Find the dry bulk density ($\rho_b$) of this sample.**

Solution:
First find the total volume ($V_t$) of the sample. This is done by using the formula for the volume of a cylinder ($V = \pi r^2 h$). The radius (r) is half of the diameter, or r = $\frac{1}{2}$(5.7 cm) = 2.85 cm. The total volume is therefore (3.14) (2.85 cm)$^2$ (6 cm) = 153 cm$^3$ = $V_{total}$. Bulk density is therefore 230 g/153 cm$^3$ = 1.50 g/cm$^3$.

**Exercise 7-2. Calculate dry bulk density of a soil sample taken with a core sampler with an inner diameter of 5.7 cm and a length of 3 cm. The dry weight of this sample is 107 g.**

The weight of the sample occupied by water ($M_w$) can be determined by drying the sample for at least 24 hr at 105 °C. This heat treatment removes almost all of the soil water. The weight of the sample after heating is the weight of the soil solids ($M_s$). So the weight of the soil water ($M_w$) is found by subtraction:

$M_{total} = M_s + M_w$, so $M_{total} - M_s = M_w$.

**Exercise 7-3. If a soil from a 32 cm³ volume has an oven dry weight of 43 g, what is its dry bulk density ($\rho_b$)?**

Solution:

$$\left[\frac{43\,g}{32\,cm^3}\right] = \frac{1.34\,g}{cm^3} = \rho_b$$

**Exercise 7-4. Calculate the weight of water ($M_w$) in a field-moist sample that had a total weight ($M_{total}$) of 230 g and a weight after oven-drying ($M_s$) of 192 g. The temperature of the field-moist sample is 20 °C**

Solution:
To find $M_w$, we use the difference between $M_{total}$ and $M_s$:

$M_w = 230\,g - 192\,g = 38\,g$

The volume of pore space contained in a sample ($V_w + V_a$) may be found by saturating the sample with water, to get rid of the air, weighing it ($M_{wsat}$), oven drying it again for atleast 24 hr at 105 °C, then reweighing it to get $M_s$ (Foth, 1984). The difference between these two weights is the weight of water that occupied all the pore space in the sample ($M_{wsat}$).

**Exercise 7-5. Find the volume of pore space in a sample that when saturated with water at 20 °C, weighed 250 g ($M_{wsat}$) and when dried, weighed 130 g ($M_s$).**

Solution:
The difference between $M_{wsat}$ and $M_s$ is 250 g - 130 g = 120 g. Dividing this result by the bulk density of water at 20 °C yields:

$$120\,g\left(\frac{1\,cm^3}{0.99821\,g}\right) = 120\,cm^3.$$

This calculation uses a density of water at 20 ºC of 0.99821 g/cm³.

Once the pore space volume is measured, the volume occupied by the soil solids ($V_s$) can be calculated. This is done by subtracting the pore volume from the total volume of the soil sample.

$V_{total} = V_s + V_w + V_a$, and since pore volume = $V_w + V_a$,
$V_{total} - (V_w + V_a) = V_s$.

Particle density ($\rho_s$) is the weight of the soil solids ($M_s$) divided by the volume of the soil solids ($V_s$):

$$\rho_s = \frac{M_s}{V_s}$$

Particle density generally ranges from 2.5 to 2.7 g/cm³.

Finally, porosity (f) is defined as the ratio, expressed as a percentage, of the pore volume ($V_w + V_a$) to the total volume of the sample:

$$f = \frac{V_w + V_a}{V_{total}} \times 100\%$$

Porosity can be expressed as a function of dry bulk density ($\rho_b$) and particle density. To find this function, we first recognize that pore volume is that portion of the total volume ($V_{total}$) not occupied by soil solids, or $V_{total} - V_s = V_w + V_a$. This allows us to express f as:

$$f = \frac{V_{total} - V_s}{V_{total}} \times 100\%$$

Next, we break apart the numerator into two separate fractions:

$$f = \left(\frac{V_{total}}{V_{total}} - \frac{V_s}{V_{total}}\right) \times 100\% \text{ or } \left(1 - \frac{V_s}{V_{total}}\right) \times 100\%$$

Next, we multiply $\frac{V_s}{V_{total}}$ by an expression equivalent to 1: $\frac{M_s}{M_s}$

This gives

$$f = \left[1 - \left(\frac{V_s}{V_{total}}\right)\left(\frac{M_s}{M_s}\right)\right] \times 100\%$$

Using the commutative property of multiplication, we get

$$f = \left[1 - \left(\frac{M_s}{V_{total}}\right)\left(\frac{V_s}{M_s}\right)\right] \times 100\%$$

Next, we recognize that

$$\frac{M_s}{V_{total}} = \rho_b \text{ and } \frac{V_s}{M_s} = \frac{1}{\rho_s}$$

Therefore,

$$f = \left(1 - \frac{\rho_b}{\rho_s}\right) \times 100\%$$

**Exercise 7-6. A soil has a dry bulk density ($\rho_b$) of 1.25 g/cm$^3$ and an oven-dry weight ($M_s$) of 10 g. What is the total soil volume ($V_{total}$)?**

Solution:

$$\rho_b = \frac{M_s}{V_{total}}$$

Multiplying both sides by $V_{total}$ gives

$$V_{total}\,\rho_b = M_s$$

Dividing both sides by $\rho_b$ gives the expression for $V_{total}$:

$$V_{total} = \frac{M_s}{\rho_b}$$

So $V_{total} = 10\text{ g}\left(\frac{\text{cm}^3}{1.25\text{ g}}\right) = 8\text{ cm}^3$

**Exercise 7-7. If a soil has a dry bulk density ($\rho_b$) of 1.25 g/cm$^3$ and a particle density ($\rho_s$) of 2.65 g/cm$^3$, what is its porosity (f)?**

Solution:

$$f = \left[1 - \frac{\rho_b}{\rho_s}\right]100\%$$

$$= \left[1 - \left(\frac{1.25\text{ g}}{\text{cm}^3}\right)\left(\frac{\text{cm}^3}{2.65\text{ g}}\right)\right]100\%$$

$$= [1 - 0.472]100\%$$

$$= 52.8\%$$

**Exercise 7-8. Total volume ($V_t$) is the sum of the volume of pores ($V_p$) + volume of the solids ($V_s$) or $V_t = V_p + V_s$. Determine $V_s$ given the following information:**

a. Total soil volume = 140 cm$^3$
b. Pore volume = 60 cm$^3$
c. Oven-dried soil weight = 130 g

**Exercise 7-9. Calculate the particle density, bulk density, and % pore space using the information above.**

**Exercise 7-10. A can that measures 5.25 cm high with a diameter of 7.65 cm is used to collect a soil sample for bulk density. The can weight is 60.75 grams. The mass of the oven dry soil and can is 335.05 g. What is the dry bulk density of the soil?**

**Exercise 7-11. What is the particle density ($\rho_s$) if porosity (f) is 45% and dry bulk density ($\rho_b$) is 1.32 g/cm$^3$?**

## Determining Weight of Soil in an Acre

Often in soil science, the weight of an acre or hectare of soil is needed to perform calculations, such as converting parts per million to pounds per acre or kilograms per hectare. To calculate the weight of the soil contained in an acre, the depth and bulk density are needed.

An often used estimate is that an acre of soil weighs 2,000,000 (2 million) lb. This estimate is based on a $\rho_b$ = 1.33 g/cm$^3$ and that the sampling depth is 6 $^2/_3$ in.

To see how this calculation is done, we note that to determine the weight of a known volume of soil ($M_s$), we multiply the volume, $V_{total}$, by the dry bulk density ($\rho_b$):

To calculate weight, the $M_s = V_{total}\rho_b$ is used.

The key to solving this equation is ensuring the volume in $V_{total}$ and in $\rho_b$ have matching units (cm$^3$).

We first need to determine the volume ($V_{total} = L \times w \times h$) of soil contained in 1 acre that is 6.67 in. deep.

$$L \times w = 1\text{ acre}\left(\frac{43,560.174\text{ ft}^2}{1\text{ acre}}\right)\left(\frac{144\text{ in.}^2}{1\text{ ft}^2}\right) = 6,272,665\text{ in.}^2$$

$h = 6.6667$ in., so

$$V_{total} = L \times w \times h = 6272665\text{ in.}^2 \times 6.6667\text{ in.} = 41817975.76\text{ in.}^3$$

Because $\rho_b$ uses cm$^3$, we next convert in.$^3$ to cm$^3$

$$V_{total} = 41817975.76\text{ in.}^3\left(\frac{16.387\text{ cm}^3}{1\text{ in.}^3}\right) = 685271168.8\text{ cm}^3$$

Now that both $\rho_b$ and $V_{total}$ are in units of cm$^3$, we can perform the calculation:

$$M_s = V_{total}\rho_b = 685271168.8\text{ cm}^3\left(\frac{1.33\text{ g}}{\text{cm}^3}\right) = 911410654.5\text{ g}$$

We now convert the units of $M_s$ from g to lb:

$$911410654.5\text{ g}\left(\frac{1\text{ lb}}{454\text{ g}}\right) = 2,007,512\text{ lb}$$

Since $\rho_b$ had only 3 significant digits (1.33 g/cm$^3$), we round the weight to 2,000,000 lb.

From this calculation, we see that the adage that an acre of soil weighs 2,000,000 lb is based on a depth of 6 $^2/_3$ in. and a $\rho_b$ of 1.33 g/cm$^3$. Altering either $\rho_b$ or the depth of the soil considered will change this estimate. An alternative approach is to calculate the weight in kg ha$^{-1}$ and then convert to lb/acre.

**Exercise 7-12. What is the weight (pounds) of one acre of soil to a depth of 6 in. if its $\rho_b$ = 1.25 g/cm$^3$?**

**Exercise 7-13. What is the weight (pounds) of one acre of soil to a depth of 8 in. if its $\rho_b$ = 1.25 g/cm$^3$?**

## Determining the Amount of Water Contained in Soil

In irrigated systems, water can be applied to reduce yield losses due to various stresses. The amount of water that should be applied is dependent on the amount of water contained in the soil and the potential plant demand. The checkbook approach is used to estimate water storage. This approach requires a rain gauge and estimated evaporative losses.

The soil serves as a water storage reservoir for the plant. The amount of water retained in the soil following rainfall or irrigation is dependent on the soil texture and organic matter content. Not all water following a rainfall is available to the plant. Soil has properties similar to a sponge (**Box 7-2**). When the sponge is placed in a bucket full of water all its pores are filled with water. When the sponge is removed from the bucket, water drips out of the sponge. Soil has these same characteristics. Following a rainfall that fills all soil pores with water, water is lost from the soil through drainage until it reaches a point where drainage stops. The soil water content at this point is called field capacity.

**Box 7-2.** Range of plant water availability in soils.

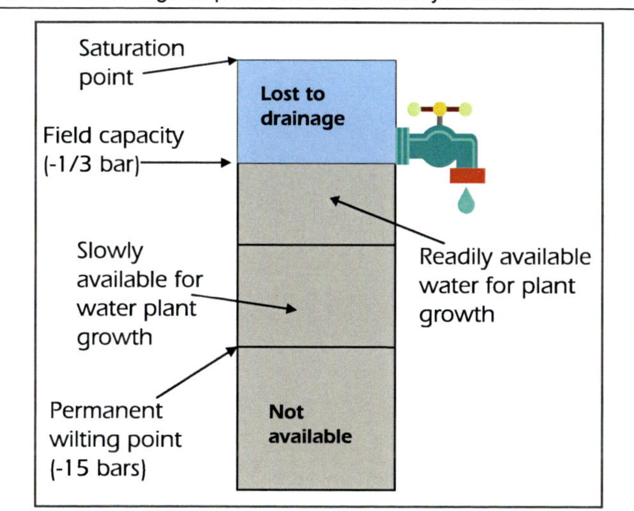


Water content can continue to decrease through plant uptake until the permanent wilting point is reached (**Box 7-2**). The water contained in the soil between field capacity and the permanent wilting point is called available water. The amount of available water is influenced by soil texture (**Box 7-3**). As a soil dries, water becomes increasingly difficult to extract. The energy needed to extract water is expressed as the water potential. The water potentials at field capacity and the permanent wilting point are approximately -1/3 and -15 bars. Soil moisture can be estimated by hand feel or measured using soil samples or sensors.

**Box 7-3.** The relationship between plant available water and soil texture.

| Soil texture | Plant available water (in./ft soil) |
|---|---|
| Fine sand | 0.7-1.0 |
| Loamy sand | 0.9-1.5 |
| Sandy loam | 1.3-1.8 |
| Loam | 1.8-2.5 |
| Silt loam | 1.8-2.6 |
| Clay loam | 1.8-2.5 |

## Gravimetric Water, Volumetric Water, and Amount of Available Water

The amount of water contained in the soil can be reported on a weight, or gravimetric basis ($\theta_g$), or on a volumetric basis ($\theta_v$). Gravimetric water content is determined by weighing a moist soil sample and then drying it in an oven at 105 °C for 24 to 48 hr until a constant dry weight is obtained. The difference between the wet and dry weights is the weight of water. When this weight is divided by the dry weight of soil, $\theta_g$ is expressed as a ratio (Jury et al., 1991):

$$\theta_g = \frac{\text{wet soil weight - dry soil weight}}{\text{dry soil weight}} = \frac{M_w}{M_s}$$

To express this ratio as a percentage, it can be multiplied by 100%:

$$\theta_g\,(\%) = \frac{M_w}{M_s} \bullet 100\%$$

**Exercise 7-14. What is $\theta_g$ if the dry soil weight is 10 g and the wet soil weight is 13 g?**

Solution:

$$\theta_g = \frac{13\text{ g} - 10\text{ g}}{10\text{ g}} \bullet 100\% = 30\%$$

**Exercise 7-15. What is $\theta_g$ if the dry soil weight is 20 g and the wet soil weight is 25 g?**

Volumetric water content ($\theta_v$) is the volume of liquid water per volume of soil:

$$\theta_v = \frac{V_w}{V_{total}}$$

If the dry bulk density ($\rho_b$) and $\theta_g$ have been determined, then $\theta_v$ can be calculated by using the equation $\theta_v = \rho_b \bullet \theta_g$. This equation is derived below:

We start with the expression for $\theta_v$:

$$\theta_v = \frac{V_w}{V_{total}}$$

We now define $P_w$ as the density of water, which is the mass of water per unit volume, or:

$$\rho_w = \frac{M_w}{V_w}$$

By multiplying both sides of this equation by $V_w$ then dividing both sides by $P_w$, we get the following equation:

$$V_w = \frac{M_w}{\rho_w}$$

$V_w$ is substituted into the equation for $\theta_v$:

$$\theta_v = \frac{\frac{M_w}{\rho_w}}{V_{total}}$$

Next, we multiply both sides of the equation for $\rho_b$ by $V_{total}$,

$$\rho_b(V_{total}) = \frac{M_s}{V_{total}} \cdot V_{total} = M_s$$

and divide both sides by $\rho_b$:

$$\frac{\rho_b(V_{total})}{\rho_b} = \frac{M_s}{\rho_b}, \text{ or } V_{total} = \frac{M_s}{\rho_b}$$

and substitute this expression for $V_{total}$ into the equation for $\theta_v$:

$$\theta_v = \frac{\frac{M_w}{\rho_w}}{V_{total}} = \frac{\frac{M_w}{\rho_w}}{\frac{M_s}{\rho_b}} = \left(\frac{M_w}{\rho_w}\right)\left(\frac{\rho_b}{M_s}\right)$$

We use the commutative property of multiplication to re-arrange this equation. The resulting equation is:

$$\theta_v = \left(\frac{\rho_b}{\rho_w}\right)\left(\frac{M_w}{M_s}\right)$$

Now note that $\left(\frac{M_w}{M_s}\right) = \theta_g$, so that the equation for $\theta_v$ becomes,

$$\theta_v = \frac{\rho_b \theta_g}{\rho_w}$$

To convert to a decimal form, multiply both sides by 100%. If the density of water is 1 g/cm$^3$, then the equation simplifies to $\theta_v = \rho_b \cdot \theta_g$.

**Exercise 7-16. If $\rho_b$ is 1.25 g/cm$^3$ and $\rho_w$ = 0.9982071 g/cm$^3$, what is the $\theta_v$ if $\theta_g$ is 30%?**

Solution

1. Convert gravimetric water $\theta_g$(%) to decimal form:
$$\frac{30\%}{100\%} = 0.30$$
2. Calculate $\theta_v$:
$$\theta_v = \frac{\rho_b \theta_g}{\rho_w} = \left(\frac{1.25\,g}{cm^3}\right)(0.30)\left(\frac{cm^3}{0.9982071\,g}\right) = 0.376$$

or expressed as %: 37.6%

If $\rho_w = 1$, then $\theta_v = \frac{1.25\,g}{cm^3} \cdot (0.30) = 0.375$ or 37.5%

**Exercise 7-17. What is the $\theta_v$ if $\theta_g$(%) is 25%, and $\rho_b$ is 1.25 g/cm$^3$ and $\rho_w$ = 0.9982071 g/cm$^3$?**

The amount of available water is often expressed in units of inches of water per foot of soil. Using a length measure for available water is done by convention and is often part of a water budget used to schedule irrigation. For instance, if a soil is known to be capable of providing 5 in. of water in the upper 3 ft. of soil, but the current moisture level is only 2 in., then up to 3 in. more water can be added through irrigation. Water budgeting can get quite complex when several factors like crop water usage rate, evaporation, and other factors are considered.

**Exercise 7-18. How many inches of water are in the surface 6 in. if $\theta_v$ is 37.5% moisture?**

Solution:

(6 in. soil)(0.375) = 2.25 in.

Determining how much available water is present in a soil sample depends on knowing how much of the water is not available to the plant. The plant available water is the difference between field capacity and permanent wilting point.

The amount of plant available water present in a soil sample is therefore found by subtracting the amount of water at the permanent wilting point from the amount of water present in the soil.

**Exercise 7-19. How much available water is contained in a soil sample with $\theta_v = 0.305$ if the permanent wilting point is $\theta_{vwp} = 0.153$?**

Solution:
Water content of the sample = (6 in.)(0.305) = 1.83 in. water permanent wilting point = (6 in.)(0.153) = 0.918 in. water. So available water = 1.83 in. – 0.918 in. = 0.912 in.

**Exercise 7-20. How much available water is in the surface 6 in. of soil that has a $\theta_g = 0.30$, a permanent wilting point with $\theta_{gwp} = 0.10$, a $\rho_b = 1.25$ g/cm$^3$, and $\rho_w = 1$ g/cm$^3$?**

## Additional Information

Available water in a soil layer is the difference between the amount of water at field capacity (-1/3 bar) and the permanent wilting point (-15 bar). Once $\theta_v$ is known at both moisture levels, available water can be found by:

available water =

$$\text{soil depth} \cdot \left[\left(\theta_v \text{ at } -\frac{1}{3} \text{ bar}\right) - \left(\theta_v \text{ at } -15 \text{ bar}\right)\right]$$

Using the associative property of multiplication, available water =

$$(\text{soil depth})\left(\theta_v \text{ at } -\frac{1}{3} \text{ bar}\right) - (\text{soil depth})(\theta_v \text{ at } -15 \text{ bar})$$

This equation can be rewritten into the equation,

available water =

$$(\text{soil depth})\left[\frac{\rho_b \theta_g(-\frac{1}{3}\text{bar})}{\rho_w} - \frac{\rho_b \theta_g(-15 \text{ bar})}{\rho_w}\right]$$

if $\rho_w = 1$, then available water =

$$(\text{soil depth})\left[\rho_b \theta_g\left(-\tfrac{1}{3}\text{bar}\right) - \rho_b\theta_g(-15 \text{ bar})\right] \text{ or}$$

$$(\text{soil depth})\rho_b\left[\theta_z\left(-\tfrac{1}{3}\text{bar}\right) - \theta_g(-15 \text{ bar})\right]$$

**Exercise 7-21. Calculate the total amount of available water (in inches) across all depths for the following data.**

| Depth, inches | inches in segment | $\theta_g$(%) at -1/3 bar | $\theta_g$(%) at -15 bar | $\rho_b$ g/cm$^3$ |
|---|---|---|---|---|
| 0-12 | 12 | 25 | 8 | 1.1 |
| 12-36 | 24 | 25 | 10 | 1.3 |

Note: $\rho_w$ = 0.9982071 g/cm$^3$

Solution:
To use the equation we derived above, $\theta_g$ rather than $\theta_g$ (%) is used. Next, for each depth, we proceed with solving the equation:

$$\text{available water} = \frac{(\text{soil depth})(\rho_b)\left(\theta_g(-\frac{1}{3}\text{bar}) - \theta_g(-15 \text{ bar})\right)}{\rho_w}$$

For the 0 to 12 in. increment:

available water =

$$12 \text{ in.}\left(\frac{1.1 \text{ g}}{\text{cm}^3}\right)(0.25 - 0.08)\left(\frac{\text{cm}^3}{0.9982071 \text{ g}}\right)$$

= 2.25 in.

For the 12 to 36 in. increment:

available water =

$$24 \text{ in.}\left(\frac{1.3 \text{ g}}{\text{cm}^3}\right)(0.25 - 0.10)\left(\frac{\text{cm}^3}{0.9982071 \text{ g}}\right)$$

= 4.69 in.

So the total available water is 2.25 in. + 4.69 in. = 6.94 in.

**Exercise 7-22. Calculate the total amount of plant available water (in meters) for the following data across all soil depths.**

| Depth (m) | $\theta_v$(%) at -1/3 bar | $\theta_v$(%) at -15 bar |
|---|---|---|
| 0-1.0 | 25 | 10 |
| 1.0-1.5 | 30 | 12 |
| 1.5-2.0 | 33 | 15 |

**Exercise 7-23. How much water (inches) needs to be added to increase the soil water content of the surface 2 ft from $\theta_v = 0.20$ to field capacity? The water content at - $^1/_3$ bar (field capacity) is $\theta_v = 0.33$.**

For planning and management purposes, it is important to understand how to convert gallons to acre inches. One acre inch is the water contained within a volume having the dimension of 1 acre (43,560 $ft^2$) by one inch. This can be easily converted to gallons or visa versa.

**Exercise 7-24. Convert 1 acre inch to gal. For this calculation it is important to know that 7.481 gal are contained in 1 $ft^3$.**

Solution:

$$\left(1.0 \text{ acre in. water} \cdot \frac{1 \text{ ft}}{12 \text{ in.}} \cdot \frac{43{,}560 \text{ ft}^2}{1 \text{ acre}} \cdot \frac{7.481 \text{ gal}}{1 \text{ ft}^3}\right) = 27{,}156 \text{ gal}$$

**Exercise 7-25. Convert 17,000 gal to acre inch.**

## Time Needed to Irrigate Fields and Gardens

Once the number of gallons or inches of water needed are known, irrigation pumping rates can be calculated.

The length of time to apply a prescribed amount of water (usually given in acre in.) can be calculated by dividing the desired amount of water to apply by the pumping rate:

$$\text{Time needed} = \frac{\text{volume of water needed}}{\text{pumping rate}}$$

**Exercise 7-26. The irrigation system pumps at a rate of 840 gal/minute. How long will it take to irrigate 132 acres with 1.25 acre inch per acre?**

Solution:
First convert acre in. to gal:

$$\left(\frac{1.25 \text{ acre in.}}{\text{acre}}\right)\left(\frac{1 \text{ ft}}{12 \text{ in.}}\right)\left(\frac{43{,}560 \text{ ft}^2}{1 \text{ acre}}\right)\left(\frac{7.481 \text{ gal}}{1 \text{ ft}^3}\right) = \frac{33{,}945 \text{ gal}}{\text{acre}}$$

This is the quantity of water needed to provide 1.25 acre in. of water. There are 132 acres that require this amount of water, so the total amount of water needed is:

$$132 \text{ acre}\left(\frac{33{,}945 \text{ gal}}{\text{acre}}\right) = 4{,}480{,}740 \text{ gal}$$

Next, this total volume is divided by the pumping rate:

$$4{,}480{,}740 \text{ gal}\left(\frac{\text{min}}{840 \text{ gal}}\right) = 5{,}334 \text{ min}$$

This time can be expressed in hours as:

$$5{,}334 \text{ min}\left(\frac{1 \text{ hr}}{60 \text{ min}}\right) = 89 \text{ hr}$$

**Exercise 7-27. How long will it take to apply 2.0 acre in. of water to 40.5 acres? The irrigation system pumps at a rate of 840 gal/min.**

Solution:
First, we find that 2 acre in./acre is:

$$\frac{2 \text{ acre in.}}{\text{acre}}\left(\frac{43{,}560 \text{ ft}^2}{1 \text{ acre}}\right)\left(\frac{1 \text{ ft}}{12 \text{ in.}}\right)\left(\frac{7.481 \text{ gal}}{1 \text{ ft}^3}\right) = \frac{54{,}312 \text{ gal}}{\text{acre}}$$

Next, we multiply by the number of acres and divide by the pumping rate:

$$40.5 \text{ acre}\left(\frac{54{,}312 \text{ gal}}{\text{acre}}\right)\left(\frac{1 \text{ min}}{840 \text{ gal}}\right)\left(\frac{1 \text{ hr}}{60 \text{ min}}\right) = 43.6 \text{ hr}$$

**Exercise 7-28. How long will it take to apply 1 in. of water with a garden hose to a 40 ft by 60 ft garden? The hose will fill a 5 gal bucket in 1 min 42 sec.**

Solution:
First, the flow rate is determined. We convert 1 min 42 sec to decimal minutes.

$$1 \text{ min} + 42 \text{ sec}\left(\frac{\text{min}}{60 \text{ sec}}\right) = 1 \text{ min} + 0.7 \text{ min} = 1.7 \text{ min}$$

So the flow rate is:

$$\frac{5 \text{ gal}}{1.7 \text{ min}} = \frac{2.9 \text{ gal}}{\text{min}}$$

Next we find the total volume of water required. This is found by multiplying the area (60 ft x 40 ft) by the depth of water desired (1 in.), making sure units match, and converting to gal:

Volume of water =

$$(40\text{ ft})(60\text{ ft})(1\text{ in.})\left(\frac{1\text{ ft}}{12\text{ in.}}\right)\left(\frac{7.481\text{ gal}}{\text{ft}^3}\right) = 1{,}496\text{ gal}$$

Dividing the total volume by the pumping rate gives:

Time needed =

$$1{,}496\text{ gal}\left(\frac{1\text{ min}}{2.9\text{ gal}}\right)\left(\frac{1\text{ hr}}{60\text{ min}}\right) = 8.6\text{ hr}$$

We can convert 8.6 hr to hr and min as follows:

$$8\text{ hr} + 0.6\text{ hr}\left(\frac{60\text{ min}}{1\text{ hr}}\right) = 8\text{ hr} + 36\text{ min}$$

**Exercise 7-29. How long will it take to apply 1 in. of water to a 20 ft by 30 ft garden with a garden hose that delivers 5 gal /min?**

How efficiently a crop utilizes water can be defined a number of ways. One method is to compare crop yield with how much water was available to the crop during the season. This ratio is called the water use efficiency. Water availability can be estimated by adding the rainfall amount to the difference between the amount of water in the soil at the beginning and end of the season.

**Exercise 7-30. What is the water use efficiency of a 200 bu/acre corn crop that during the season received 20 in. of rainfall? In addition to this precipitation, $\theta_v$ of the upper 3 ft of soil at the beginning of the season was 0.31. At harvest, $\theta_v$ of this soil layer was 0.15. Report results as both bu grain/in. water and kg grain/cm water.**

Solution:
First, the amount of water provided by the upper 3 ft of soil during the season is estimated from the change in $\theta_v$:

$$3\text{ ft}(0.31 - 0.15)\left(\frac{12\text{ in.}}{1\text{ ft}}\right) = 5.76\text{ in.}$$

This amount is then added to the amount of rainfall to provide the total amount of available water:

$$5.76\text{ in.} + 20\text{ in.} = 25.76\text{ in.}$$

Since the yield was 200 bu/acre, the water use efficiency is

$$\left(\frac{200\text{ bu}}{\text{acre}}\right)\left(\frac{1}{25.76\text{ in.}}\right) = \frac{7.76\text{ bu}}{\text{acre in water}}$$

We can convert this efficiency parameter to metric units using the bushel weight of corn grain (56 lb/bu), the conversion of lb to kg, the conversion of in. to cm, and the conversion of acre to ha:

$$\left(\frac{7.76\text{ bu}}{\text{acre in.}}\right)\left(\frac{56\text{ lb}}{1\text{ bu}}\right)\left(\frac{0.454\text{ kg}}{1\text{ lb}}\right)\left(\frac{1\text{ in.}}{2.54\text{ cm}}\right)\left(\frac{1\text{ acre}}{0.4047\text{ ha}}\right) = \frac{192\text{ kg grain}}{\text{ha cm water}}$$

This can also be converted to gal water/bu

$$\frac{1\text{ acre in.}}{7.76\text{ bu}} \bullet \frac{27{,}156\text{ gal}}{\text{acre in.}} = \frac{3{,}500\text{ gal}}{\text{bu}}$$

## Water Flow Measurements

Water moves from areas of higher potential energy (wet) to areas of lower potential energy (dry). Such potential energy gradients determine the direction water will move in a soil, which can be up, down, or any other direction. A calculation that is often performed is how quickly water will move downward when ponded at the surface. Like all such calculations, a special set of conditions apply. In this case, we consider a soil that is rigid and already saturated with water. We are interested in the rate of water movement downward through this saturated soil when a known depth of free water is ponded above the soil surface, as illustrated in **Box 7-4.** We consider a cylinder of soil with a known length and radius. Under this set of conditions, the volume of water that will move downward over time is described by Darcy's Law:

$$Q = \frac{K_s A H}{z}$$

Where Q = volume of water flowing per unit time ($cm^3$/hr), $K_s$ = saturated hydraulic conductivity (a constant determined experimentally and is in units of cm/s or cm/hr), A = cross sectional area of the soil cylinder ($A = \pi r^2$ and has $cm^2$ units), z = length (cm) of the saturated soil column, and H = the sum of the lengths (cm) of the saturated soil (Z) and the free water above the soil surface (p) (**Box 7-4**).

**Box 7-4.** Range of plant water availability.

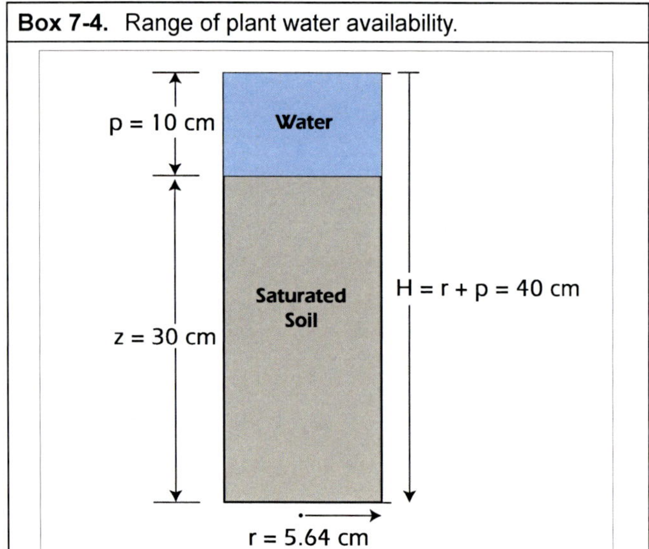


**Exercise 7-31. In Box 7-4, what is the rate of saturated flow (Q) for the soil cylinder with radius r = 5.64 cm, length z = 30 cm, a depth of free water p = 10 cm, and $K_s$ = 20 cm/hr?**

Solution:
First, find the cross sectional area, A:

$$A = \pi r^2 = 3.14159(5.64\text{ cm})^2 = 100\text{ cm}^2$$

Next find H:

$$H = p + z = 10\text{ cm} + 30\text{ cm} = 40\text{ cm}$$

Finally, use Darcy's Law to find Q:

$$Q = \left(\frac{20\text{ cm}}{\text{hr}}\right)(100\text{ cm}^2)(40\text{ cm})\left(\frac{1}{30\text{ cm}}\right) = \left(\frac{2{,}667\text{ cm}^3}{\text{hr}}\right)$$

Another way of expressing water movement is volume of water (Q) per unit of cross sectional area (A). This expression is termed flux ($J_w$) and is expressed mathematically as:

$$J_w = \frac{Q}{A}$$

In this expression, $J_w$ has units of cm/hr, as shown by the cancellation of units in the expression for $J_w$:

$$\frac{\text{cm}}{\text{hr}} = \left(\frac{\text{cm}^{\not 3}}{\text{hr}}\right)\left(\frac{1}{\not{\text{cm}}^{\not 2}}\right)$$

**Exercise 7-32. Find the saturated water flux ($J_w$) for the situation described in Box 7-4.**

Solution:
From Exercise 7-30 we know that A = 100 cm² and Q = 2667 cm³/hr.
Therefore:

$$J_w = \frac{Q}{A} = \left(\frac{2{,}667\text{ cm}^3}{\text{hr}}\right)\left(\frac{1}{100\text{ cm}^2}\right) = \frac{26.67\text{ cm}}{\text{hr}}$$

Box 7-5 provides saturated conductivity values for three soil conditions.

**Box 7-5.** Saturated conductivity values for three soil conditions.

| Soil | $K_s$, cm/sec | $K_s$, cm/hr |
|---|---|---|
| Sand | $5 \times 10^{-3}$ | 18 |
| Well drained | $5 \times 10^{-4}$ | 1.8 |
| Impermeable | $5 \times 10^{-8}$ | $3.6 \times 10^{-5}$ |

**Exercise 7-33. Using Darcy's law, determine Q, if p = 10 cm, A = 200 cm², $K_s$ = 10 cm/hr, and z = 20 cm.**

Soil pores act like capillaries that direct the movement of water. A simple model of this movement considers a soil pore to behave like a cylindrical glass capillary. In this model, the glass capillary is typically considered to be in a vertical position. Of interest in this model is how far water will rise up capillary tubes that have different radii. Capillary rise is caused by a difference in pressure between the soil air and the soil water in the capillary tube.

When water enters a capillary tube, it spreads over its inside walls, which creates a curved surface over the water. Water is higher along the walls of the tube than it is in the middle. This curved water surface creates a low pressure area above the water. Water moves up in the tube in response to this pressure difference. This movement continues until the upward force of the pressure gradient equals the downward force exerted by gravity.

When water completely wets the inside of the capillary tube, capillary rise may be expressed as:

$$H = \frac{2\sigma}{\rho_w g r}$$

Where H = height of rise (cm); $\sigma$ = surface tension which at 20 °C equals 72.7 g/s$^2$; $\rho_w$ is the density of water = 0.9982071 g/cm$^3$ at 20 °C; g is the gravitational force = 980 cm/s$^2$; and r is the inner radius of the capillary tube (cm). If we keep the temperature constant at 20 °C, the equation for H simplifies to:

$$H = 2\left(\frac{72.7\,g}{s^2}\right)\left(\frac{cm^3}{0.9982071\,g}\right)\left(\frac{s^2}{980\,cm}\right)\left(\frac{1}{r\,cm}\right)$$

$$H = \frac{0.1486\,cm^2}{r\,cm}\ at\ 20\,°C$$

**Exercise 7-34. How much higher will water rise in a loam soil with r = 0.0019 cm than in a sandy soil with r = 0.0037 cm?**

## Soil Texture and Surface Area

The soil texture represents the relative amount (mass balance) of sand, silt, and clay in a soil. These soil separates vary in their diameters and surface areas.

**Box 7-6.** Diameter and surface area of sand, silt, and clay sized particles.

| Soil separate | Diameter, mm | Specific surface area, cm$^2$/g |
|---|---|---|
| Sand | 2.0-0.05 | 30 |
| Silt | 0.05-0.002 | 1,500 |
| Clay | <0.002 | 8,000,000 |

Soil texture does not consider materials greater than 2 mm in diameter, nor organic matter. An approximate surface area can be calculated from the decimal fractions of sand, silt, and clay contained in a sample using the equation,

$$\text{Surface area} = f_{sand}SA_{sand} + f_{silt}SA_{silt} + f_{clay}SA_{clay}$$

Where f is the decimal fraction and SA the specific surface area of these separates.

Consider a soil that contains 30% sand, 40% silt and 30% clay. We want to find the surface area contained in 1g of this soil. The first step is to convert each percentage into a decimal fraction. This is done by recognizing that the percentages represent the weight of a given fraction in 100 units of soil. Since specific surface area is in units of cm$^2$/g, we choose g for the units in which to represent the percentages:

$$30\%\ sand = \frac{30\,g}{100\,g} = \frac{0.30\,g}{1\,g} = 0.30$$

$$40\%\ silt = \frac{40\,g}{100\,g} = \frac{0.40\,g}{1\,g} = 0.40$$

$$30\%\ clay = \frac{30\,g}{100\,g} = \frac{0.30\,g}{1\,g} = 0.30$$

These decimal fractions can then be multiplied by their respective specific surface areas (listed in **Box 7-6)** to come up with a specific surface area for the soil:

$$\left(\frac{30\,cm^2}{1\,g\,soil}\right)(0.30) + \left(\frac{1,500\,cm^2}{1\,g\,soil}\right)(0.40) + \left(\frac{8,000,000\,cm^2}{1\,g\,soil}\right)(0.30) = \frac{2,400,609\,cm^2}{1\,g\,soil}$$

This specific surface area is then multiplied by the weight of the soil sample to come up with the surface area present:

$$1\,g\,soil\left(\frac{2,400,609\,cm^2}{1\,g\,soil}\right) = 2,400,609\,cm^2$$

**Exercise 7-35. What is the surface area of a 2 g sample of soil containing 20% sand, 50% clay, and 30% silt?**

Soil textural class is a convention used to describe the physical properties of soil. This classification system is based on percent sand, silt, and clay. The three separates must total 100%. So if the percentages of any two of the three are known, the third can be calculated by subtraction from 100%. The use of only three size classes and their restriction that they always add to 100% allow soil textural class to be conveyed and calculated using a triangle, termed the "textural triangle" (see **Box 7-7**).

For these calculations, only material less than 2 mm are considered. Soil texture does not consider organic matter. Materials greater than 2 mm are sieved out of the sample, which are grouped into gravel (2-75 mm), cobbles (75-250 mm), stones (250-600 mm), and boulders (> 600 mm). The percent of material > 2 mm exceeds 15%, or is between 15-35%, 35-60%, and 60-95%, then the soil name includes moderately, very, or extremely respectively. Soils with > 95% 2 mm material are termed gravel.

As an example, consider a soil that contains 40% silt and 40% sand. From **Box 7-7**, we see that this soil is classified as a loam and must contain 20% clay.

**Box 7-7.** Soil textural triangle showing the sand textural classes (http://www.uwsp.edu/geo/faculty/ritter/glossary/s_u/soil_texture_triangle.html).

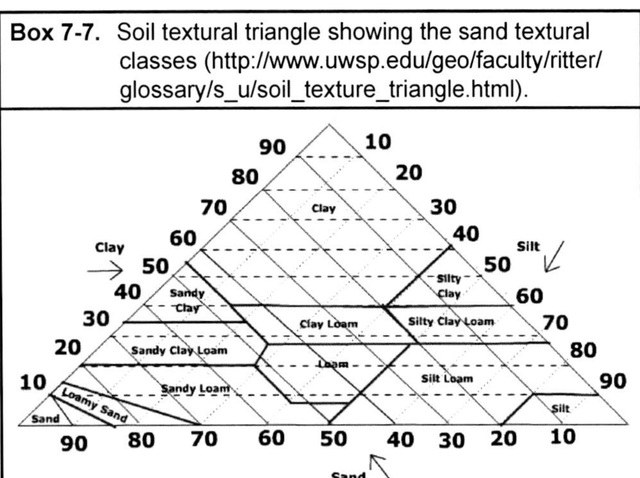


**Exercise 7-36. What is the texture of a soil if it contains 50% sand, 20% clay, and 30% silt?**

For soil texture determinations, only materials less than 2 mm are considered. Soil texture does not consider organic matter. Materials greater than 2 mm are sieved out of the sample. Materials greater than 2 mm are grouped into gravel (2-75 mm), cobbles (75-250 mm), stones (250-600 mm), and boulders (>600 mm). The textural classification is modified to account for the percentage, based upon weight, of total material greater than 2 mm **(Box 7-8).**

**Box 7-8.** Modification of textural name based on percentage of materials >2 mm.

| percent of total | Modification |
|---|---|
| <15% | Textural name only |
| 15-35% | >2 mm material type + textural name |
| 35-60% | Very+>2 mm material type + textural name |
| 60-95% | Extremely + type + textural name |
| >95% | > 2 mm material type only |

**Exercise 7-37. If a soil contains 25 g of material between 2-75 mm, 50 g of clay, 25 g of silt, and 25 g of sand, what is its textural classification?**

Solution:
Calculate the total weight of the sample:
25g + 50g + 25g + 25g = 125g

Next calculate the percent, by weight, of the gravel size fraction:
gravel: (25 g / 125 g) x 100% = 20%

Because this percentage is in the 15-25% range, the term "gravelly" will be placed before the textural name.

To find the textural name, we first calculate the weight of the sample when the gravel size fraction is omitted:
50 g + 25 g + 25 g = 100 g

Calculate the weight percent of the 100 g sample of each size fraction:
clay: (50 g / 100 g) x 100% = 50%
silt: (25 g / 100 g) x 100% = 25%
sand: (25 g / 100 g) x 100% = 25%

Next find the textural name, using the textural triangle in **Box 7-7:**
50% clay, 25 % sand, 25% silt = clay

So the complete name is "gravelly clay"

## Additional Information


Brady, N.C. and R.R. Weil. 2008. The Nature and Properties of Soil. 14th edition, Pearson Prentice Hall.

Foth, H.D. 1984. Fundamentals of soil science. 7th ed. p. 38-39. John Wiley & Sons, New York, NY.

Jury, W.A., W.R. Gardner, and W.H. Gardner. 1991. Soil physics. 5th ed. John Wiley & Sons, Inc., New York, NY.

Logsdon, S.D., O.S. Mbuya, and T. Tsegaye. 2008. Bulk density and soil moisture sensors, p. 211-220. *In* S. Logsdon et al. (ed.) Soil science: Step by step field analysis. SSSA, Madison, WI.

# MOLARITY, CONCENTRATIONS, AND STABLE ISOTOPES

KEY PROBLEMS: How to use parts per million (ppm) and parts per billion (ppb) in problem solving.

MATHEMATICAL SKILLS: Review of multiplication, addition, and equations.

CHAPTER CONCEPTS: This chapter demonstrates how to calculate molarity, ppm, and ppb in solutions. These units are used in many different types of natural resource applications.

**Box 8.1.** Definitions of molarity, parts per million, and parts per billion.

**Definitions**

**Molarity (M) = mol/L.** This is a measure of the concentration of a solute in a solution, or of any molecular, ionic, or atomic species in a given volume. The chemical literature traditionally uses mol/L or moles of a substance per liter (L) of solution. A mole is the amount of pure substance containing the same number of molecules ($6.023 \cdot 10^{23}$). For example, a mole of carbon contains $6.023 \cdot 10^{23}$ carbon atoms that weight 12.0107 g.

**Parts per million (ppm).** This is number of units of interest per 1,000,000 units of sample. For example, if a soil sample contains 5.2 ppm Ca, this means that it contains 5.2 lb of Ca per 1,000,000 lb of soil, or 5.2 kg of Ca per 1,000,000 kg of soil.

**Parts per billion (ppb).** This is number of units of material per 1,000,000,000 units of sample. For example, if a soil sample contains 5.2 ppb Ca, this means that it contains 5.2 lb of Ca per 1,000,000,000 lb of soil.

Many fertilizer recommendations and environmental quality guidelines are based on concentration, which may be reported in parts per million (ppm). One way to explain ppm and molarity is to provide a series of examples.

## Molarity

Molarity is used in many different natural resource applications. The ability to calculate molarity is necessary to produce **accurate** and **precise** laboratory results. Accuracy is defined as the capacity of providing a correct readings, while precision is the ability to obtain repeatable values.

**Box 8.2.** Atomic mass of selected elements.

| Element | Symbol | Atomic mass | Element | Symbol | Atomic mass |
|---|---|---|---|---|---|
| Aluminum | Al | 26.97 | Magnesium | Mg | 24.32 |
| Bromide | Br | 79.92 | Mercury | Hg | 200.61 |
| Cadmium | Cd | 112.41 | Nitrogen | N | 14.01 |
| Calcium | Ca | 40.08 | Oxygen | O | 16 |
| Carbon | C | 12.01 | Phosphorus | P | 30.98 |
| Chlorine | Cl | 35.45 | Potassium | K | 39.1 |
| Copper | Cu | 63.57 | Selenium | Se | 78.96 |
| Hydrogen | H | 1.01 | Sodium | Na | 23 |
| Iron | Fe | 55.84 | Sulfur | S | 32.064 |
| Lead | Pb | 207.21 | | | |

**Exercise 8-1. If we have 74.56 g KCl and we mix it with water until the solution contains exactly 1.00 L. What is the molarity of this solution?**

**Solution:**
Molecular weight KCl = 39.1 + 35.46 = 74.56 g

$$\left(\frac{74.56\,\text{g}}{1\text{L}}\right)\cdot\left(\frac{1\,\text{mole}}{74.56\,\text{g}}\right)=1.0\,\text{M}$$

Atomic masses are shown in **Box 8.2.** The weight of KCl (74.56) is determined by adding the atom mass of K (39.1 g/mole) with Cl (35.6 g/mole). Notice that the units of molarity are mole divided by liters. Neither unit cancels. A replacement for mol/L is capital M. So the answer is 1.00 M.

**Exercise 8.2. What is the molarity of a solution that has 3.00 moles of solute dissolved in 6 L of solution?**

**Exercise 8-3. If 58.45 grams of NaCl are dissolved in exactly 4.00 L of solution, what would be the molarity of the solution? To solve this exercise, the grams of NaCl need to be converted to moles and the weight of one mole of NaCl must be determined. Grams are converted to moles as shown below. One mole of NaCl weighs 58.45 g (23 + 35.45).**

**Exercise 8-4. What is the molarity of 10.0 g of acetic acid ($CH_3COOH$) dissolved in 500.0 mL of solution? 1 mole of acetic acid weighs 60.06 g.**

Solution:

$$\left(\frac{10\text{ g acetic acid}}{500\text{ mL}}\right)\cdot\left(\frac{1\text{ mole acetic acid}}{60.06\text{ g}}\right)\cdot\left(\frac{1{,}000\text{ mL}}{\text{L}}\right)=$$

$$\frac{0.33\text{ moles}}{\text{L}}=0.33\text{ M}$$

**Exercise 8-5. To produce a 0.1 M solution of $CaCl_2$, how many grams of $CaCl_2$ are needed per liter of solution? 1 mole of $CaCl_2$ weighs 110.98 g (40.08 + 70.9).**

**Exercise 8-6. To make a 2.0 M solution of KCl, how much KCl should be added to 15 L?**

**Exercise 8-7. To make 1 L of 0.1 M solution of $NH_4Cl$, how much $NH_4Cl$ should be added into a 1 L volumetric flask?**

**Exercise 8-8. Sea water contains roughly 28.0 g of NaCl per liter. What is the molarity of sodium chloride in sea water?**

**Exercise 8-9. How many moles of $Na_2CO_3$ are in 10.0 mL of a 2.0 M solution?**

In many situations, normality is used rather than molarity. **Normality** is defined as the number of equivalents divided by the liters of solution. There is a relationship between normality and molarity. Normality can only be calculated when we know the reactions, because normality is a function of equivalents. For more information on normality, see Barbalance (1999). In cation exchange reactions, $Ca^{2+}$ has 2 equivalents, while $K^{+}$ has 1 equivalent.

**Exercise 8.10. Convert 100 g $P_2O_5$ to P.**

Solution:

$$100\text{ g }P_2O_5\left(\frac{61.96\text{ g P}}{141.96\text{ g }P_2O_5}\right)=43.65\text{ g P}$$

The 43.65 value is often used to convert $P_2O_5$ values to P values. To make this conversion, convert 43.65 to a decimal form (0.4365) and multiply this value times the $P_2O_5$. This conversion may be necessary to conduct a nutrient budget (Chapter 6).

**Exercise 8.11. Convert 100 g $K_2O$ to K.**

**Exercise 8-12. Convert 100 g nitrate to g nitrate-N.**

Solution:
100 g nitrate represents the weight of the N atom and 3 oxygen atoms ($NO_3$), while nitrate-N only considers the N atom.

$$100\ \text{g}\ NO_3 \cdot \left(\frac{14\ \text{g}}{62\ \text{g}\ NO_3}\right) = 22.58\ \text{g N}$$

Soil laboratories typically report values in ppm $NO_3$-N rather than $NO_3$.

**Exercise 8-13. Convert 325 g nitrate to nitrate-N.**

## Parts Per Notation

The parts per notations are dimensionless quantities that do not have units associated with them. They are defined as some amount of something in some specified amount of something else. A part per hundred is represented by the % sign. A part per hundred indicates 1 part per 100 parts. One part per thousand should be spelled out, and in some situations it is identified by the ‰ notation. Part per ten thousand is denoted by the ‱ symbol. The part per million and parts per billion are identified by the ppm and ppb notation.

**Exercise 8-14. Rewrite $\left(\frac{5\ \text{kg P}}{1{,}000{,}000\ \text{kg soil}}\right)$ in ppm and ppb.**

Solution:
$\left(\frac{5\ \text{kg P}}{1{,}000{,}000\ \text{kg soil}}\right)$ represents 5 kg P per 1,000,000 kg of soil and therefore is 5 ppm.
$\left(\frac{5\ \text{kg P}}{1{,}000{,}000\ \text{kg soil}}\right)$ is converted to ppb by multiplying the ppm value by a number that will result in 1,000,000 in the denominator.

$$\left(\frac{5\ \text{kg P}}{1{,}000{,}000\ \text{kg soil}}\right) \cdot \left(\frac{1{,}000}{1{,}000}\right) = \frac{5{,}000\ \text{kg P}}{1{,}000{,}000{,}000\ \text{kg soil}} = 5{,}000\ \text{ppb}$$

**Exercise 8-15. Convert 15 ppm P in soil to lb P per lb soil.**

Solution:

$$15\ \text{ppm P} = \left(\frac{15\ \text{lb P}}{1{,}000{,}000\ \text{lb soil}}\right)$$

**Exercise 8-16. Convert 15 ppm P in soil to kg P per kg of soil.**

**Exercise 8-17. Convert 0.03 kg chlorosulfuron/ha to ppm in the surface 5 cm if the bulk density is 1.1 g/cm$^3$.**

Solution:

1. Calculate amount of soil

$$5\text{ cm}\cdot\left(\frac{1.1\text{ g}}{\text{cm}^3}\right)\cdot\left(\frac{10,000\text{ m}^2}{\text{m}^2}\right)\cdot\left(\frac{10,000\text{ m}^2}{\text{ha}}\right)\cdot\left(\frac{\text{kg}}{1,000\text{ g}}\right)=$$

550,000 kg soil

2. Determine ppm

$$\left(\frac{0.03\text{ kg chlorsulfuron}}{550,000\text{ soil}}\right)\cdot\frac{2}{2}=\left(\frac{0.06\text{ kg chlorsulfuron}}{1,100,000\text{ soil}}\right)\cdot\frac{\left(\frac{1}{1.1}\right)}{\left(\frac{1}{1.1}\right)}$$

$$=\frac{0.0545\text{ kg}}{1,000,000}=0.0545\text{ ppm}$$

Note: To calculate ppm, the denominator must equal 1,000,000.

**Exercise 8-18. Convert 10 g nitrate-N per kg of water to ppm.**

**Exercise 8-19. Convert 3 ppm nitrate-N to g per kg of water.**

Solution:
When solving these exercises, an understanding of what nitrate-N represents is needed. Nitrate-N represents only the N in the solution, rather than the amount of nitrate in the solution. These values are different because N has a molecular weight of 14 g while nitrate has a molecular weight of 62 g.

First, rewrite what 3 ppm represents

$$\frac{3\text{ g nitrate-N}}{1,000,000\text{ g of water}}$$

Second, change the units to match the desired units in the answer.

$$\frac{3\text{ g nitrate-N}}{1,000,000\text{ g of water}}\cdot\frac{1,000\text{ g}}{1\text{kg}}$$

Third, cancel and multiply

$$\frac{3\text{ g nitrate-N}}{1,000,000\text{ g of water}}\cdot\frac{1,000\text{ g}}{1\text{kg}}=0.003\text{ g nitrate-N/kg}$$

**Exercise 8-20. A soil sample from the surface 6 in. of soil is collected and analyzed using appropriate techniques. The sample contains 10 ppm nitrate-N. How many lb of nitrate are contained in the soil if an acre that is 6 in. thick contains 2.3 million lb?**

**Exercise 8-21. Soil samples from 0 to 6 and 6 to 24 in. soil depths are collected. The 0 to 6 and 6 to 24 in. depths contain 5 ppm and 7 ppm $NO_3$-N, respectively. How much nitrate is in the soil if each 6 in. segment contains 2,000,000 lb?**

**Exercise 8-22. You want to make a solution containing 1,000 ppm $NO_3$-N. How much $Ca(NO_3)_2$ should be weighed into a 1 L volumetric flask?**

Solution:

1,000 ppm can be rewritten as $\frac{1,000 \text{ mg}}{\text{L}}$ or $\frac{1,000 \text{ mg}}{\text{g}}$

$$1\text{L} \cdot \frac{1,000 \text{ mg } NO_3\text{-N}}{\text{L}} \cdot \frac{164 \text{ g } Ca(NO_3)_2}{28 \text{ g } NO_3\text{-N}} \cdot \frac{\text{g}}{1,000 \text{ mg}}$$
$$= 5.857 \text{ g } Ca(NO_3)_2$$

**Exercise 8-23. You want to make 2 L of solution containing 100 ppm $NO_3$-N. How much $Ca(NO_3)_2$ should be weighed into a 2 L volumetric flask?**

## Converting ppm to the Amount of Chemical in a Substance

Many environmental regulations are based on ppm. For example, EPA has drinking water standards for mercury (0.002 ppm), nitrate-N (10 ppm), and atrazine (3 ppm).

**Exercise 8-24. A producer wants to know the atrazine (herbicide) carryover in a field. A soil sample from the surface 6 in. contains 0.1 ppm. If atrazine was applied at 1 lb/acre, what percent of the atrazine remains? Assume that the surface 6 in. of soil weighs 2,000,000 lb/acre.**

Solution:

$$\left(\frac{0.1 \text{ lb atrazine}}{1,000,000 \text{ lb soil}}\right)\left(\frac{2,000,000 \text{ lb soil}}{\text{acre (6 in.)}}\right) = \frac{0.2 \text{ lb atrazine}}{\text{acre (6 in.)}}$$

$$\% \text{ remaining} = 100\%\left(\frac{0.2}{1.0}\right) = 20\%$$

**Exercise 8-25. A fish sample contains 0.01 ppb mercury. If you eat a fish that weighs 2 lb, how many µg of Hg have you eaten? Note: ppb is parts per billion; 1,000 ppb = 1 ppm; 1,000,000 µg = 1 g.**

**Exercise 8-26. A soil sample contains 10 ppm P. How many lb P/acre are contained in an acre of soil that is 6 in. deep? Assume that the soil weighs 2,000,000 lb.**

**Exercise 8-27. A 6 in. soil sample contains 40 ppm K. How many lb K/acre are contained in this soil? Assume that the soil weighs 2,000,000 lb.**

Solution:

$$\left(\frac{40 \text{ lb K}}{1,000,000 \text{ lb soil}}\right)\left(\frac{2,000,000 \text{ lb soil}}{\text{acre (6 in.)}}\right) = \frac{80 \text{ lb K}}{\text{acre (6 in.)}}$$

This calculation is used for an example. In many fertilizer recommendations, the concentration in ppm is used rather than the amount of nutrient.

**Exercise 8-28. Determine how many pounds/acre of inorganic N are contained in a soil if the samples from the 0 to 6 and 6 to 24 in. depths contain 5 and 15 ppm inorganic N, respectively. Assume each 6 in. segment of soil weighs 2,000,000 lb.**

## Stable Isotopes

There are many stable isotopes in nature **(Box 8-3)**. The stable isotope $^{15}N$ contains one more neutron than $^{14}N$. These isotopes can be used for source tracking.

**Box 8.3.** Relative abundance of stable isotopes found in soil.

| Element | Weight, g/mole | Abundance, % |
|---|---|---|
| H | 1 | 99.985 |
| | 2 | 0.015 |
| C | 12 | 98.89 |
| | 13 | 1.11 |
| N | 14 | 99.63 |
| | 15 | 0.37 |
| O | 16 | 99.759 |
| | 17 | 0.037 |
| | 18 | 0.204 |
| S | 32 | 95 |
| | 33 | 0.76 |
| | 34 | 4.22 |
| | 36 | 0.014 |

Different units are used to report the concentration of stable isotopes. Several of these are shown below.

$$At\%^{15}N = \left(\frac{^{15}N}{^{14}N + ^{15}N}\right) \cdot (100\ At\%)$$
$$At\%^{15}N + At\%^{14}N = 100$$

$$At\%^{13}C = \left(\frac{^{13}C}{^{12}C + ^{13}C + ^{14}C}\right) \cdot (100\ At\%)$$
$$At\%^{13}C + At\%^{12}C + At\%^{14}C = 100$$

Studies examining stable isotopes at or near natural abundance levels often report values as delta values. These values can be reported in parts per thousand or per mil ("o/oo"). Delta values are not absolute isotope abundances, but differences between sample readings and one or another of the widely used natural abundance standards. Delta calculations rely on the absolute isotope ratios (R). These values are defined below,

$$\delta^{15}N = \left(\frac{R_{sample} - Rstd}{Rstd}\right) \cdot (1{,}000\delta t‰)$$
$$R = \left(\frac{At\%^{15}N}{At\%^{14}N}\right);\ Atom\% = \left[\frac{R}{(1+R)}\right] \cdot 100$$
$$\delta^{13}C = \left(\frac{R_{sample} - Rstd}{Rstd}\right) \cdot (1{,}000‰)$$
$$R = \left(\frac{At\%^{13}C}{At\%^{12}C}\right);\ Atom\% = \left[\frac{R}{(1+R)}\right] \cdot 100$$
$$\Delta = \left(\frac{\delta^{13}C_{PDBofair} - \delta^{13}C_{PDBsample}}{1 + \delta^{13}C_{PDBsample}/1{,}000}\right)$$

For N, the Rstd is air (delta = 0, At%$^{15}$N = 0.3663033, R = 0.0036436) and for C, Rstd is Pee Dee Belemnite (At%$^{13}$C = 1.1112328, R = 0.0112372).

Different photosynthetic systems have different amounts of C discrimination. $C_4$ plants often have $^{13}C$ discrimination values ($\Delta$) that range from 2 to 4 ‰ ($\delta^{13}C$ values ranging from -12 to -14 ‰) while $C_3$ plants have $^{13}C$ discrimination values that range from 16 to 18 ‰ ($\delta^{13}C$ values range from -24 to -26 ‰). These differences provide the opportunity to track the fate of carbon added to soil. In short-term experiments, the relative proportion of the new carbon remaining (F) in the soil can be determined using the equation,

$$F = \left(\frac{\left|\delta^{13}C_{soil\ sample} - \delta^{13}C_{new\ carbon}\right|}{\left|\delta^{13}C_{new\ carbon} - \delta^{13}C_{old\ carbon}\right|}\right)$$

This equation should not be used for long-term experiments because $^{13}C$ enrichment of SOC can bias the results. The | | symbols mean absolute value. Details for including $^{13}C$ isotopic discrimination in turn over calculations is available in Clay et al. (2007).

**Exercise 8-29. Determine the proportion of new carbon remaining in soil if the δ¹³C value of a soil sample collected after 1 year is -15‰, the δ¹³C value of the old carbon is -16‰ (soil before the experiment was started), and the δ¹³C value of the new carbon is -12‰.**

Solution:

$$F = \left(\frac{\delta^{13}C_{soil\ sample} - \delta^{13}C_{new\ carbon}}{\delta^{13}C_{new\ carbon} - \delta^{13}C_{old\ carbon}}\right)$$
$$F = \left(\frac{-15-(-16)}{-12-(-16)}\right) = \frac{1}{4} = 25\%$$

A similar equation can be used to calculate N fixation in legume plants and N in new plants derived from fertilizer. These calculations are based on $\delta^{15}N$ in fixed N and fertilizer N being < 0‰ and $\delta^{15}N$ in soil, N being > 0‰.

**Exercise 8-30. Calculate N fixed based on the following measurements $\delta^{15}N$ of fixed N = -1‰, $\delta^{15}N$ of non-fixed N = 4‰. Total N in plant = 200/bu/acre plant sample had a $\delta^{15}N$ value of 0.5‰.**

Solution:

$$\%N\ fixed = 100\left(\frac{4 - 0.5}{4 - (-1)}\right)$$

$$Total\ N\ fixed = (0.7)\left(\frac{200 - lb\ N}{acre}\right)$$

$$= \frac{140\ lb\ N}{acre}$$

**Exercise 8-31. A sample has a R value of 0.003715. What is its $d^{15}N$ value?**

Solution:

Atom $\%^{15}N$ air = 0.3663033

Atom $\%^{14}N$ = 100 - Atom $\%^{15}N$

Rair = $N^{15}/N^{14}$ = 0.003676506

$d^{15}N$ = [(0.003715-0.003677)/(0.003677)] • 1,000 = 10.33‰

**Exercise 8-32. A sample has a $d^{13}C$ value of -14.3. What is its $^{13}C$ isotopic discrimination value (△)?**

Solution:

△ = [-8 – (-14.3)]/[1+(-14.3/1,000)] = 6.39‰

**Exercise 8-33. If $R_{PDB}$ is 0.0112372, calculate the atom % $^{13}C$ if the sample has a $d^{13}C$ value of -55.32.**

Solution:

$R_{PDP}$ is the standard and therefore its value replaced $R_{STN}$ in the delta equation

-55.32 = 1,000 • (Rsam-$R_{PDP}$)/($R_{PDP}$)

Rsample = 0.010616

**Exercise 8-34. A plant sample has a $d^{13}C$ value of -13‰. What is its $^{13}C$ isotopic discrimination value?**

Solution:

$$\triangle = \left(\frac{\dfrac{(d^{13}C_{PDB}\ of\ air - d^{13}C_{PDB}\ sample)}{(1 + d^{13}C_{PDB}\ sample)}}{1,000}\right)$$

Where

$d^{13}C$ air is -8‰

△ = [-8 – (-13)] / [1+(-13/1,000)] = 4.93‰

**Exercise 8-35. Fertilizer has a $d^{15}N$ value of -1‰. Corn grain harvested from the field has a $d^{15}N$ value of 1‰. Corn collected from unfertilized areas has a $d^{15}N$ value of 5‰. What is the percentage of N that was derived from fertilizer?**

Solution:

$$\%\ Fertilizer = \frac{[Sample - Unfertilized]}{[Fertilizer - Unfertilized]} \times 100$$

$$= \frac{|1-5|}{|-1-5|} \times 100 = \left(\frac{4}{6}\right)100 = 66\%\ from\ Fertilizer$$

This same calculation can be used to determine N fixed by legumes. More complex methods are available for determining yield losses due to N and water stress in plants.

## Additional Information

Barbalace, R.C. 1999. Molarity, Molality and Normality. EnvironmentalChemistry.com. 1999. Accessed on-line: 7/9/2008 http://EnvironmentalChemistry.com/yogi/chemistry/Molarity MolalityNormality.html

Clay, D.E., C.E. Clapp, C. Reese, Z. Liu, C.G. Carlson, H. Woodard, and A. Bly. 2007. $^{13}C$ fractionation of relic soil organic C during mineralization effects calculated half-lives. Soil Sci. Soc. Am. J. 71:1003-1009.

Clay, D.E., C.G. Carlson, S.A. Clay, C. Reese, Z. Liv, and M.M. Ellsbury, 2006. Theoretical derivation of new, stable and non-isotopic approaches for assessing soil organic turnover. Agron. J. 98:443-450.

Clay, D.E., Ki-In Kim, J. Chang, S.A. Clay, and K. Dalsted. 2006. Characterizing water and N stress in corn using remote sensing. Agron. J. 98:579-587.

# UNIT CONVERSIONS
# PROBLEMS INVOLVING PESTICIDES

**Key problems:** Calibrate sprayer output. Determine the amount of pesticide that should be added to a sprayer to apply the desired rate of product. Determine the application rate from a dry (granular) applicator.

**Mathematical skills:** Review addition, multiplication, percent, and multistep conversions.

**Chapter concepts:** Many problems in natural resource management are solved by a combination of applying an equation and unit conversions. Solving these problems should be done in a step by step process. These problems link many individual steps discussed in previous chapters and also incorporate specific measurements and equations.

## Determining Sprayer Application Rates

Many pest problems require the use of chemicals for cost effective control. To ensure that labeled rates are followed, it is important to calibrate the sprayer and calculate how much chemical should be placed in each tank. Companies that produce and label agrichemicals spend hundreds of millions of dollars over multiple years to develop safe, effective products for producers. Applying the compound at the labeled rate is the legal obligation of the user. Due to calibration errors, application equipment often does not apply the desired rate (Hofman and Solseng, 2004).

The application rate can be changed by modifying applicator speed, nozzle output, spray pressure, and distance between nozzles. There are several calculation steps to correctly determine the output of an applicator and the amount of product that should be added to a mixture.

## Calibration for Liquid Solutions

### Step 1. Determining nozzle and boom output.

The sprayer nozzle is an integral portion of the spray system. The nozzle regulates flow, atomizes the solution to droplets, and disperses the solution in a desired pattern. There are many guides available to help determine which type of nozzle to use (Hofman and Solveng, 2004). Prior to calibration each nozzle on the boom should be checked to make sure that all nozzles (and screens) are of the same type. The output of each nozzle should be similar. Nozzles that have outputs 10% below the average should be checked to make sure the orifice (or screen) is not plugged. Nozzles that are 10% above average should be checked for damage and wear. Replace nozzles as needed with the same nozzle type and recheck individual output.

The next step to correctly calibrate a sprayer is to measure the output in volume per unit time. In these examples, the output of all nozzles on the boom is assumed to be within 10% of the average boom output. Output from the boom will be calculated in the following exercises.

**Exercise 9-1a. Calculate the output (gal/min) of a sprayer that has an average nozzle output of 21 fluid oz in 30 sec. The number of nozzles on the boom is 50 and nozzles are spaced every 20 in. (20 in. on center).**

Solution:

$$\text{Sprayer output per nozzle} = \left(\frac{21\text{ oz}}{30\text{ sec}} \cdot \frac{1\text{ gal}}{128\text{ oz}} \cdot \frac{60\text{ sec}}{1\text{ min}}\right) =$$

0.328 gal/min/nozzle

Sprayer output over the boom = 0.328 gal/min/nozzle • 50 nozzles = 16.4 gal/min

**Exercise 9-1b. Calculate the output of the sprayer in Exercise 9-1a in liters/min.**

**Exercise 9-2a. Calculate the output (gal/min) of a sprayer that has 25 nozzles on the boom and an average nozzle output of 145 mL/15 sec.**

**Exercise 9-2b. Calculate the output (gal/min) of a sprayer that has 40 nozzles on the boom and an average nozzle output of 190 ml/20 sec.**

### Step 2. Speed of applicator.

The applicator speed influences the amount of chemical applied to a given area. Tire slippage due to nonuniform terrain conditions can result in actual speeds that differ from those shown by the speedometer, and therefore speeds need to be checked on a terrain similar to that of the application area (Hofman and Solseng, 2004). The time to drive the course is measured and the speed of the equipment is calculated with the equation,

$$\text{Tractor speed} = \left(\frac{\text{length of course}}{\text{time over course}} \cdot \text{conversion}\right)$$

**Exercise 9-3a. The course length is 200 ft and the time to drive the course is 25 sec. Calculate the tractor speed in a) ft/min, and b) miles per hour (mph).**

Solution:

$$\text{a. Tractor speed} = \left(\frac{200\text{ ft}}{25\text{ sec}} \cdot \frac{60\text{ sec}}{1\text{ min}}\right) = 480\text{ ft/min}$$

$$\text{b. Tractor speed} = \left(\frac{200\text{ ft}}{25\text{ sec}} \cdot \frac{1\text{ mile}}{5{,}280\text{ ft}} \cdot \frac{60\text{ sec}}{1\text{ min}} \cdot \frac{60\text{ min}}{1\text{ hr}}\right) = 5.45\text{ mph}$$

**Exercise 9-3b. The course length is 150 ft and the time to drive the course is 14 sec. Calculate the tractor speed in a) ft/min, and b) miles per hour (mph).**

### Step 3. Determining the area covered per unit of time.

Based on the speed of the applicator and the width of the boom the amount of acres sprayed in a given time can be calculated.

**Exercise 9-4a. Calculate the number of acres that are sprayed in 1 min if the sprayer has 50 nozzles on 20 in. center spacing, and the tractor speed is 480 ft/min.**

Solution:

Area = Length • boom width

$$\text{Boom width} = \left(\frac{20\text{ in.}}{1\text{nozzle}} \cdot \frac{1\text{ ft}}{12\text{ in.}}\right) \cdot 50\text{ nozzles} = 83\text{ ft}$$

$$\text{Area covered} = \left(\frac{480\text{ ft}}{1\text{ min}} \cdot 83\text{ ft}\right) = 39,840\text{ ft}^2/\text{min}$$

$$\text{Area covered} = \left(\frac{39,840\text{ ft}^2}{1\text{ min}} \cdot \frac{1\text{ acre}}{43,560\text{ ft}^2}\right) = 0.915\text{ acre/min}$$

**Exercise 9-4b. Calculate the number of acres that can be sprayed in 1 min if the tractor speed is 642 ft/min, and the spray boom has 25 nozzles on 30 in. center spacing.**

## Step 4. Determine the volume output of the sprayer per unit area (i.e. gal/acre).

By calculating the output per unit time and the area covered per unit time, the output/area can be determined by multiplying (output/unit time) • inverse of (area/unit time):

$$\text{Output/area} = \frac{\text{output}}{\text{min}} \cdot \frac{\text{min}}{\text{area}}$$

**Note that the time increment must be in the same units.**

**Exercise 9-5a. Calculate the gal/acre output if the sprayer output is 16.4 gal/min and the area is sprayed at a rate of 0.915 acre/min.**

Solution:

$$\text{Sprayer output} = \left(\frac{16.4\text{ gal}}{1\text{ min}} \cdot \frac{1\text{ min}}{0.915\text{ acre}}\right) = 17.9\text{ gal/acre}$$

**Exercise 9-5b. Calculate the output of a sprayer in gal/acre if the tractor is traveling 642 ft/min, the boom width is 62.5 ft, and the output of the sprayer is 6 gal/min.**

## Alternative method: Step 4.

If the output (in gallons per nozzle,) speed of the equipment (in miles per hour), and nozzle spacing (in inches) on the boom have been measured, the sprayer output (in gallon per acre) can be calculated using an alternative approach using the equation below,

$$\text{gpa} = \left(\frac{\text{gpm} \cdot 5940}{\text{mph} \cdot \text{w}}\right)$$

or

$$\frac{1\text{ gal}}{1\text{ acre}} = \frac{1\text{ gal}}{1\text{ min}} \cdot \frac{1}{1\text{ miles inch}/1\text{ hour}} \cdot \frac{12\text{ inch}}{1\text{ ft}} \cdot \frac{1\text{ mile}}{5,280\text{ ft}} \cdot \frac{43,560\text{ ft}^2}{1\text{ acre}} \cdot \frac{60\text{ min}}{1\text{ hour}}$$

where gpa = gallons per acre;
gpm = gallon per minute PER NOZZLE;
mph = miles per hour of the applicator;
w = nozzle spacing width in INCHES and
5,940 is a conversion constant.

The conversion constant 5,940 that is used to make the units cancel is derived from the following equation,

$$5,940 = \left(\frac{12\text{ in.}}{1\text{ ft}}\right)\left(\frac{1\text{ mile}}{5,280\text{ ft}}\right)\left(\frac{43,560\text{ ft}^2}{1\text{ acre}}\right)\left(\frac{60\text{ min}}{1\text{ hr}}\right)$$

**Note: The units on the constant 5,940 are inches • mile • min/a • hr. The units on this number are often not shown, making it imperative to either 1) memorize the formula exactly or 2) have the formula written somewhere it can be easily referenced.**

**Exercise 9-6a. What is the application rate gal/acre of a sprayer that is traveling at 5.4 mile/hr, nozzles spaced 20 in. apart, and an average nozzle output of 21 oz/30 sec?**

Solution:

$$\text{gpm} = \left(\frac{21\,\text{oz} \cdot 1\,\text{gal} \cdot 60\,\text{sec}}{30\,\text{sec} \cdot 128\,\text{oz} \cdot 1\,\text{min}}\right) = 0.328\,\text{gpm}$$

$$\text{gpa} = \left(\frac{0.328\,\text{gpm} \cdot 5940}{5.4\,\text{mpg} \cdot 20\,\text{in.}}\right) = 18.04\,\text{gal per acre}$$

**Exercise 9-6b. What is the application rate (gal/acre) of a sprayer that is traveling at 9.8 mile/hr, nozzles spaced 30 in. apart, and with an average nozzle output of 29 oz/32 sec?**

## Step 5. Determining the Amount of Pesticide to Add to a Spray Tank

A common question that many resource managers ask is how much chemical should be added in each tankload of mix. To solve these problems, check the label to determine the appropriate rate. Land managers typically treat hundreds of acres rather than single acre areas. Pesticides are available in many formulations including liquids, dry flowables, aerosols, and granules. Examples of calculations for liquid formulations will be discussed first.

Liquid formulations include water soluble (WS), emulsifiable concentrates (EC), and flowables (F). The application rates typically are expressed in fluid ounces, pints, or quarts per unit area. The general equation to solve these problems is:

Pesticide amount to put into tank = recommendation • number of acres to be covered

**Exercise 9-7a. The recommendation for a liquid formulation is 36 fl oz per acre. How many gallons of formulation are needed if a tankful of solution that will treat 120 acres?**

Solution:

$$\text{Formulations needed} = \left(\frac{36\,\text{fl oz}}{1\,\text{acre}} \cdot 120\,\text{acre} \cdot \frac{1\,\text{gal}}{128\,\text{fl oz}}\right) = 33.75\,\text{gal}$$

**Exercise 9-7b. The recommendation for a liquid formulation is 10 fl oz per acre. How many gallons of formulation need to be added to a tank if 60 acres will be treated?**

**Exercise 9-8a. How much pesticide should be mixed into a spray tank solution if the tank holds 500 gal, the desired chemical rate is 1.25 pt/acre, and the sprayer is calibrated to apply 20 gal/acre?**

**Exercise 9-8b. How much pesticide should be mixed into a spray tank solution if the tank holds 200 gal, the desired chemical rate is 0.8 pt/acre, and the sprayer is calibrated to apply 6 gal/acre?**

**Exercise 9-9a. A spray tank holds 1,000 gal, the intended application rate of solution is 15 gal/acre, and the pesticide product rate is 2.5 pt/acre.**
**1) How many acres can be treated with a tankful of solution?**
**2) How many gal of product need to be added to the tank to get the correct application rate?**

First determine how many acres will be sprayed with 1,000 gallons of mix.

$$\text{Acres} = \left(\frac{1{,}000\text{ gal}}{1\text{ Tank Load}}\right)\left(\frac{1\text{ acre}}{15\text{ gal}}\right) = \left(\frac{66.7\text{ acres}}{1\text{ Tank Load}}\right)$$

Total Pints of Product/tank =

$$\left(\frac{\text{Pt Pesticide}}{1\text{ acre}}\right)\left(\frac{1\text{ acre}}{1\text{ Tank Load}}\right) = \left(\frac{2.5\text{ pt}}{1\text{ acre}}\right)\cdot\left(\frac{66.7\text{ acres}}{1\text{ Tank Load}}\right) = \frac{166.7\text{ pt}}{1\text{ Tank Load}}$$

In most cases, *pints* are inconvenient for measurement. Changing from *pints* to *gallons* can be done by multiplying by 1 (1 *gal* = 8 *pt*). The calculations are,

$$\left(\frac{166.7\text{ pt}}{1\text{ Tank Load}}\right)\left(\frac{1\text{ gal}}{8\text{ pt}}\right) = 20.8\text{ gal product}$$

**Exercise 9-9b. A spray tank holds 350 gal, the intended application rate of solution is 12 gal/acre, and the pesticide product rate is 0.33 pt/acre.**
**1) How many acres can be treated with a tankful of solution?**
**2) How many pints of product need to be added to the tank to get the correct application rate?**

Dry formulations of pesticides are also available. The amount of product used is typically expressed in ounces or pounds of product. Dry formulations can be applied dry or mixed with water. When applying dry materials the spreaders also need to be calibration. See the spreader instruction manuals for calibration information.

**Exercise 9-10a. A spreader travels 300 ft. The spreader output is 5 oz of dry formulation and gives a 14 in. coverage. What is the output of the spreader in pounds per acre?**

Solution

1. Calculate the area covered.

$$\text{Area covered} = (300\text{ ft})(14\text{ in.})\left(\frac{1\text{ ft}}{12\text{ in.}}\right)\left(\frac{1\text{ acre}}{43{,}650\text{ ft}^2}\right) = 0.008\text{ acres}$$

2. Calculate the pounds of material spread

$$\text{Pounds spread} = (5\text{ oz})\left(\frac{1\text{ lb}}{16\text{ oz}}\right) = 0.312\text{ lb}$$

3. Calculate the output in pounds/acre

$$\text{Pounds spread per acre} = \left(\frac{0.312\text{ lb}}{0.008\text{ acre}}\right) = 39\text{ lb/acre}$$

**Exercise 9-10b. Small differences in output over a calibration course can lead to large differences in output per acre. In this example, the same spreader as in exercise 9.10a is used. The spreader travels 300 ft. but the new spreader output is 7 oz of dry formulation with the same 14 in. coverage. What is the new output of the spreader in pounds per acre?**

**Exercise 9-10c. A spreader travels 200 ft. The spreader output is 0.3 lb of dry formulation and gives a 20 in. coverage. What is the output of the spreader in pounds per acre?**

Some dry formulations need to be added to water, such as water dispersible granules (WDG), dry flowables (DF), or wettable powders (WP). Typically, WDG and DF formulations have a high concentration of active ingredient(s) and use rates can be less than 5 oz/acre. Small miscalculations with some of these products can lead to very large problems including crop injury, high carryover with rotational crop injury, crop residue concerns, or other environmental problems. Care must be taken when mixing and applying these concentrated products. Check and recheck calculations prior to mixing and applying. NOTE: It is important that fluid ounces (volume, 128 fluid ounces per gallon) not be confused with dry ounces (weight, 16 ounces per pound).

**Exercise 9-11a. The recommendation for a water dispersible granule product is 0.8 oz/acre. How many pounds of formulation are needed if the solution in the tank can treat 55 acres?**

Solution:
Pounds needed =

$$\left(\frac{0.8\text{ oz}}{1\text{ acre}} \cdot 55\text{ acres} \cdot \frac{1\text{ lb}}{16\text{ oz}}\right) = 2.75\text{ lb of formulation per tank}$$

**Exercise 9-11b. The recommendation for a dry flowable product is 0.38 oz/acre. How many pounds of formulation are needed if the solution in the tank can treat 30 acres?**

## Step 6. Calculating oil, surfactants, and other adjuvant amounts

Other ingredients or adjuvants such as crop oils, nonionic surfactants (NIS), and nitrogen fertilizers (AMS) are often recommended to be included in the spray tank mixture. The amount of additive is usually given on a volume per volume (*v/v*) format. For example, if the label states that crop oil concentrate (COC) should be added at 1% *v/v,* this means that 1 gal of COC is needed for every 100 gal of solution (water + pesticide + COC).

The amounts of adjuvants, crop oils, and surfactants that need to be added to a tank depend on the amount of liquid in the tank and NOT on the application rate per acre. If the recommendation is to add 0.25% COC, then 0.375 gal of COC should be added and then filled to 150 gal.

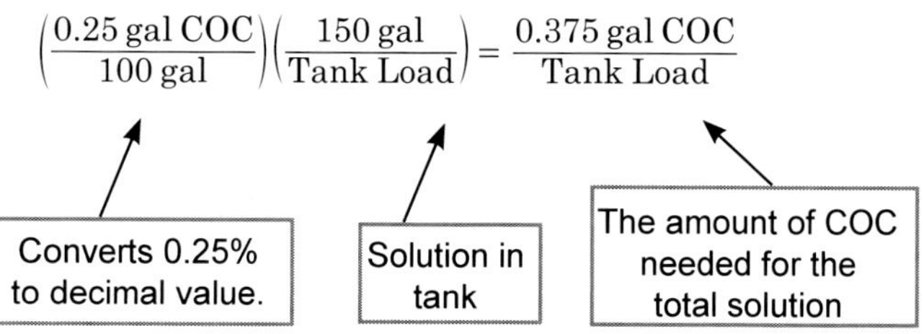


**Exercise 9-12a. If the recommendation is to add 0.25% oil, how much oil should be added to a 1,000 gallon tank if the sprayer is calibrated for 15 gal/acre?**

Solution:

$$\left(\frac{0.25\text{ gal COC}}{100\text{ gal}}\right)\left(\frac{1{,}000\text{ gal}}{\text{Tank Load}}\right) = \frac{2.5\text{ gal COC}}{\text{Tank Load}}$$

Add 2.5 gal to the tank and then fill to 1,000 gal.

**Exercise 9-12b. If the recommendation is to add 0.35% of COC, how much should be added to 750 gal of spray solution?**

Often, an entire area cannot be treated with a single tank load of chemical. The last load may only be a partial tank of solution. The amount of solution in the tank should not exceed the amount of acres to be sprayed, as this will increase cost and may be difficult to use in labeled manner. The next calculation is an example of one of the many ways of determining a partial load.

**Exercise 9-13a. A spray tank holds 750 gal of solution. The intended application rate of solution is 18.5 gal/acre, and the pesticide product rate is 12 fluid oz/acre. NIS is recommended at 0.25% (v/v). There are 60 acres that are being treated. 1) How many acres can be treated with a tankful of solution? And 2) how many pints of pesticide and NIS need to be added to the tankful to get the correct application rate? 3) How many acres will be treated with the second tank of solution? How many gallons of total solution are needed? How much pesticide and NIS should be added to this solution?**

Solution:

First determine how many acres will be sprayed with 750 gal of solution and how much of each chemical needs to be added to the first load.

$$\text{Acres} = \left(\frac{750\text{ gal}}{\text{Tank Load}}\right)\left(\frac{1\text{ acre}}{18.5\text{ gal}}\right) = \frac{40.5\text{ acres}}{\text{Tank Load}}$$

Total Product/750 gal tank =

$$\left(\frac{12\text{ fl oz}}{1\text{ acre}}\right)\left(\frac{40.5\text{ acres}}{1\text{ Tank}}\right)\left(\frac{1\text{ gal}}{128\text{ fl oz}}\right) = 3.8\text{ gal}$$

$$\text{Total NIS} = \left(\frac{750\text{ gal}}{\text{Tank Load}}\right)\left(\frac{0.25\text{ gal}}{100\text{ gal}}\right) = \frac{1.875\text{ gal}}{1\text{ Tank Load}}$$

Next determine how many acres remain to be treated.

Acres to be covered by second tank = 60 acres – 40.5 acres = 19.5 acres

Next determine the amount of water and chemical that needs to be added to finish the job.

$$\text{Volume of solution for tank 2} = 19.5\text{ acres}\left(\frac{18.5\text{ gal}}{1\text{ acre}}\right) = 361\text{ gal}$$

$$\text{Product needed for tank 2} = \left(\frac{12\text{ fl oz}}{1\text{ acre}}\right)\left(\frac{1\text{ gal}}{128\text{ fl oz}}\right)(19.5) = 1.83\text{ gal}$$

**Exercise 9-13b. A spray tank holds 500 gal of solution. The intended application rate of solution is 9 gallons/acre, and the pesticide product rate is 0.75 pt/acre. AMS is recommended at 17 lb/100 gal. There are 160 acres that are being treated. 1) How many acres can be treated with a tankful of solution?, and 2) How many gal of pesticide and pounds of AMS need to be added to the tankful to get the correct application rate? 3) How many acres will be treated with the second and third tanks of solution? How many gallons of total solution are needed? How much pesticide and AMS should be added to this solution?**

### Step 7. Determining the total cost of a pesticide treatment.

For management purposes, it is important to estimate costs of proposed treatments. These costs are determined by summing the costs of several different materials. These problems are solved by the equation:

Cost = amount needed • cost per unit amount • conversions

**Exercise 9-14. What is the cost of 5 gal of pesticide if each gallon costs $10.**

Solution:

Cost = 5 gal • $10/gal = $50.00

**Exercise 9-15a. What will the total chemical cost be for 320 acres if: 1) the product costs $33.75/oz; 2) surfactant costs $6.75/gal; 3) spray rate is 20 gal/acre; and 3) the surfactant rate is 1.4% (v/v); 4) the treatment rate is 0.7oz/acre; and 5) and the tank holds 500 gal?**

Solution:

The cost of the treatment is determined by multiplying the amount of material needed by the cost of the materials. In these calculations the amount of pesticide product and surfactant for the entire area must be determined.

Pesticide: $\left(\frac{0.7\text{ oz}}{1\text{ acre}}\right)\left(\frac{\$33.75}{1\text{ oz}}\right)\left(\frac{320\text{ acre}}{1\text{ job}}\right)=\frac{\$7{,}560\text{ product}}{1\text{ job}}$

Surfactant:

$$\left(\frac{0.014\text{ gal surfactant}}{1\text{ gal solution}}\right)\left(\frac{20\text{ gal solution}}{1\text{ acre}}\right)\left(\frac{320\text{ acre}}{1\text{ job}}\right)\left(\frac{\$675}{1\text{ gal surfactant}}\right)=\frac{\$604.8}{1\text{ job}}$$

Total chemical cost for the 320 acres = $7,560 + $605 = $8,165.

Note: This is the chemical cost only. Application cost (either custom rate or sprayer costs) are not included.

**Exercise 9-15b. What will the total chemical cost be for 720 acres if:**
**1) the product costs $2.60/oz;**
**2) the product treatment is 6 oz/acre;**
**3) the surfactant costs $4.75/gal;**
**4) the surfactant rate is 2.5% v/v;**
**5) the application rate is 12 gal/acre;**
**and**
**6) the tank can hold 450 gal.**

## Purchasing Herbicide Mixtures

Pesticides can be bought as a single ingredient or premixed formulations. When more than one active ingredient is needed, the question is, should they be purchased separately and mixed together in the tank (tank mix) or purchased as a premix? Because the premix may not have the ratio of chemicals that are desired to be applied, additional chemical may be needed. As you can tell by the examples, chemical applications can be costly and saving a few dollars per acre can quickly add up. The next set of exercises will be used to determine if a premix combination of herbicides is the most cost effective control method, or if single ingredient formulations should be purchased and tank-mixed.

**Exercise 9-16a. A product contains 4 lb active ingredient/gal and sells for $100/gal. If the application rate is 1.5 lb of active ingredient/acre, determine the amount of product needed per acre and cost of product per acre.**

Solution:

$$\frac{1.5\text{ lb ai}}{1\text{ acre}}\cdot\frac{1\text{ gal}}{4\text{ lb ai}}=\frac{0.375\text{ gal}}{1\text{ acre}}$$

$$\frac{0.375\text{ gal}}{1\text{ acre}}\cdot\frac{\$100}{1\text{ gal}}=\frac{\$37.50}{1\text{ acre}}$$

**Exercise 9-16b. A product contains 2.5 lb active ingredient/gal and sells for $151/gal. If the application rate is 0.75 lb of active ingredient/acre, determine the amount of product needed per acre and cost of product per acre.**

**Exercise 9-17a. A herbicide application guide states that product A (7.6 lb ai/gal) may be applied at 0.8 to 1.7 pt/acre and may be tank mixed with product B (4 lb ai/gal) at 0.5 to 1 pt/acre. A premix product C can be applied at 1 to 2.5 pt/acre and contains 6.3 lb/gal of the active ingredient of product A + 1.5 lb of the active ingredient of product B. Based on the weed pressure and species present in the field the recommendation is that 1.7 pt/acre of A and 1 pt/acre of B should be applied. Costs are:**

- **Product A, $110/gal, that contains 7.6 lb of active ingredient per gallon.**
- **Product B, $115/gal, that contains 4 lb of active ingredient per gallon.**
- **Product C, $105/gal, that contains 6.3 lb/gal of active ingredient A and 1.5 lb/gal of active ingredient B.**

**Should the chemicals be purchased as the premix or individually and tank mixed for the lowest cost?**

Solution:
First determine the cost of the tank mix by determining the amount of each product needed and the cost for the tank mix.

Cost of purchasing the individual products and tank-mixing.

Product A: Determine the amount of product A active ingredient per acre

$$\left(\frac{1.7\text{ pt}}{1\text{ acre}}\right)\left(\frac{7.6\text{ lb active ingredient}}{1\text{ gal}}\right)\left(\frac{1\text{ gal}}{8\text{ pt}}\right)=\frac{1.6\text{ lb A}}{1\text{ acre}}$$

Determine cost of A per acre

$$\left(\frac{1.7\text{ pt}}{1\text{ acre}}\right)\left(\frac{1\text{ gal}}{8\text{ pt}}\right)\left(\frac{\$110}{1\text{ gal}}\right)=\$23.38$$

Product B: Determine the amount of product B active ingredient per acre

$$\left(\frac{1.0\text{ pt}}{1\text{ acre}}\right)\left(\frac{4.0\text{ lb active ingredient}}{1\text{ gal}}\right)\left(\frac{1\text{ gal}}{8\text{ pt}}\right)=\frac{0.5\text{ lb active ingredient B}}{1\text{ acre}}$$

Determine cost of B per acre

$$\left(\frac{1.0\text{ pt}}{1\text{ acre}}\right)\left(\frac{1\text{ gal}}{8\text{ pt}}\right)\left(\frac{\$115}{1\text{ gal}}\right)=\$14.38$$

Total cost per acre = $23.38 + $14.38 = $37.76

**Exercise 9-17b. Should a premix (product C) be purchased?**

Solution:
Based on the calculations above it was determined that the desired amount of active ingredient for product A was 1.6 lb ai/acre and for product B 0.5 lb ai/acre was needed. Based on the label information, each gallon of premix C, contains 6.3 lb ai of product A and 1.5 lb ai of product B. To determine how much of premix is needed, the amount premix needed to get the desired rate of each chemical should be calculated.

**Step1:** Calculate the amount of premix C required to provide the recommended amount of product A.

$$\left(\frac{1.6\text{ lb active ingredient A}}{1\text{ acre}}\right)\left(\frac{1\text{ gal premix C}}{6.3\text{ lb active ingredient A}}\right)=\frac{0.254\text{ gal premix C}}{1\text{ acre}}$$

**Step 2:** Calculate the amount of premix C required to provide the correct amount of product B.

$$\left(\frac{0.5\text{ lb active ingredient B}}{1\text{ acre}}\right)\left(\frac{1\text{ gal premix C}}{1.5\text{ lb active ingredient B}}\right)=\frac{0.33\text{ gal premix C}}{1\text{ acre}}$$

Note that there is a difference in the amounts of premix C to use per acre when the individual active ingredient of A and B is calculated. If you use the correct amount of premix C to get the active ingredient in product A then product B will be under applied. If premix C is used to get the correct amount of B then product A will be over applied. To reconcile this, use the LESSER amount of premix C and ADD enough of the single formulation to enhance the mix so that the desired active ingredient amount of each is used.

In this case, the lesser amount of premix C is when the correct amount of product A is used (0.254 gal of premix C).

How much active ingredient of product B is in this amount of premix C? How much more of product B is needed to get the desired amount of active ingredient of B?

**Step 3:** Determine the amount of active ingredient of B in premix C when the premix rate is based on product A (0.254gal/acre).

$$\left(\frac{0.254\text{ gal premix C}}{1\text{ acre}}\right)\left(\frac{1.5\text{ lb active ingredient B}}{1\text{ gal premix C}}\right)=\frac{0.38\text{ lb active ingredient B}}{1\text{ acre}}$$

**Step 4:** Determine the difference between the amount of product B applied with the premix and the amount required. This is determined by subtracting the two values.

Solution:
Additional B needed = (0.5 lb B desired) – (0.38 lb B applied with premix) = 0.12 lb active in B

*...continued on page 72*

**Step 5**: Determine the gallons of product B that is required

$$\left(\frac{0.12\text{ lb active ingredient B}}{1\text{ acre}}\right)\left(\frac{1\text{ gal product B}}{4\text{ lb active ingredient B}}\right) = \frac{0.03\text{ gal product B}}{1\text{ acre}}$$

Step 6: Determine the cost of Premix C plus the additional amount of Product B per acre?

Cost of product C =

$$\left(\frac{0.254\text{ gal premix C}}{1\text{ acre}}\right)\left(\frac{\$105}{1\text{ gal}}\right) = \frac{\$26.67\text{ premix C}}{1\text{ acre}}$$

Cost of product B =

$$\left(\frac{0.03\text{ gal B}}{1\text{ acre}}\right)\left(\frac{\$115}{1\text{ gal product B}}\right) = \frac{\$3.45\text{ product B}}{1\text{ acre}}$$

Total cost = $26.67 + $3.45 = $30.12

The total cost of mixing product A and product B to get the desired application amount for each was $37.76/acre. Using premix C and adding enough of product B to make up the active ingredient difference was $30.12/acre. Therefore, using the premix C and adding product B will save $7.64/acre ($37.76 - $30.12). Over a 160 acre field the savings by using the premix C plus product B will be $1,222.40 ($7.64/acre * 160 acres), a substantial savings for a few minutes of calculation. (even if it took an hour to work through this problem, your wage would be $1,222/hr).

**Exercise 9-17c. You determine that the application rate for a desirable tank mix is 0.6 pt/acre of herbicide A that contains 3.2 lb ai/gal and 10 fluid oz of product B that contains 1.2 lb ai/gal. A premix, product C, contains 5.2 lb ai of product A + 0.5 lb ai of product B. Costs are,**

- **Product A, $201.60/gal**
- **Product B, $104.23/gal**
- **Product C, $438/gal**

**1) Determine the cost of the tank mix per acre.**

**2) Determine the amount of product C per acre to give the desired amount of A and B.**

**3) Determine the correct amount of product C to use and which (if any) product needs to be added to make up the difference in rate.**

**4) Determine the total cost of premix + any additional chemical.**

**5) Determine the cost difference per acre of using the tank mix vs the premix combination.**

## Additional information


Renz, M.J. 2006. Extension Pesticide Applicator Training Series-#5: Calculating pesticide amounts for broadcast applications. University of New Mexico GA- 614. Available at http://www.cahe.nmsu.edu:16080/pubs/_a/A-614.pdf.

Hofman, V. and E. Solseng. 2004. Spray Equipment and Calibration. NSDU Extension Service AE-73 (Revised). North Dakota State Universtiy, Fargo, ND.

# ESTIMATING SEEDING RATES, PLANT POPULATIONS, CORN AND SOYBEAN YIELDS, AND YIELD LOSSES DURING COMBINING

**Key problems:** Estimating seeding rates and potential yields.

**Mathematical skills:** Review of addition, subtraction, and multiplication.

**Chapter concepts:** To maximize profitability, input costs and losses must be kept to a minimum. This chapter provides information on how to calculate these values.

## Estimating Planting Rates

The desired seeding rate is a function of the desired plant stand, seed size, germination rate, and expected seedling mortality. Different plants have different size seeds (**Box 10-1**). To obtain the desired stand population, the number of seeds per pound and the germination rate must be considered. The number of seeds per pound can be determined by weighing 1,000 seeds in grams and then converting these units to seeds per pound.

**Exercise 10-1. What is the number of seeds per pound if 1,000 seeds weigh 80 g?**

Solution:

$$\frac{1{,}000 \text{ seeds}}{80 \text{ g}} \cdot \frac{454 \text{ g}}{1 \text{ lb}} = \frac{5{,}675 \text{ seeds}}{1 \text{ lb}}$$

**Exercise 10-2. What is the number of seeds per pound if 500 seeds weigh 60 g?**

The number of seeds per pound can be subsequently combined with the germination test information to determine planting rates and cost of seed per acre.

Box 10-1. Range of seeds per pound for selected crops.

| Plant | Seeds/lb |
|---|---|
| Alfalfa | 200,000-250,000 |
| Barley | 12,000-15,000 |
| Corn grain | 1,000-1,500 |
| Flax | 70,000-90,000 |
| Kentucky bluegrass | 2,000,000-2,500,000 |
| Pea | 1,500-3,600 |
| Reed Canarygrass | 500,000-5,500,000 |
| Soybean | 2,500-3,000 |
| Wheat (hard red) | 12,000-15,000 |

**Exercise 10-3. What is the cost and planting rate in lb/acre if the desired population is 10 plants/ft$^2$, a pound of seed contains 200,000 seeds, the seed has a germination of 30%, and it costs \$5/lb?**

Solution:
To solve this exercise convert the germination rate to the decimal form, 1 seed/0.3 plants, and solve the problem.

$$\frac{10 \text{ plants}}{1 \text{ ft}^2} \cdot \frac{43{,}560 \text{ ft}^2}{1 \text{ acre}} \cdot \frac{1 \text{ seed}}{0.3 \text{ plants}} \cdot \frac{1 \text{ lb}}{200{,}000 \text{ seeds}} = \frac{7.26 \text{ lb}}{1 \text{ acre}}$$

$$\frac{7.26 \text{ lb}}{1 \text{ acre}} \cdot \frac{\$5.00}{1 \text{ lb}} = \frac{\$36.3}{1 \text{ acre}}$$

**Exercise 10-4. What is the cost and planting rate in lb/acre if the desired population is 0.8 plants/ft², a pound of seed contains 1,500 seeds, the seed has a germination rate of 95%, and it costs $2.68/lb?**

## Estimating Plant Populations

Plant population information is needed to estimate yields and germination rates. Estimating population involves measuring the number of plants in a given area and converting this value to the number of plants per acre.

**Exercise 10-5. If 29 corn plants are measured in 20 ft of row and the rows are spaced 30 in. apart, what is the population?**

Solution

The population is $\frac{29 \text{ plants}}{20 \text{ ft} \cdot 30 \text{ in.}}$. However, this answer is not in the usual format of plants/acre and therefore the 20 ft x 30 in. need to be converted to acres. This is accomplished by first converting the value 20 ft x 30 in. to ft², and then converting ft² to acres.

$$\frac{29 \text{ plants}}{20 \text{ ft} \cdot 30 \text{ in.}} \cdot \frac{12 \text{ in.}}{1 \text{ ft}} \cdot \frac{43{,}560 \text{ ft}^2}{1 \text{ acre}} = \frac{25{,}265 \text{ plants}}{1 \text{ acre}}$$

**Exercise 10-6. If 30 corn plants are measured in 25 ft of row and the rows are spaced 21 in. apart, what is the population?**

**Exercise 10-7. If 20 corn plants are measured in 20 ft of row and the rows are spaced 21 in. apart, what is the population?**

## Estimating Yields

Estimating corn yields involves determining the number of kernels per ear, and the number of ears in a sampling area, and then converting these numbers to bushels per acre. The calculation involves multiple conversions. Crop yield estimates are very useful for marketing and planning purposes. Protocols have been developed for estimating both corn and soybean yields.

## Develop a Protocol for Estimating Corn Yields

The approach described below is one of many that can be used to estimate corn yields.

1. Determine the sampled area. For fields with rows spaced at 30 in., it is convenient to sample an area one row (30 in.) wide by 17 ft 5 in. long. The area is determined by multiplying length times the width:

   $$\text{Sampled Area} = \text{length} \cdot \text{width} =$$

   $$\left(17 \text{ ft} \cdot \frac{12 \text{ in.}}{1 \text{ ft}} + 5 \text{ in.}\right) \cdot 30 \text{ in.} \cdot \frac{1 \text{ ft}^2}{144 \text{ in}^2} = 43.55 \text{ ft}^2$$

   This area represents approximately 1/1,000 (0.001) of an acre. (1 acre = 43,560 ft².)

2. Determine the ear population of a given area. This value is determined by counting the number of average sized ears in a 17 ft 5 in. long row (for 30 in. rows). In this example, there are 27 plants with ears in the row.

3. Determine the average number of kernels on an ear. Select a representative ear, count the number of rows and the number of kernels in a row. Note: care must be used in selecting this ear. If the largest ear is selected, the estimates

will over-estimate yields. Several ears can also be selected from which an average yield can be determined.

4. The number of kernels on the ear is calculated by multiplying the number of kernels in a row times the number of rows. In this example, there are 16 rows on an ear with 35 kernels per row. The number of kernels in the sampling area is calculated below:

$$\frac{\text{kernels}}{\text{1 ear}} = \frac{\text{kernels}}{\text{1 row}} \cdot \frac{\text{rows}}{\text{1 ear}} = \frac{\text{35 kernels}}{\text{1 row}} \cdot \frac{\text{16 rows}}{\text{1 ear}} = \frac{\text{560 kernels}}{\text{1 ear}}$$

$$\frac{\text{kernels}}{\text{sampling area}} = \frac{\text{kernels}}{\text{1 ear}} \cdot \frac{\text{ears}}{\text{sampling area}} =$$

$$\frac{\text{560 kernels}}{\text{1 ear}} \cdot \frac{\text{27 plants}}{\text{sampling area}} = \frac{\text{15,120 kernels}}{\text{sampling area}}$$

6. Convert the value of kernels/sampling area to bushels per acre. For this calculation, assume that a bushel contains approximately 80,000 kernels. If the kernels appear small, 90,000 kernels/bu or 100,000 kernels/bu may be more appropriate.

$$\frac{\text{bu}}{\text{acre}} = \frac{\text{\# kernels}}{\text{sampling area}} \cdot \frac{\text{1 bu}}{\text{\# kernels}} =$$

$$\frac{\text{15,120 kernels}}{\text{0.001 acre}} \cdot \frac{\text{bu}}{\text{80,000 kernels}} = \frac{\text{189 bu}}{\text{acre}}$$

**Exercise 10-8. What length of row is needed to produce 1/1,000 (0.001) of an acre if the row spacing is 22 in.?**

Solution:

$$43.56\ \text{ft}^2 = \text{length} \cdot \left(22\ \text{in.} \cdot \frac{1\ \text{ft}}{12\ \text{in.}}\right)$$

Solve for length

$$\text{Length} = \frac{43.56\ \text{ft}^2}{1.833\ \text{ft}} = 23.76\ \text{ft} = 23\ \text{ft} + 9.12\ \text{in.}$$

**Exercise 10-9. If a corn ear contains 40 kernels/row and has 16 rows, what is its estimated yield if there are 25 plants in 1/1,000 (0.001) of an acre?**

Solution:

$$\frac{\text{kernels}}{\text{ear}} = \frac{\text{40 kernels}}{\text{1 rows}} \cdot \frac{\text{16 rows}}{\text{1 ear}} = \frac{\text{640 kernels}}{\text{1 ear}}$$

$$\frac{\text{kernels}}{\text{sampling area}} = \frac{\text{640 kernels}}{\text{1 ear}} \cdot \frac{\text{25 plants}}{\text{1 sampling area}} = \frac{\text{16,000 kernels}}{\text{sampling area}}$$

$$\frac{\text{bu}}{\text{acre}} = \frac{\text{16,000 kernels}}{\text{0.001 acre}} \cdot \frac{\text{bu}}{\text{80,000 kernels}} = \frac{\text{200 bu}}{\text{acre}}$$

Note: this calculation assumes that a bushel contains 80,000 kernels. This estimate could be improved by weighing the kernels.

**Exercise 10-10. Develop a protocol for estimating soybean yields.**

Solution:

Research has shown that soybean yield estimate accuracy improves with decreasing time until maturity. An eight step process is used to estimate yield.

1. Determine the number of feet of row to produce 1/1,000 of an acre (see above or appendix 1 for a table of values);
2. Count the number of plants in this length of row at several locations (more the better);
3. Determine the average of these counts;
4. Count the number of pods on 10 randomly selected plants;
5. Calculate the number of pods/acre by multiplying the plants/acre by the number of pods/plant;
6. Open up 10 different pods and calculate the average number of seeds/pod. If each pod contains on average 2.5 seeds then the seeds/acre is determined by multiplying pods/acre by 2.5 seeds/pod;
7. Calculate lb/acre by dividing seed/acre by 2,900 seeds/lb.
8. Estimate yield in bushels/acre by dividing lb/acre by 60 lb/bu.

**Exercise 10-11. What is the estimated yield if 150 soybean plants are counted in a row length of 17 ft 5 in. if the distance between the rows is 30 in., each pod contains 2.5 seeds, and each plant contains 15 pods?**

**Exercise 10-12. What is the estimated yield if 200 soybean plants are counted in a row length of 17 ft 5 in. if the distance between the rows is 30 in., each pod contains 3.0 seeds, and each plant contains 12 pods?**

## Estimating Corn Yield Losses During Combining

A goal when harvesting is to leave as little grain in the field as is possible. Under good field conditions, new combines leave less than 0.5 bu/acre in the field. Making an estimate of how much grain is left behind in a harvested field can be accomplished very simply. Place a 1 ft by 1 ft (inside dimension) box on the ground and count the kernels, beans, or seeds found within the box. It is recommended that at least three locations behind the combine be used (one behind the left side of the header, one behind the center of the combine, and one behind the right side of the combine). In soybeans, you must consider those beans that are left on the main stem below the height of cut. The average number of individual beans, kernels of corn, or wheat seeds found in 1 ft$^2$ must be determined. Conversion to yield losses vary among crops.

| Box 10-2. Seeds on ground in 1 ft$^2$ area representing 1 bushel of grain lost. | |
|---|---|
| Crops | Seeds/bu |
| Corn | 2 |
| Soybeans | 5 |
| Wheat | 22 |

If four corn kernels are found per ft$^2$, then 2 bu/acre were lost, whereas if 22 wheat seeds were found then only 1 bu/acre was lost.

1) **Preharvest loss**. First calculate the pre and post harvest losses using **Box 10-2**. The preharvest loss is determined in an unharvested area of the field, usually before harvest and at physiological grain maturity. The loss attributed to the combine is the difference between the pre and post harvest values. In corn, each ¾ of an ear in 1/100 of an acre represents approximately 1 bu/acre. See **Box 10-3** below to estimate 1/100 of an acre. If most of the total losses are from preharvest damage, there is little or nothing that can be done to increase harvest efficiency.

2) **Header loss.** If preharvest losses are small, the agronomist must next check if there are significant losses from the header. When a combine pass is being made in the field, stop in the row and back up. Use the method described above to take 3 or 4, 1-ft$^2$ measurements in the area between the uncut grain and the area where chaff was spread. Compare this to the 1-ft$^2$ measurements taken from behind the combine. This analysis will determine if the loss is from the combine header. If these losses are a significant part of the total harvest losses, check with your operators manual or dealer to learn how adjustments can be accomplished to reduce header yield losses. New major machinery manufacturer and after market headers now on the market can lower losses appreciably. If one of these headers will save you a bushel or more per acre, the annualized capital cost of the header attachment per acre should be compared with the loss of yield.

3) **Corn ear loss.** As you walk down a harvested row, whole ears of corn may be found. Assuming a typical ear to be 18 rows by 35 kernels, or 80,000 kernels/bu corn, 1 ear/.01 acre amounts to about 0.75 bu/acre. Below is a table helpful for determining the distance for 0.01 or 1/100 of an acre.

| Box 10-3. Row width and length to determine 1/100 or 0.01 acres. | |
|---|---|
| Row width | Length (ft) for 1/100 acre |
| 20 in. | 261.36 |
| 30 in. | 174.2 |
| 40 in. | 130.68 |

4) **Separation loss**. If preharvest and header losses are insignificant, then you will conclude that your losses result from less than perfect threshing/separation of grain within the combine. If this is the

case, there are a number of adjustments that can be made that will have significant impact upon losses from within the combine. Check your operators manual and/or your dealer to determine the corrective adjustments or other action that should be made. Again, it is important to balance the annualized capital cost for combine upgrade against the yield loss.

**Exercise 10-13. What is the yield loss if 13 corn kernels are found in a 1 ft² area behind a combine?**

Solution:

$$\text{Yield Loss} = \left(\frac{13\text{ kernels}}{1\text{ ft}^2}\right)\left(\frac{43,560\text{ ft}^2}{1\text{ acre}}\right)\left(\frac{1\text{ bu}}{80,000\text{ kernels}}\right) = \frac{7.1\text{ bu}}{1\text{ acre}}$$

Note this calculation assumes that a bushel of corn contains 80,000 kernels.

**Exercise 10-14. What is the yield loss if 10 kernels are found in a 0.5 ft² area behind a combine?**

Solution:

$$\text{Yield Loss} = \left(\frac{10\text{ kernels}}{0.5\text{ ft}^2}\right)\left(\frac{43,560\text{ ft}^2}{\text{acre}}\right)\left(\frac{1\text{ bu}}{80,000\text{ kernels}}\right) = \frac{10.9\text{ bu}}{\text{acre}}$$

## Yield Monitor Data

The use of yield monitors to assist in on-farm decisions continues to increase. Analyzing these data is complicated and requires appropriate software. Yield monitor data processing software is available at "GIS Applications in Agriculture" Eds Francis J. Pierce and David Clay; available from CRC Press at http://www.crcpress.com. One use of yield monitor data is to define management zones. Programs that can assist in defining management zones are available at http://www.ars.usda.gov/SP2UserFiles/Place/36221500/briefs/Brief-MZA.pdf, and http://zonemap.umac.org/. In the management zone approach, a large field is separated into smaller areas that can be managed uniformly. Data that have been used to classify management zones include soil survey, EC, yield monitor, elevation, and remote sensing.

Remote sensing data have been very useful for defining crop health. A number of different remote sensing indexes are being used. Two indexes that are widely used are NDVI and GNDVI. NDVI is calculated using the equation, NDVI = (NIR – red)/(NIR + red). GNDVI is calculated with the equation GNDVI = (NIR – green)/(NIR + green). Free Landsat data and assistance in using remote sensing are available at http://www.americaview.org/currentmembers.htm and http://eros.usgs.gov/. Remote sensors that can be used for management purposes are being commercially manufactured. Information about two sensors being used for a variety of purposes is available at http://www.hollandscientific.com/ and http://www.ntechindustries.com/.

## Additional Information


Carlson, C.G., T. Doerge, and D.E. Clay. 2002. Estimating corn yield losses from unevenly spaced corn. SSMG 37. Clay et al. (Ed) *Site-Specific Management Guidelines.* Potash & Phosphate Institute. Norcross, GA. Available at http://www.ipni.net/ssmg.

Lyon, D.J. and R.N. Klein. 2001. Estimating winter wheat grain yields. Nebraska Extension Service, Neb Guide G1429. Available at http://www.ianrpubs.unl.edu/epublic/live/g1429/build/g1429.pdf.

Nielsen, R.L. 2004. Estimating corn grain yields prior to harvest. Corny New Network Articles. Purdue University. Available at http://www.agry.purdue.edu/ext/corn/news/articles.04/YieldEst-0718.html.

# FORAGE AND GRAIN YIELDS — MOISTURE AND SHRINKAGE

**Key problems:** Determining crop and hay yields; moisture content; estimating grain yields; yield losses during combining; determining seeding rates based on germination tests and desired populations.

**Mathematical skills:** Review of equation solving, algebra, addition, and subtraction.

**Chapter concepts:** The ability to calculate and estimate yields is critical for solving natural resource management problems. This chapter shows many yield calculation examples.

## Forage Yields

Commodities are sold using different measurement units. To ensure that an adequate payment is received, an understanding of the commodity unit measurement is required. Some commodities are traded on a weight basis, while others are based on a volume basis.

In forages, yield can be based on nutritional content or dry matter (DM) yield. Dry matter yield is traditionally estimated by measuring the height of the standing forage. Heights are converted to dry matter with conversion factors that range from 100 lb of dry matter per inch for forages in fair condition to over 400 lb/in. for forages in excellent condition. Conversion factors are based on actual measurements that include clipping, weighing, and drying forages from an area of 1 $m^2$. Forage dry matter and moisture percentages are calculated using the equations:

$$\text{Dry matter percentage} = \frac{\text{dry weight} \cdot 100\%}{\text{wet weight}}$$

$$\text{Moisture percentage} = 100 - \text{dry matter percentage}$$

**Note:** to avoid confusion, remember that the definition of moisture content differs from soil to plant material. Recall that the gravimetric moisture content of soil was given as:

$$\%\text{ soil moisture} = \frac{\text{water weight}}{\text{dry soil weight}}$$

The gravimetric moisture content of plant and/or grain is defined as:

$$\%\text{ moisture in plant material or } (\%\text{ moisture in grain}) = \frac{\text{water weight}}{\text{water weight} + \text{dry plant material weight}}$$

$$\text{OR water weight} = \frac{\left(\frac{\text{moisture }\%}{100}\right)(\text{grain weight})}{\left(1 - \frac{\text{moisture }\%}{100}\right)}$$

**Exercise 11-1. Estimate dry matter yield in a good stand of mixed pasture grass that is 4 in. taller than cutting height. Each inch is estimated to contain 300 lb of dry matter (DM) per acre.**

Solution:

$$\left(4\text{ in. forage} \cdot \frac{300\text{ lb DM}}{1\text{ in. forage} \cdot \text{acre}}\right) = \frac{1{,}200\text{ lb DM}}{\text{acre}}$$

**Exercise 11-2. A pasture contains 9.0 in. of forage. Each inch contains 350 lb/acre of forage dry matter. If a cow eats 118 lb of dry matter per day, how many cow-days of food are in 100 acres of that pasture?**

**Exercise 11-3. Calculate the dry matter percentage if the dry and wet weight of a forage sample are 30 and 89 g, respectively.**

## Grain Yields

Many grain-based (corn, soybean, wheat) commodities are sold as a volume measure (bushels). A grain species-specific standard bushel is defined by a specified weight and moisture content. The values in **Box 11-1** are typically used to define a standard bushel and are used to determine yield by converting weight of grain per acre to standard bushels per acre.

**Box 11-1.** Standard bushel weight at indicated moisture.

| Grain | lb/bu | % moisture | lb/bu at 0% moisture |
|---|---|---|---|
| Corn | 56.00 | 15.5% | 47.32 |
| Wheat | 60.00 | 13.5% | 51.90 |
| Soybeans | 60.00 | 13.0% | 52.20 |
| Barley | 48.00 | 14.5% | 41.04 |
| Oats | 32.00 | 14.0% | 27.52 |
| Rye | 56.00 | 14.0% | 48.16 |

For example, a standard bushel of soybeans has a moisture content of 13% and weighs 60 lb. A standard bushel of wheat has a moisture content of 13.5% and weighs 60 lb.

**Exercise 11-4. A weigh wagon contains 4,500 lb of corn at 15.5% moisture. How many bushels of corn are in the weigh wagon?**

Solution:
Corn grain at 15.5% moisture weighs 56 lb/bu. Based on these numbers, the answer is

$$bu_{corn} = \left(4,500 \text{ lb} \cdot \frac{1 \text{ bu corn}}{56 \text{ lb}}\right) = 80 \text{ bu of corn}$$

**Exercise 11-5. The weigh wagon contains 4,500 lb of soybeans at 13% moisture. How many bushels of soybeans are in the weigh wagon?**

**Exercise 11-6. A weigh wagon full of wheat at 13.5% moisture contains 4,500 lb. How many bushels are in the wagon?**

## Grain Moisture Percentage

To convert grain weights to standard bushels, the moisture percent of the grain must be known. Grains with different moisture percentages have different weights of dry grain per bushel.

The percentage of dry matter (DM) in a bushel is calculated with the equation

$$\% \text{ DM} = (100 - \% \text{ moisture})$$

and the amount of dry grain (DG) in a bushel is

$$\text{DG} = (\text{standard weight})\left(\frac{\% \text{ DM}}{100}\right)$$

For corn at 15.5% moisture

$$\% \text{ DM} = (100 - 15.5\%) = 84.5\%$$

DG in 1 bu of corn

$$\left(\frac{56 \text{ lb corn}}{\text{standard bu}}\right)(0.845) = \frac{47.32 \text{ lb dry corn}}{\text{bu}}$$

**Exercise 11-7. Calculate the amount of dry corn in 1,000 lb of corn at 20% moisture.**

**Exercise 11-8. Calculate the amount of dry wheat in 50,000 lb at 22% moisture.**

**Exercise 11-9. Calculate the amount of dry barley in a standard bushel at 14.5% moisture.**

**Exercise 11-10. Determine the amount of dry corn contained in a standard bushel of corn at 0% moisture.**

Solution:
Grain moisture contents are reported on a wet grain basis. The amount of dry corn in a standard bushel should be the same for corn at 20, 10, and 5% moisture. Therefore, the amount of dry corn contained in a standard bushel can be determined by solving for Dry Grain in the equation,

$$\% \text{ moisture} = \left(\frac{\text{water weight}}{\text{water weight} + \text{dry grain weight}}\right)100\%$$

However, this equation can not be solved because it contains two unknowns (water weight and dry corn weight). To solve this equation, more information is needed. We assume that the standard weight of corn is 56 lb and that the standard moisture content is 15.5%, then we can substitute these into the equation above and determine the weight of water,

$$15.5\% = \left(\frac{\text{water weight}}{56 \text{ lb}}\right)100\%$$

This equation is rearranged

Water weight = 0.155 x 56 lb, so water weight = 8.68 lb

And it follows that since water weight + dry grain weight = 56, then

Dry grain weight = 56 - 8.68 = 47.32 lb

Therefore, the amount of dry corn in 1 standard bushel (56 lb of corn at 15.5% moisture) is 47.32 lb.

**Exercise 11-11. Determine the amount of dry matter contained in a standard bushel (60 lb at 13.5% moisture) of wheat at 0% moisture.**

## Estimating Grain Test Weight (TW)

Test weight (TW) is a measure of grain density, and is determined by weighing a measured amount of grain, usually a dry quart, and converting that value to pounds per bushel. Grain terminals use more precise measurements and equipment, but a set of kitchen measuring cups and a postal scale can provide a good estimate of test weight at a fraction of the cost. It is important to consider that a dry quart (1 dry quart = 67.200625 in.$^3$) is not equal to 1 liquid quart (1 liquid quart = 57.75 in.$^3$). Calibrated dry quart containers are available, but $4\frac{2}{3}$ cups will approximate 1 dry quart (there are 32 dry quarts in a bushel).

**Exercise 11-12. Determine the test weight of a dry quart ($4\frac{2}{3}$ cups) of corn that weighs 29 oz.**

Solution:

$$\left(\frac{\text{test weight lb}}{\text{bu}}\right) = \left(\frac{\text{measured weight oz}}{\text{dry qt}}\right) \times \left(\frac{32 \text{ dry qt}}{1 \text{ bu}}\right) \times \left(\frac{1 \text{ lb}}{16 \text{ oz}}\right)$$

$$\text{Test weight} = \left(\frac{29 \text{ oz}}{1 \text{ dry qt}}\right) \times \left(\frac{32 \text{ qt}}{1 \text{ bu}}\right) \times \left(\frac{1 \text{ lb}}{16 \text{ oz}}\right) = \frac{58 \text{ lb}}{\text{bu}}$$

**Exercise 11-13. Determine the wet weight of a bushel of corn that contains 10% moisture.**

Solution:
To solve this exercise, the water weight in the % moisture equation must be determined. As before, the dry corn weight is 47.32 lb which is substituted into the equation and solved for water weight as follows,

$$10\% = \left(\frac{\text{water weight}}{\text{water weight} + 47.32\ \text{lb}}\right)100\%$$

Rearrangement results in

0.10 (water weight + 47.32 lb) = (water weight)

Rearrangement results in

water weight - 0.10 • water weight = 0.10 • 47.32 lb

water weight • (1 - 0.1) = 4.732 lb

0.90 • water weight = 4.732 lb

$$\text{water weight} = \frac{4.732\ \text{lb}}{0.90} = 5.26\ \text{lb}$$

Based on these calculations, a standard bushel of corn (47.32 lb of dry corn) at 10% moisture contains 5.26 lb of water, so (5.26 + 47.32) = 52.58 lb.

**Exercise 11-14. A combine contains 2,300 lb of corn grain at 20% moisture. How many standard bushels of corn are contained in the combine if a standard bushel of corn has a moisture content of 15.5% and weighs 56 lb?**

## Grain Shrinkage

Grain is bought and sold based on rules developed by our forefathers. Before large scale weighing capability was available, grain was sold by volume, thus the bushel became the basic unit of grain commerce. A bushel (United States dry measure) equals 2150.42 in.[3] When wet corn was traded, it was found that as the grain dried, it lost volume which increased test weight by 0.25 to 0.5 lb/bu point. One moisture percentage point is commonly abbreviated in grain trade discussions and is referred to as a point, thus the term shrinkage was used to describe volume loss. Today, because of our ability to accurately weigh a delivered load, we don't measure the volume. Therefore, the word shrinkage is associated with the weight loss rather than volume loss due to drying.

Grain traders often use a total shrink factor (TS) to compensate for the excess moisture that will be removed from grain that exceeds the specifications of a standard bushel. The excess moisture will be lost when the wet grain is dried to the secondary buyer's specification. The total shrink factor (TS) is the sum of the moisture shrink factor (MS) and handling shrink factor (HS) calculated as:

$$TS = MS + HS$$

**Box 11-2.** Theoretically correct % shrinkage loss per point for varied final moisture contents.

| Final % moisture contents | % shrink/pt |
|---|---|
| 15.5 | 1.18343195 |
| 15 | 1.17647059 |
| 14.5 | 1.16959064 |
| 14 | 1.1627907 |
| 13.5 | 1.15606936 |
| 13 | 1.14942529 |
| 12.5 | 1.14285714 |
| 12 | 1.13636364 |
| 11.5 | 1.1299435 |
| 11 | 1.12359551 |
| 10.5 | 1.11737844 |
| 10 | 1.11111111 |

The % shrink/point column in **Box 11-2** shows the theoretically correct factors to use for the MS factor. Handling shrinkage (HS) varies from buyer to buyer. The actual amount of handling shrink has extreme variations. In most cases, grain buyers use a TS shrink factor that is slightly larger than the theoretical moisture shrink (MS) factor (1.1834319% for 15.5% basis corn) to allow for their losses in handling grain. Typical total shrink factors (TS) used by grain buyers range from 1.25 to 1.4% per point of moisture content.

If the moisture content of grain offered for sale is less than the standard moisture content, then there may be "hidden dockage". Again, bushels are usually determined by dividing the total weight by standard bushel weight (for corn this is 56 lb). If the corn delivered is drier than the standard, seldom are the number of bushels adjusted upward to account for the fact that more dry grain is being delivered. This means that when grain is over dried (the grain moisture content is less than standard moisture content) the seller will get paid for fewer bushels.

These calculations suggest that the economic consequences of selling corn drier than the standard moisture percentage rest on the seller. This consequence must be balanced with the need to store the grain safely. The length that corn can be stored is related to temperature. Wet corn must be used or dried before long-term storage. A rule of thumb is that corn should not be harvested until the moisture content is less than 30% and it should be dried to 14% if storage is planned for 12 months or less and 13% if storage is planned for longer than one year. The relationship between storage lengths, grain moisture, and storage temperature is shown in **Box 11-3**.

**Exercise 11-15. An elevator is purchasing corn at 15.5% moisture for $4.50/bu. Your corn is at 22% moisture. What is the selling price based on the theoretical shrinking factors?**

Solution:

The following steps should be followed to solve this problem:

1. Calculate points

   22 points - 15.5 points = 6.5 points

2. Calculate discount

   $$(6.5 \text{ points})(1.18343)\left(\frac{1}{100\%}\right)\left(\frac{\$4.50}{\text{bu}}\right) = \$0.346$$

3. Calculate selling price

   $$\$4.50 - \$0.346 = \frac{\$4.15}{\text{bu}}$$

**Exercise 11-16. A buyer wants to purchase corn at 13% moisture. The corn is currently at 16% moisture. The selling price is $5.00/bu. What is the discounted price if it is assumed that a bushel weighs 56 lb?**

Solution:

The following steps should be followed to solve this problem:

1. Calculate points

   16 points - 13 points = 3 points

2. Calculate discount

   $$(3 \text{ points})(1.1494\%)\left(\frac{1}{100\%}\right)\left(\frac{\$5.00}{\text{bu}}\right) = \$0.17$$

3. Calculate purchase price

   $$\frac{\$5.00}{\text{bu}} - \$0.17 = \$4.83$$

**Box 11-3.** Allowable storage time as influenced by grain moisture content and storage temperature (modified from Pohl and Durland, 2002).

| Temperature, °F | Grain moisture content % | | | | | | | | | | |
|---|---|---|---|---|---|---|---|---|---|---|---|
| | 14 | 15 | 16 | 17 | 18 | 20 | 22 | 24 | 26 | 28 | 30 |
| | Allowable storage time (days) | | | | | | | | | | |
| 30 | | | 847 | 503 | 323 | 160 | 95 | 64 | 47 | 37 | 31 |
| 35 | | | 634 | 377 | 242 | 120 | 71 | 48 | 35 | 28 | 23 |
| 40 | | 879 | 474 | 282 | 181 | 90 | 53 | 36 | 26 | 21 | 17 |
| 50 | | 492 | 265 | 158 | 101 | 50 | 30 | 20 | 15 | 12 | 10 |
| 60 | 576 | 275 | 149 | 88 | 57 | 28 | 17 | 11 | 8 | 6 | 5 |

**Exercise 11-17. A load of 22.5% moisture corn is taken to the elevator. The elevator is advertising that they will pay $2.05/bu for the corn. The elevator dockage sheet indicates that they will dock $0.05 for each moisture percentage point above 15.5% moisture for shrinkage, and $0.03 for each moisture percentage point above 15.5% for drying. After dockage, you calculate that you will receive $1.49 for your 22.5% moisture corn. Is this fair?**

Solution:
**Even though it is not technically correct, many grain purchasers consider a bushel of corn to be 56 lb regardless of the moisture content or volume.** To determine what is fair, start by determining how much dry corn is contained in each bushel (56 lb) of delivered corn. To calculate the amount of dry corn in 56 lb of 22.5% moisture corn, the weight of water and dry corn must be determined.

After substitution,

$$\text{water weight in 22.5\% moisture corn} = \frac{22.5\%}{100\%} \cdot 56 \text{ lb} = 12.6 \text{ lb}$$

Therefore, dry corn weight at 22.5% moisture is 56 lb – 12.6 lb = 43.4 lb

Since from above

$$\text{water weight in 15.5\% moisture corn} = \frac{\left(\text{grain moisture \%}/_{100\%}\right)(\text{weight dry corn})}{1-\left(\text{grain moisture \%}/_{100\%}\right)} =$$

$$\frac{0.155 \cdot 43.4}{1-0.155} = 7.96094675 \text{ lb}$$

Add this weight to the weight of dry corn to determine total weight: 43.4 lb + 7.96 lb = 51.36 lb corn. This is the weight of corn after drying the 56 lb of 22.5% corn to 15.5% corn.

The weight loss relative to the initial weight is:

$$\left(\frac{56 \text{ lb} - 51.36 \text{ lb}}{56 \text{ lb}}\right) \times 100 = 8.284\%,$$

which divided by 22.5% – 15.5% = 7 points

$$= \frac{8.284\%}{7 \text{ points}} = \frac{1.1834319\%}{\text{point}}$$

We would have reached the same results by using the 1.1834% of loss/point for a base/final moisture content of 15.5% from **Box 11-2** on the previous page.

$$7 \text{ points} \cdot \frac{1.1834319\%}{\text{point}} \cdot 56 \text{ lb/bu} = 4.6390 \text{ lb}$$

$$56 \text{ lb} - 4.6391 \text{ lb} = 51.3609 \text{ lb}$$

$$7 \text{ points} \cdot \frac{1.1834319\%}{\text{point}} \cdot \$2.05/\text{bu} = \$0.17/\text{bu}$$

The elevator is charging a total shrinkage loss (TS) of $.05 • 7 points = $0.35 and $0.21 for drying, resulting in a selling price of $2.05 - 0.56 = $1.49/bu. Based on the value of dry corn, a bu is worth it. $2.05 - 0.17 = $1.88. The difference between $1.88 and $1.49 is $0.39, and therefore if drying is greater than $0.39/bu, this is a good deal.

**Exercise 11-18. If soybeans (standard bu = 60 lb at 13% moisture) are selling for $7.25/bu, theoretically, what should a bushel (60 lb) of 19% moisture beans be worth?**

**Exercise 11-19. If wheat (standard bu = 60 lb at 13.5% moisture) is selling for $12.25/bu, theoretically, what should the moisture shrinkage cost per bu (60 lb) of 18% moisture wheat be worth?**

## Additional Information

Cloud, H.A., L.D. Van Fossen, L.F. Backer, R.C. Brook, L.A. Brook, K.J. Hellevang, B.A. McKenzie, W.H. Peterson, and R.O. Pierce. 1997. Grain Drying, Handling and Storage Handbook. Midwest Plan Service, Iowa State University, Ames, Iowa 50011. Available for purchase at: https://www.mwps.org/. Accessed 06/12/2008.

Pohl, S. and G.R. Durland. 2002. Grain drying guidelines for a wet fall harvest. SDSU Cooperative Service Extension ExEx 1014. http://agbiopubs.sdstate.edu/articles/ExEx1014.pdf. Accessed 06/22/2008.

# CALCULATING GRAIN AND FORAGE STORAGE SPACE

**Key problems:** How much storage room do we need to maintain grain and forage quality with time?

**Mathematical skills:** Converting external length measurements to volumes.

**Chapter concepts:** Show student how to use equations to calculate storage capacity.

When storing grain, it is important to match the amount of grain harvested to the available storage capacity. To determine the available storage, the volume of the bin and the amount of grain that can be held within that volume must be determined. The volume for a cylinder is: $volume = \pi \cdot r^2 \cdot height$, where $\pi$ is approximately equal to 3.14159 and r is the radius ($2 \cdot r =$ diameter) of the cylinder. A bushel is equivalent to 1.244 $ft^3$. The storage capacity of a bin is calculated with the equation:

$$\frac{bu}{bin} = \frac{volume}{bin} \cdot \frac{bu}{volume}.$$

**Box 12-1.** Diagram and equation to determine the volume of a cylinder.

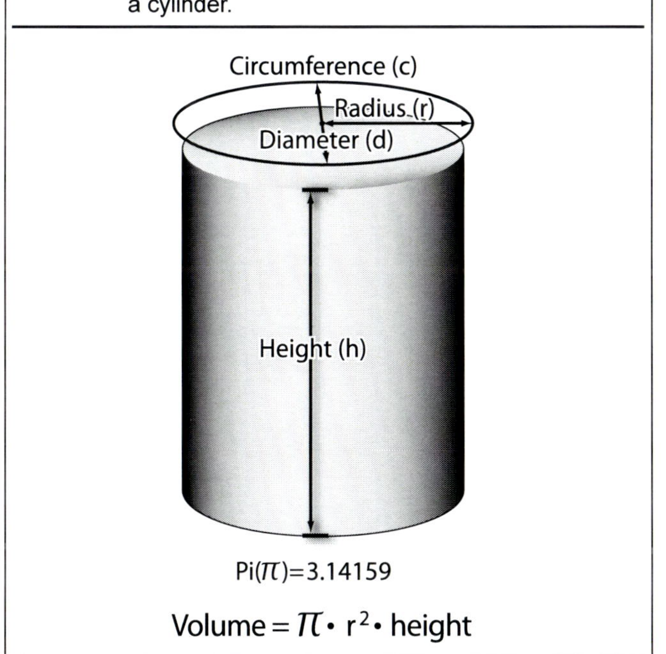


**Exercise 12-1. How many bushels of corn can be stored in a cylindrical bin that has an inside diameter of 26 ft and height of 24 ft?**

Solution:
The amount of grain that can be stored in a cylindrical bin is calculated by dividing its volume by the volume of a bushel. The volume of a bushel is 1.244 $ft^3$. By substituting a value for the radius (half the diameter) into the cylinder equation, the storage capacity of the bin is determined.

$$\frac{\pi \cdot 13^2 \cdot 24\ ft^3}{bin} \cdot \frac{1\ bu}{1.244\ ft^3} =$$

$$\frac{3.14159 \cdot 13^2 \cdot 24}{bin} \cdot \frac{1\ bu}{1.244\ ft^3} = 10,243\ bu$$

**Exercise 12-2. How many bushels of soybeans will be held in a cylindrical grain bin that has an inside diameter of 28 ft and is 30 ft tall?**

**Exercise 12-3. If a producer averages yields of 160 bu/acre, how many acres of corn production can be stored in a bin that holds 17,046 bu?**

**Exercise 12-4. How many bins that hold 17,046 bu are needed to store the production from 1,000 acres which produce 160 bu/acre?**

## Calculating Hay Storage Requirements

Producers can preserve hay quality by storing it in an appropriate facility. To determine the size of a facility, the number of hay bales of a specified size to be stored must be determined.

**Exercise 12-5. An 80-acre field has a hay yield of 5 tons/acre that will be stored in bales that are 14 in. (1.17 ft) by 18 in. (1.5 ft) by 40 in. (3.33 ft). The average bale weighs 65 lb. What is the length of storage facility needed if the width and height are 50 ft and 14 ft?**

Solution:
First, determine the volume of hay produced

$$80\text{ acres} \cdot \frac{5\text{ ton}}{\text{acre}} \cdot \frac{2{,}000\text{ lb}}{\text{ton}} \cdot \frac{1\text{ bale}}{65\text{ lb}} \frac{(1.17\text{ ft})(1.5\text{ ft})(3.33\text{ ft})}{\text{bale}} = 71{,}928\text{ ft}^3$$

Second, determine the dimensions of the building. The volume of a rectangular building is

Volume = Length • Width • Height

Solve for length = Volume/(Width • Height)

$$\text{Length} = \frac{71{,}928\text{ ft}^3}{50\text{ ft} \cdot 14\text{ ft}} = 102.8\text{ ft}$$

Note: This solution assumes perfect stacking. Based on this calculation, the building dimensions need to be at least 50 ft by 14 ft by 103 ft.

**Exercise 12-6. Calculate the height of a hay storage facility needed for 100 acres of alfalfa. The field has an average yield of 6 tons/acre, and the hay will be stored as bales that are 1 ft by 2 ft by 3 ft. The average bale weighs 70 lb. The building length and width are 128 ft and 60 ft, respectively.**

## Determining the Amount of Corn in a Pile and Using Slope to Estimate Height

The amount of grain contained in a pile can be estimated using the equation to determine the volume of a cone **(Box 12-2)**. To determine the volume, the radius and height must be measured or estimated. The radius may be determined by measuring the circumference of the pile with a flexible measuring tape. The radius is then calculated with the following equation.

$$\text{Radius (r)} = \frac{\text{circumference}}{2 \cdot \pi}$$

The height can be estimated using a variety of approaches. One approach is to use a clinometer or an Abney level. Using either instrument, the % slope is defined by the equation, $\%\text{ slope} = \frac{\text{rise}}{\text{run}} \cdot 100\%$. The run is the radius of the bottom of the pile plus the distance that you and the clinometer are from the edge of the pile.

**Box 12-2.** Diagram and equation to define the volume of a cone.

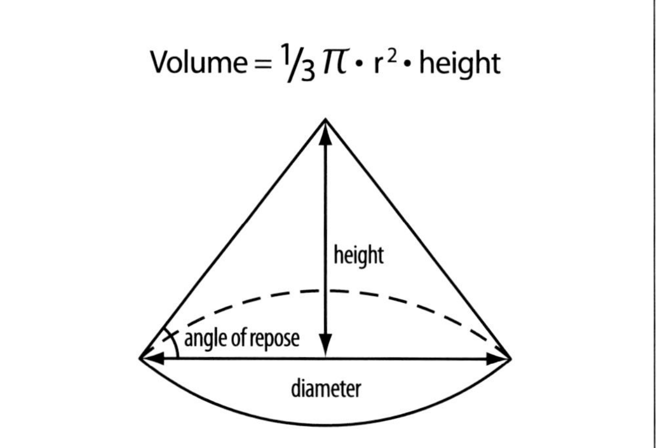

**Box 12-3.** Measuring volume of a grain pile.

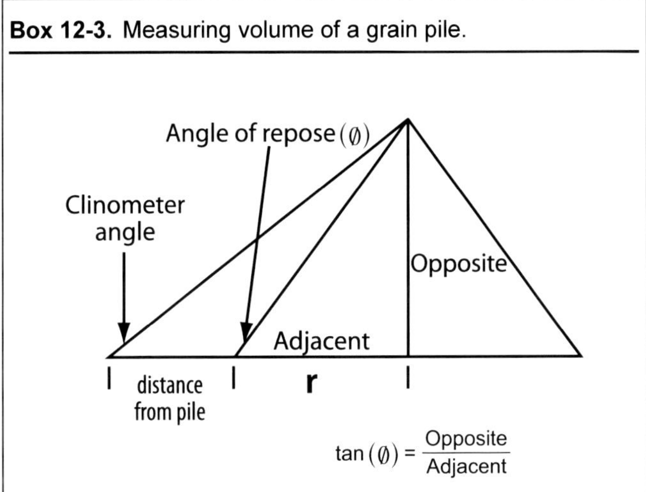


**Exercise 12-7. If the radius is 400 ft and you are 100 ft from the edge of the pile, how high is a pile if a clinometer is used to determine that the slope is 20%?**

Solution:

$$\% \text{ slope} = \frac{\text{rise}}{\text{run}} \cdot 100\%$$

$$\text{rise} = \frac{\% \text{ slope}}{100\%} \cdot \text{run}$$

$$\text{rise} = 0.2(400 \text{ ft} + 100 \text{ ft}) = 100 \text{ ft}$$

**Box 12-4.** Angles of repose that typically result from piling grain.

| Crop | Angle (AR) | Tan (AR) |
|---|---|---|
| Barley | 28 | 0.53 |
| Corn (Shelled) | 23 | 0.42 |
| Oats | 28 | 0.53 |
| Soybeans | 25 | 0.47 |
| Sunflowers | 27 | 0.51 |
| Wheat | 25 | 0.47 |

Adapted from Wilke & Wyatt, 2002 and Grain Drying, Storage, and Handling Handbook, MWPS-13.

**Exercise 12-8. What is the height and volume of a pile if the circumference is 100 ft and the angle of repose is 28°.**

Solution:

Circumference = $\pi \cdot 2r$

r = circumference/($\pi \cdot 2$) = 100 ft/(2 • 3.14) = 15.9 ft

Because tan ∅ = opposite/adjacent, Tan ∅ = height/radius

tan 28° = 0.53

0.53 • 15.9 = height

Height = 8.43 ft

Volume = $\pi \cdot r^2 \cdot$ height

Volume = 3.14 • (15.9 ft)$^2$ • 8.43 ft = 6,691.9 ft$^3$

Note that in the table in **Box 12-4**, the tangent (AR) angle of repose (AR) and tangent (tan AR) are listed. This is an estimate of the rise/run for different grain. Once the radius and height are known, the volume of the cone and number of bushels contained in the pile can be calculated using the equation,

$$\frac{\text{bushels}}{\text{pile}} = \text{cone volume} \cdot \frac{\text{bushels}}{\text{volume}}$$

**Exercise 12-9. You determine that the circumference of a cone shaped pile of wheat is 146 ft. How many bushels of wheat are in the pile?**

**Exercise 12-10. How many bushels of corn are contained in a cone shaped pile that is 23 ft high and has a circumference of 264 ft?**

Solution:

$$r = \frac{\text{circumference}}{\pi \cdot 2}$$

$$r = \frac{264\ \text{ft}}{\pi \cdot 2} = 42.0\ \text{ft}$$

$$\text{Cone Volume} = \frac{1}{3} \cdot \pi \cdot r^2 \cdot \text{height} = \frac{1}{3} \cdot 3.14 \cdot 42^2 \cdot 23 = 42{,}465\ \text{ft}^3$$

$$42{,}465\ \text{ft}^3 \cdot \frac{1\ \text{bu}}{1.244\ \text{ft}^3} = 34{,}136\ \text{bu}$$

Note: This calculation is based on each bushel having a volume of 1.244 ft$^3$.

**Exercise 12-11. A clinometer is used to determine that the slope to the top of a tree is 25%. The clinometer reading was taken 100 ft from the center of the trunk of the tree. How high is the tree?**

**Exercise 12-12. How many bushels of soybeans are contained in a cone shaped pile that when measured with a clinometer at 20 ft from the edge of the pile has a slope of 25 %, and has a circumference of 503 ft?**

## Additional Information


Wilke. B. 1998. Calculating bushels. University of Minnesota Extension Service. Available at http://www.bbe.umn.edu/extens/postharvest/bushels.html.

Wilke. W. and G. Wyatt. 2002. Grain storage tips. Factors and Formulas for crop drying, storage, and handling. University of Minnesota Extension Service. St. Paul M-1080-FS.


| **Box 12-5.** Equations used in chapter | |
|---|---|
| Circumference circle | $\pi \cdot 2r$ |
| % slope | rise/run |
| cylinder volume | $\pi \cdot r^2 \cdot$ height |
| cone volume | $^1/_3 \cdot \pi \cdot r^2 \cdot$ height |
| tan $\theta$ | opposite/adjacent |
| lbv | 1.244 ft$^3$ |

# SOIL pH, CATION EXCHANGE CAPACITY, BASE SATURATION, CALCIUM CARBONATE EQUILIBRIUM, AND LIMING

**KEY PROBLEMS:** Understanding soil pH is an essential part of understanding reactions occurring in the soil.

**MATHEMATICAL SKILLS:** Using logarithms, using equilibrium relationships.

**CHAPTER CONCEPTS:** The ability to control soil pH is important for effective management. This chapter shows a number of examples that demonstrate how pH is managed.

In many soils, the application of N fertilizers results in a gradual decrease in soil pH. To maintain the soil pH level, lime is routinely applied. A better understanding of lime requirements can be obtained by understanding soil pH, calcium carbonate equilibrium, and cation exchange capacity.

Soil pH can be separated into active, salt replaceable, and residual activity pools. Active acidity is the acidity in the soil solution, while salt replaceable acidity is the $H^+$ on the soil cation exchange sites that is replaced with an unbuffered salt. Active and salt replaceable acidity can be measured with a pH meter. When pH is measured in a soil/water mixture, the measured value represents the active acidity. When measured in a dilute salt, such as 0.01 M $CaCl_2$, the pH represents the combined active and salt-replaceable pools. Soil pH measured with water generally is higher than when measured with a dilute salt.

## Soil pH and Weak Acids

Soil pH is one of the least expensive and easiest soil chemical properties to measure. Soil pH is used as an indicator of many problems. For example, at high pH (>8) the soil may contain high sodium (Na) levels, whereas at low pH values, P availability may be reduced. In addition, some herbicides should not be applied due to carry over problems. In particular, the dissipation of atrazine and simazine, and sulfonylurea containing products (Classic, Canopy, and Peak) is reduced by high soil pH. At low pH (<6.0) the dissipation of Clomazone (Command) is reduced. Slow dissipation means that crops sensitive to those chemicals may be injured one or more years after application.

The p in pH means negative logarithm (see **Box 13-1** for logarithm operators) while the H represents the molar concentration of $H^+$ (we will assume that concentration = activity). At 25 °C, the sum of pH and pOH is 14. The equations to calculate pH and pOH are,

$$pH = -\log [H^+]$$

$$pOH = -\log [OH]$$

$$pH + pOH = 14$$

$$pOH = 14 - pH$$

**Box 13-1**. Log operator rules.

$$\log(a \times b) = \log(a) + \log(b)$$

$$\log\left(\frac{a}{b}\right) = \log(a) - \log(b)$$

$$\log(a^n) = n + \log(a)$$

$$\log(\sqrt[n]{a}) = \frac{\log(a)}{n}$$

to change from log base a to log base b

$$\log_a(x) = \frac{\log_b(x)}{\log_b(a)}$$

**Exercise 13-1. What is the pH if [H⁺] is 0.001 M? What is the pOH?**

Solution:
This exercise is solved by log transforming the $H^+$ concentration.

$$-\log(0.001) = 3$$

$$pOH = 14 - 3 = 11$$

**Exercise 13-2. What is the pOH if $OH^-$ is 0.00005 M? What is the pH? What is the molar concentration of $H^+$ in the solution?**

Many materials added to soils are weak acids. Weak acids do not completely dissociate in solution, whereas strong acids do. Examples of strong acids are sulfuric acid and hydrochloric acid, while an example of a weak acid is phosphoric acid. Weak acids can be defined by the equilibrium relationship, $HA \leftrightarrow H^+ + A^-$. Where HA is the undissociated acid, $H^+$ is the hydrogen ion, and $A^-$ is the associated anion. If the pH is less than the $pK_a$, the acid primarily exists in the undissociated form. For example, at pH 1, $H_3PO_4$ is the primary species, while at pH 5 M $H_2PO_4^{1}$ is the primary species. The $pK_a$ values of selected weak acids are shown in **Box 13-2**. Often times the $K_a$ values are reported. The $K_a$ and $pK_a$ values can be converted back and forth relatively easy using the log operators shown in **Box 13-1**.

**Box 13-2.** $pK_a$ values of weak acids important in natural resource management.

| Weak acid | $pK_a$ |
|---|---|
| $H_3PO_4 \rightarrow H^+ + H_2PO_4^-$ | 2.23 |
| $H_2PO_4^{-1} \rightarrow H^+ + HPO_4^{-2}$ | 7.2 |
| $HPO_4^{-2} \rightarrow H^+ + PO_4^{-3}$ | 12.37 |
| $H_2CO_3 \rightarrow H^+ + HCO_3^-$ | 6.37 |
| $HCO_3^- \rightarrow H^+ + CO_3^{-2}$ | 10.33 |
| $NH_4 \rightarrow H^+ + NH_3$ | 9.2 |
| $H_2BO_3 \rightarrow HBO_3^- + H^+$ | 9.3 |
| $H_2O \rightarrow H^+ + OH^-$ | 14 |

**Exercise 13-3. Convert a $pK_a$ value of 7.2 to $K_a$.**

Solution:

$\log K_a = -7.2$

$K_a = 10^{-7.2}$

$K_a = 6.3 \cdot 10^{-8}$

**Exercise 13-4. Convert the $K_a$ value of $5.5 \cdot 10^{-5}$ to $pK_a$.**

**Exercise 13-5. Will the reactants or products be the dominant species for atrazine, 2,4-D, and ammonium at pH 5?**

Solution:
The reactions and $pK_a$ values for atrazine, 2,4-D, and ammonium are shown below (Bohn et al., 2001).

| Species | $pK_a$ | Reaction |
|---|---|---|
| Atrazine (s-triazine) | 1.68 | $BH^+ \rightarrow B + H^+$ |
| 2,4-D | 2.8 | $R\text{-}COOH \rightarrow R\text{-}COO^- + H^+$ |
| $NH_4^+$ | 9.26 | $NH_4^+ \rightarrow NH_3 + H^+$ |

At pH 5, atrazine and 2,4-D will primarily be in the product forms (B and $R\text{-}COO^-$), while ammonium will exist in the reactant configuration ($NH_4^+$).

**Exercise 13-6. At pH 6.5, what will be the dominant species of phosphate?**

## Predicting Leaching Potential

The charge on the molecule can be used to predict its potential leaching in soil. Many soils in North America have negative charges. The negative charges are a result of isomorphic substitutions that occurred during the formation of clays. The negative charges are balanced with positively charged cations. Cations that are sorbed to the soil matrix are less likely to leach than those that are not sorbed to the matrix. A weak acid with a $pK_a$ that is less than the pH, will tend to form the products. For example, 2,4-D with a $pK_a$ of 2.8 primary exists as a negatively charged species ($R\text{-}COOH \rightarrow R\text{-}COO^- + H^+$) in soil with pH between 6 -8 and therefore can leach in soil.

**Exercise 13-7. What is the pH of solution of 0.025 M $HNO_3$?**

Solution:

$HNO_3$ is a strong acid and, for all practical purposes, it dissociates completely.

$HNO_3 + H_2O \rightleftharpoons NO_3^- + H_3O^+$

$[H^+] = 0.025$ M

pH = -log(0.025 M) = 1.6

**Exercise 13-8. What is the pH of a solution of 0.05 M HCl?**

## pH Values of Weak Acids

The amount that a weak acid dissociates is defined by the equation,

$$K_a = \left(\frac{(H^+)\cdot(A^-)}{(HA)}\right)$$

where $K_a$ is the dissociation constant of the weak acid. If the acid concentration and $pK_a$ are large [$K_a$ is small ($1 \cdot 10^4$)], then the pH of a solution containing the weak acid can be determined with the equation,

$$pH \approx \left(\frac{pK_a + pHA}{2}\right)$$

The error associated with this equation increases with decreasing acid concentration and increasing amount of acid that is dissociated.

**Exercise 13-9. What is the pH of 0.001 M solution of a weak acid if the $K_a$ is $10^{-6}$ M?**

Solution:
When solving equilibrium equations, the first step is to write out the equation. In this example:
$HA \longleftrightarrow H^+ + A^-$
pHA = -log (0.001) = 3
$pK_a = -\log(1 \cdot 10^{-6}) = 6$
If $K_a$ is small (i.e. if $pK_a$ is large) then the equation,

$$pH = \left(\frac{pK_a + pHA}{2}\right)$$

can be used to determine pH

$$pH = \left(\frac{6+3}{2}\right) = 4.5$$

**Exercise 13-10. What is the pH of 0.1 M solution of ammonia ($NH_3$) if the $pK_a$ is 9.2?**

**Exercise 13-12. Determine the calcium carbonate equivalent of $Ca(OH)_2$?**

## Calcium Carbonate Equivalent

The amount of calcite ($CaCO_3$), dolomite [$CaMg(CO_3)_2$], quick lime (CaO), and hydrated lime [$Ca(OH)_2$] needed to raise the pH to a desired level is dependent on the soil type, initial pH, desired pH change, and relative effectiveness of the limestone. Even though each mole of these calcium materials neutralize two moles of $H^+$, different amounts of the liming material are needed because they have different molecular weights. The calcium carbonate equivalent (CCE) is used to compare different materials. The CCE is the **grams of liming material** required to neutralize 1 mole of $H^+$ **divided by the grams of $CaCO_3$** required to neutralize one mole of $H^+$. Calcium carbonate equivalent is calculated with the equation,

$$CCE = \left( \frac{\frac{\text{molecular weight } CaCO_3}{\text{moles } H^+}}{\frac{\text{molecular weight liming material}}{\text{moles } H^+}} \right)$$

**Exercise 13-11. Determine the CCE of CaO?**

Solution:

$CaO + 2H^+ \Rightarrow Ca^{++} + H_2O$

Because each mole of CaO weights 56 g (40 g + 16 g) and will neutralize 2 moles of $H^+$, $\frac{56}{2}$, 28 g of CaO are required to neutralize 1 mole of $H^+$.

Because one mole of $CaCO_3$ weighs 100 g (40 + 12 + 16 • 3) and neutralizes 2 moles of $H^+$, 50 g are needed to neutralize 1 mole of $H^+$.

$CCE = \left(\frac{50}{28}\right) = 1.78$

A CCE value of 1.78 means that 1.78 times more $CaCO_3$ is needed than CaO to provide the same neutralizing power.

## Cation Exchange Capacity

The cation exchange capacity (CEC) of a soil is the sum total of the exchangeable cations that it can adsorb. It is influenced by soil texture, clay content, organic matter content, and soil pH. The CEC can be estimated from a soil's clay and organic mater content (**Box 13-3**) or measured in a laboratory. Many different techniques can be used to measure this value. For comparative purposes, a standard measurement approach should be used. If measured, the CEC is the summation of cations that balance the negative charges in the soil. The soil CEC is reported in units of $cmol_c/kg$. The units, $cmol_c/kg$, are equal to values reported as meq/100 g (milli equivalents per 100 g soil). The units for CEC ($cmol_c/kg$) represents the centimoles of positive charges per kg of soil required to balance all the negative charges in the soil.

**Box 13-3.** The influence of soil phase on the approximate cation exchange capacity.

| Soil phase | Approximate CEC, $cmol_c/kg$ |
|---|---|
| Organic matter | 200 |
| **Clays** | |
| Vermiculite | 150 |
| Smectite | 100 |
| Mica | 30 |
| Kaolinite | 9 |
| Gibbsite | 4 |

**Exercise 13-13. Estimate the soil CEC if a soil contains 20% smectite clay and 5% organic matter.**

Solution:

$$\frac{100\ \text{cmol}_c}{\text{kg smectite}} \cdot \frac{20\ \text{kg smectite}}{100\ \text{kg soil}} + \frac{200\ \text{cmol}_c}{\text{kg organic matter}} \cdot \frac{5\ \text{kg organic matter}}{100\ \text{kg soil}} = \frac{30\ \text{cmol}_c}{\text{kg soil}}$$

**Exercise 13-14. Estimate the soil CEC if a soil contains 10% mica and 3% organic matter.**

## Base Saturation

The cation exchange capacity is also used to determine the percentage base saturation (%BS) which is defined by the equations,

$$\%\text{BS} = 100\% \cdot \frac{\text{CEC - Acidic Cation}}{\text{CEC}}$$

Where, the acidic cations are $H^+$ and $Al^{3+}$. The %BS is used as a relative measure of weathering and in soil classification. Highly weathered soils tend to have low %BS values.

**Exercise 13-15. What is the CEC and percent base saturation (%BS) of a soil if the soils' negative charges are balanced by 15.1, 0.3, 2.3, 1.5, and 4.0 $\text{cmol}_c$/kg of $Ca^{2+}$, $Mg^{2+}$, $K^+$, $Na^+$, and $H^+$, respectively?**

Solution:

CEC = summation of all cations

CEC = 15.1 + 0.3 + 2.3 + 1.5 + 4.0 $\text{cmol}_c$/kg = 23.2 $\text{cmol}_c$/kg

Acidic cations are $H^+$ or $Al^{+3}$ = 0 + 4 $\text{cmol}_c$/kg

$$\%\text{BS} = 100\% \cdot \frac{\text{CEC} - \text{Acidic Cation}}{\text{CEC}} = \frac{23.2 - 4}{23.2} \cdot 100\% = 82.8\%$$

**Exercise 13-16. What is the %BS if the CEC is 25 $\text{cmol}_c$/kg and the acidic cations are 10 $\text{cmol}_c$/kg?**

## Lime Recommendations

A critical question when managing soil pH is the use of lime to increase pH. Lime recommendations are dependent on soil and location and are based on laboratory and field calibration studies. Plant species and soils vary in their response to lime. For example, alfalfa has much lower yields at low pH values than wheat (**Box 13-4**). Based on interactions between plants and pH, one of the first questions that must be asked is: What plant is being grown?

**Box-13-4.** Relative yields of different crops at five soil pH values (modified from Bohn et al. (2001).

| | pH | | | | |
|---|---|---|---|---|---|
| Crop | 4.7 | 5 | 5.7 | 6.8 | 7.5 |
| | (average relative yield) | | | | |
| Sweet clover | 0 | 2 | 49 | 89 | 100 |
| Alfalfa | 2 | 9 | 42 | 100 | 100 |
| Corn | 34 | 73 | 80 | 100 | 93 |
| Wheat | 68 | 78 | 89 | 100 | 99 |
| Oats | 77 | 93 | 99 | 98 | 100 |

Once a decision is made that lime is needed (adding lime will increase yield), the question is: How much lime? To answer this question, laboratory and field calibration studies should be conducted. In the laboratory, titrations (mixing soil with base) are conducted by adding a quantity of base and then measuring the change in soil pH. Both active and salt replaceable acidity are neutralized during a titration. To be realistic, these reactions must be slow enough for the base to react with the soil. The problem with the titration approach is that the titration of individual soil samples is impractical for soil-testing procedures. This problem is often solved by using a buffer solution to quantify the relationship between pH change and the amount of base added. General recommendations based on buffer pH change are not available because the relationship between buffer pH change and yield response are site-specific. When lime is added to field soils, the reactions are generally incomplete because the reactions require considerable time. The following examples are used to demonstrate why liming recommendations are site-specific.

**Exercise 13-17. How much $CaCO_3$ should be added to increase the %BS from 40 to 80% of a sandy soil (1,600,000 kg soil/ha) if the CEC is 5 $cmol_c$/kg?**

Solution:

To solve this exercise, the molecular weight and reaction of $CaCO_3$ are needed. $CaCO_3$ has a molecular weight of 100 g, and therefore 1 cmol weights 1 gram. The reaction of $CaCO_3$ in soil, $CaCO_3 + 2H^+ \rightarrow Ca^{2+} + H_2O + CO_2$, shows that 1 cmol of $CaCO_3$ will neutralize two moles of $H^+$ (1 cmol $CaCO_3$ = 2 $cmol_c$ ).

$$(0.8-0.4)\frac{5\ cmol_c}{kg\ soil}\cdot\frac{1,600,000\ kg\ soil}{ha}\cdot\frac{1\ cmol\ CaCO_3}{2\ cmol_c}\cdot\frac{1\ g\ CaCO_3}{1\ cmol\ CaCO_3}\cdot\frac{1\ kg\ CaCO_3}{1,000\ g}=\frac{1,600\ kg\ CaCO_3}{ha}$$

**Exercise 13-18. How much $CaCO_3$ should be added to increase the %BS from 40 to 80% of a silty clay loam soil (1,600,000 kg soil/ha) if the CEC is 25 $cmol_c$/kg?**

## Effective Liming Rate

The reactivity of lime is related to its fineness and purity. The amount of lime that needs to be applied should be adjusted based on both these factors. The approach to adjust purity is best shown through an example,

$$1,000\ kg\ CaCO_3\cdot\frac{1\ kg\ lime}{0.85\ kg\ CaCO_3}=1,176\ kg\ lime$$

The rates should also be adjusted for size of the liming material. If the material does not pass through a 10 mesh screen (10 wires/in.) then it will have little impact on soil pH. If the liming material passes through a 10 mesh but not a 50 mesh screen (50 wire/in.), then the amount of liming material should be doubled (**Box 13-5**).

| **Box. 13-5.** Mesh size and liming effectiveness. | |
|---|---|
| Lime passes through mesh size | Relative liming effectiveness |
| >50 | 100 |
| 10-50 | 50 |
| <10 | 0 |

**Exercise 13-19. If calculations show that 1,000 kg of $CaCO_3$ are needed to increase the pH, how much lime should be applied if analysis shows that it is 90% pure and 25% passes through a 50 mesh screen and 75% passes through a 10-50 mesh screen?**

Solution:

Purity

$$1{,}000\ \text{kg CaCO}_3 \cdot \frac{1\ \text{kg lime}}{0.90\ \text{kg CaCO}_3} = 1{,}111\ \text{kg lime}$$

Effective

$$0.25\ \text{lime} \cdot \frac{1\ \text{effective lime}}{1\ \text{actual lime}} + 0.75 \cdot \frac{0.5\ \text{effective lime}}{1\ \text{actual lime}}$$

$$= 0.625\ \text{effective lime}$$

Lime needed

$$1{,}111\ \text{kg CaCO}_3 \cdot \frac{1\ \text{kg lime}}{0.625\ \text{effective lime}} = 1{,}778\ \text{kg lime}$$

In most situations, the amount of lime needed is based on calibration studies that compare soil pH in water with the pH in a buffer solution. These curves are soil dependent. It is advisable to check with local soil specialists to develop recommendations for your soils. Sample recommendations based on soil pH in water and a buffer solution are available at: http://ohioline.osu.edu/agf-fact/0505.html.

## References and Additional Information

Bohn, H.L., B.L. McNeal, and G.A. O'Connor. 2001. Soil Chemistry 2nd Edition. John Wiley and Son, Inc. New York.

| **Box 13-6**. Equations used in chapter |
|---|
| pH = -log H |
| 14 = pOH + pH |
| weak acid pH = ($pk_a$ + pHA)/2 |
| % Base Saturation = 100 • (CEC - acid cations)/CEC |
| Lime required = (calculated lime requirement)/ (relative effectiveness) |

# SOIL SALINITY, SODICITY, AND ELECTRICAL CONDUCTIVITY

**Key problems:** Soil electrical conductivity (EC), and the soil's exchangeable sodium percentage (ESP) are used to assess salt hazards. In arid and semi-arid regions, mismanagement of salts is a primary factor leading to a loss of productivity.

**Mathematical skills:** Review

**Chapter concepts:** The ability to put EC and ESP into relevant values is important for effective salt management. This chapter shows a number of examples that demonstrate how these values are calculated.

In the United States, about one-third of the arid and semi-arid soil regions are affected by salinity. Soils with salinity problems typically contain high concentrations of $CaSO_4$, $MgCO_3$, $Na_2CO_3$, $CaCl_2$, and NaCl. Salinity problems result from water not washing (leaching) the salts out of the profile, fossil deposits, inadequate drainage capacity, improper irrigation management, summer fallow, and/or the replacement of native deep rooted perennials with annual crops.

Many different approaches are used to measure salinity. It can be measured in the field with an EM meter (Geonics) or a Veris Soil EC Mapping System. In the laboratory, EC is typically measured using a saturated paste extraction or in a soil/water suspension. These two approaches produce different results (**Box 14-1**). A saturated paste is made by adding water to soil until it glistens and flows slightly if jarred. After allowing it to stand overnight, the soil water solution is extracted by suction filtration. The electrical conductivity is measured with a meter and is reported in the units in dS/m. Generally, the meter reading increases with ions in the solution. A soil is considered saline if the saturated paste EC is greater than 4 dS/m (**Box 14-1**). Many soil testing laboratories measure soil EC using a 1:1 water to soil ratio. This measurement approach will result in slightly lower EC values.

The electrical conductivity of the saturated paste solution can be used to estimate total dissolved solids (TDS), soil osmotic potential, and $mmol_c/L$, total cations (+ charged ions) or total anions (- charged ions) using the equations,

640 mg Na/L = 1 dS/m

800 mg Ca/L = 1 dS/m

0.36 bars Osmotic Potential = 1 dS/m

10 $mmol_c/L$ total cations = 1 dS/m

Because the dominant cations are $Ca^{2+}$, $Mg^{2+}$, and $Na^+$, total cations are often replaced with $mmol_c$ Ca/L + $mmol_c$ Mg/L + $mmol_c$ Na/L. When using this notation, it is important to understand what $mmol_c/L$ represents. The $mmol_c$ represents milli moles of charge. Different cations have different valances (charges). Calcium and magnesium have a valance of +2 (plus two), while Na, K, and $NH_4$ have a valance of +1. To convert mmol of Ca to $mmol_c$ of Ca, the mmol of Ca must be multiplied by its valance (2). For example, 2 mmol Ca is equivalent to 4 $mmol_c$ Ca.

**Box 14-1.** The electrical conductivity of silt-loam to clay loam soil using different soil to water ratios.

| Soil salinity | Saturated paste | 1:1 soil to water ratio |
|---|---|---|
| | dS/m | |
| Non-saline | 0-2 | 0-1.3 |
| Slightly | 2.1-4.0 | 1.4-2.5 |
| Moderate | 4.1-8.0 | 2.6-5.0 |
| Strongly | 8.1-16.0 | 5.1-10.0 |
| Very | >16.0 | >10.0 |

**Exercise 14-1. Convert 10 mmol of $Ca^{2+}$ and 10 mmol of $Na^+$ to $mmol_c$.**

Solution:

2 $mmol_c$ of Ca are equal to 1 mmol of Ca because Ca has a positive charge of 2. 1 $mmol_c$ of Na equals one mmol of Na because Na has a charge of 1.

$$10\ \text{mmol Ca} \cdot \frac{2\ \text{mmol}_c\ \text{Ca}}{1\ \text{mmol Ca}} = 20\ \text{mmol}_c\ \text{Ca}$$

$$10\ \text{mmol Na} \cdot \frac{1\ \text{mmol}_c\ \text{Na}}{1\ \text{mmol Na}} = 10\ \text{mmol}_c\ \text{Na}$$

**Exercise 14-2. Convert 5 mmol of $Mg^{2+}$ and 10 mmol of $K^+$ to $mmol_c$.**

**Exercise 14-3. Estimate the $mmol_c/L$ of Ca, Mg, and Na if the EC is 5.2 dS/m.**

Solution:

$$EC(\text{dS/m}) \cdot 10 = \left[\frac{\text{mmol}_c\ \text{Ca}}{\text{L}} + \frac{\text{mmol}_c\ \text{Mg}}{\text{L}} + \frac{\text{mmol}_c\ \text{Na}}{\text{L}}\right]$$

$$5.2\ \text{dS/m} \cdot 10 = \frac{52\ \text{mmol}_c}{\text{L}}$$

**Exercise 14-4. Estimate the $mmol_c/L$ of Ca and Mg if the EC of a saturated paste is 4.2 (dS/m) and the Na content is 5 $mmol_c/L$.**

## Sodic Soils (ESP and SAR)

Worldwide, high Na content of soil is one of the most common factors responsible for a loss of productivity. Sodium is a problem because high Na concentrations can cause soil aggregates to disperse. Sodic soils typically have very slow water infiltration rates, poor aeration, and poor plant growth. A soil is considered sodic if the exchangeable Na percent (ESP) exceeds 15%. The ESP is the relative amount of Na on the soil exchange sites and is calculated with the equation,

$$ESP = 100\% \cdot \frac{\text{cmol}_c\ \text{Na}/\text{kg Soil}}{\text{CEC}}$$

where CEC is the cation exchange capacity. Soils with a ESP > 15% are considered sodic soils.

The ESP is difficult and expensive to measure and therefore the Na adsorption ratio (SAR) of the soil water extraction is often used to characterize sodic problems. Soil solution extracts with a SAR greater than 13 are characterized as sodic. For SAR determinations, a soil water extract is analyzed for Ca, Mg, and Na. The SAR is calculated with the equation,

$$SAR = \frac{\frac{\text{mmol}_c\ \text{Na}}{\text{L}}}{\left(\frac{[\text{mmol}_c\ \text{Ca/L} + \text{mmol}_c\ \text{Mg/L}]}{2}\right)^{0.5}}$$

When calculating SAR values the correct units must be used. If the laboratory reports Ca and Mg in mmol/L, then these values must be converted to $mmol_c/L$. Each mmol of Ca or Mg contains 2 $mmol_c$, and therefore mmols of Ca or Mg are converted to $mmol_c/L$ by multiplying the number of mmols by 2. For Na, mmol/L and $mmol_c/L$ are identical. If the Ca and Mg are reported as meq/L, then these values replace the $mmol_c/L$ value in the equation above.

If we make the assumption that $Ca^{2+} + Mg^{2+} + Na^+$ make up the vast majority of the positive ions in a soil, then

10 $mmol_c$ total cations/L = 1 dS/m so

10 $mmol_c$ $(Ca^{2+} + Mg^{2+} + Na^+)/L$ = 1 dS/m

**Exercise 14-5. Determine the SAR of a soil water extract if it contains 10 $mmol_c$ Ca/L, 5 $mmol_c$ Mg/L, and 1 $mmol_c$ Na/L.**

Solution:

$$SAR \approx \frac{\frac{1\ mmol_c Na}{L}}{\left(\frac{[10\ mmol_c\ Ca + 5\ mmol_c\ Mg]}{2}\right)^{0.5}} = 0.36$$

**Exercise 14-6. Determine the SAR of a soil water extract if it contains 15 $mmol_c$ Ca/L, 5 $mmol_c$ Mg/L, and 12 $mmol_c$ Na/L.**

**Exercise 14-7. A solution has an EC of 3.9 and 7 $mmol_c$ Na/L. What is its SAR?**

Solution:

Step 1. Calculate the amount of Ca + Mg + Na
3.9 • 10 = 39 $mmol_c$/L

Step 2. Calculate the amount of Ca + Mg
39 $mmol_c$/L - 7 $mmol_c$ Na/L = 32 $mmol_c$ (Ca, Mg)/L

Step 3. Determine SAR
SAR = 7 $mmol_c$ /L/[(32 $mmol_c$ /L)/2]$^{0.5}$ = 1.75

1.75 is less than 13, therefore the soil is classified as non-sodic.

**Exercise 14-8. Determine the SAR if a saturated paste water solution has an EC of 3.8 dS/m and a Na concentration of 5 $mmol_c$/L.**

Extreme care must be used in the management of sodic soils. We recommend that you contact a local Cooperative Extension Service representative to help develop management plans for these soils.

## Leaching Requirement

A common source of salt in soil is derived from irrigation water. After irrigation, the water is used by the crop and the salt from the water is left behind. All irrigation waters contain some salts. If this salt is not removed, it will accumulate in soil. The process of salt accumulation is called salinization. Salinization is responsible for the loss of some of the most productive lands on the earth. The leaching requirement (LR) equation,

$$LR = \frac{\text{Irrigation water EC (dS/m)}}{\text{Acceptable deep drainage EC (dS/m)}},$$

can be used to estimate the amount of water needed to remove excess salts.

**Exercise 14-9. Determine the leaching requirement if irrigation water EC is 2dS/m and the acceptable deep drainage EC value is 6 dS/m (50% yield reduction).**

Solution:

$$LR = \frac{2\ dS/m}{6\ dS/m} = 0.33$$

A leaching requirement of 0.33 means that 33% more water is needed than the amount required to meet the plants' requirements. For example, if 3 in. of water are needed by the plant, then the amount of water needed to meet the needs of the plant and wash the salts out of the profile is 4 in. of water (= 3 + 3 • 0.33) x = 4 inches.

**Exercise 14-10. What is the leaching requirement if irrigation water EC is 1 dS/m and the acceptable deep drainage EC value is 4 dS/m?**

When Na problems occur in a soil profile, resolving them requires that some of the $Na^+$ on the soil exchange complex be replaced with $Ca^{2+}$. The most commonly used reclamation amendment is gypsum, $CaSO_4$. To determine the amount of gypsum needed to reclaim a soil, it is necessary to know the soil CEC, the depth of reclamation, the initial concentration of Na on the exchange complex, $\left(cmol_c Na^{+1}/kg\ soil\right)$, and the target (desired) concentration of Na on the exchange complex $\left(cmol_c Na^{+1}/kg\ soil\right)$. Gypsum is usually $CaSO_4 \cdot 2H_2O$, 172 g/mol.

**Exercise 14-11. The ESP of a soil (CEC 30) is 20%. How many kg/ha of gypsum will it take to bring the top 15 cm of soil down to an ESP of 5%?**

Solution:

Gypsum is $CaSO_4$ and has a molecular weight of 136 g (40.08 + 32 + 16 • 4)

$$ESP = 100\% \cdot \frac{cmol_c Na^{+1}/kg\ soil}{30\ (CEC)} = 20 \text{ and } ESP = 100\% \cdot \frac{cmol_c Na^{+1}/kg\ soil}{30\ (CEC)} = 5$$

$$Na^{+1} = 6\ cmol_c Na/kg\ soil \text{ and } Na^{+1} = 1.5\ cmol_c Na/kg\ soil$$

Na that must be replaced =

$$so = 6\ cmol_c Na^{+1}/kg\ soil - 1.5\ cmol_c Na^{+1}/kg\ soil = 4.5\ cmol_c Na^{+1}/kg$$

$$\left(4.5\ cmol_c Na^{+1}/kg\ soil\right)\left(2,000,000\ kg\ soil/ha\ 15\ cm\ soil\right)\left(0.00086^*\ kg\ gypsum/cmol_c\right)$$

$$= 7,740\ kg\ gypsum/ha\ 15\ cm\ soil$$

This assumes exchange of 100% efficiency which is too optimistic, so we typically increase the amount by 125%. That equals 9,675 kg/ha, or about 9.7 t/ha.

* 0.00086 is calculated by converting gypsum mole weight (172) to kg/$cmole_c$

**Exercise 14-12. The ESP of a soil (CEC 18) is 12%. How many kg/ha of gypsum will it take to bring the top 15 cm of soil down to an ESP of 5%? Assume 2,000,000 kg soil per ha to a 15 cm depth.**

## Management of Saline and Sodic Soils

In managing saline and sodic soils, care must be used to prevent further degradation. To prevent further degradation: 1) plant appropriate plants; 2) collect soil and water samples to identify the scope and magnitude of the problem; 3) eliminate source of salt or balance salt additions with salt losses; 4) apply chemicals such as gypsum to sodic soils if needed; 5) apply crop residues to improve water infiltration, and 6) apply irrigation water to leach salts from the soil.

In many areas, salts typically concentrate near the soil surface in areas with high water tables. In these areas, water and salts dissolved in the water rise through capillary movement from the water table to the surface. Water that evaporates at the soil surface is replaced by more water from the water table. The net result is an accumulation of salts. Capillary action provides a continuous transport of salts to the soil surface. One of the most common approaches for managing salts is to install tile drainage. However, in many situations installing tile drainage is illegal or physically impossible.

If tile lines are not an option, the producer should eliminate bare or black summer fallow and consider planting late-maturing, deep-rooted plants with high salt tolerance. Late-maturing plants are beneficial for several reasons. First, they mulch the soil, which reduces the potential for surface evaporation. Second, they reduce the potential for capillary movement of salts to the surface. The most important management consideration for these areas is to maximize transpiration and minimize evaporation (Franzen, 2007).

For sodic soils, extreme care must be used. High Na has the added problem that it can greatly reduce water infiltration. If a problem with Na is suspected, a soil sample should be collected and a water extract from that soil analyzed for Na, Ca, and Mg. Based on this value, the SAR should be calculated. If Na is a problem, the long-term goal should be to prevent further degradation and reduce new Na additions. This can be accomplished by providing drainage, adding low Na manure or gypsum, or lowering the pH (if the soil pH is high) with elemental S. If gypsum is present at deeper soil depths, tillage may help. If drainage and soil amendments are not possible, consider placing the field into pasture and planting it with grasses tolerant of salt and Na.

## Additional Information

Cardon, G.E., J.G. Davis, T.A. Bauder, and R.M. Wascum. 2007. Managing saline soils. #503 Colorado Extension, Colorado State University. Available at http://www.ext.colostate.edu/pubs/crops/00503.pdf.

Franzen, D. 2007. Managing saline soils in North Dakota. SF1087. North Dakota State University. Available at http://www.ag.ndsu.edu/pubs/plantsci/soilfert/sf1087.pdf.

# SCIENCE, DISCOVERY, AND DECISION MAKING

**Key terms:** Recommendations, scientific method, problem definition, hypothesis, inference space, experimental design, data collection, sample analysis, interpretation and modeling, conceptual model, relational diagram, mathematical model, testing, implementation.

**Mathematical skills:** Addition, subtraction, multiplication, division (including fractions)

**Chapter concepts:** This chapter will discuss the different components of the scientific method and how conceptual diagrams and flow charts can be used to develop a better understanding of the problem. The goals of the scientific method are to design studies and measure outcomes in order to make dependable prediction of future events.

## The Role of Science in Decision Making

### Defining the role of science

The decisions that we make are improved by learning from our mistakes, watching how our friends or families react to various circumstances, and/or trying to emulate people we admire. We often end up comparing where we are at to where we want to be. When such introspection identifies that the two are not the same, change is needed. But exactly which changes to make may elude us. To figure out where to start, consult with experts for guidance.

Making good natural resource and agronomic management decisions follows the same general flow as described above. We set goals for our management, compare them to our current practices, determine if change is needed, and then look to experts to provide us with probable outcomes of different scenarios. The importance of integrating science into the process is that it improves the probability that a specific outcome will be achieved.

### Differentiating recommendations from science

In natural resource management, university and government scientists have traditionally provided recommendations. Recommendations have been in the form of fact sheets and guidelines. While recommendations utilize the outcomes of scientific investigations (probability and magnitude), the recommendations themselves are not pure science. It is important to recognize the difference between recommendations and science. Recommendations take scientific findings into consideration, but they are the result of interpreting that science with their associated biases. Recommendations can contain biases.

Biases occur when data collected at one location is used to make recommendations at a different location. For example, experiments conducted in one area or climate zone are used to make recommendations in a different area. The recommendations will be applicable only if the areas are similar in climate, Growing Degree Days (GDD), or daylength. The more dissimilar the areas, the less applicable the findings are to making good management decisions. A second bias occurs when all possible impacts are not considered. For example, a cost analysis does not consider the impact of nitrate runoff or leaching losses on hypoxia in the Gulf of Mexico.

Bias may also consider human preferences. For instance, the risk of having a weedy field is not acceptable even though the weeds may not reduce yields. Biases generally are not explicitly stated. Distrust can occur if the biases of the agronomist and managers do not match. Consequently, recommendations must clearly state the assumptions and biases with which they are formulated. If assumptions and biases are clearly stated, those using the recommendations will be better able to tailor the recommendation to their needs.

## Policy and recommendations

Readers who are or will be involved in public policy must remember the difference between recommendations and science. In the USA, university Extension recommendations are currently considered a regulatory standard by some agencies. Where policymakers have recognized the biases and assumptions built into university recommendations, language has been included in regulations to recognize the legitimacy of localized practices if they are founded on documented science pertinent to the area. Without such caveats, generalized suggestions based upon the biases of particular individuals become widely implemented and maintained rigidly as standards, which may or may not fit local conditions or meet the goals that policy was trying to reach when crafted. Consequently, responsibility is on the shoulders of those creating policy and those wanting the freedom to find local solutions to problems. Policymakers have the responsibility to include language that allows local variations in management to be made. Those wishing to tailor recommendations to their specific situations have the responsibility to use scientific approaches when evaluating and testing new practices. Creators of recommendations, whether at the university or in the field office, have the responsibility to clearly state the biases and assumptions they made when crafting their guidelines.

# The Scientific Method

The **scientific method** is "a body of techniques for investigating phenomena and acquiring new knowledge, as well as for correcting and integrating previous knowledge" (Wikipedia contributors, 2007). The scientific method offers a structured, reproducible, and defensible process for generating new information that addresses specific questions.

A conceptual view of the scientific method is shown in **Box 15-1**. In this illustration, several components of the scientific method are shown. The components flow in logical order and start with problem definition.

**Box 15-1**. Components of the scientific method conceptualized as a cycle of discovery that leads toward progress.

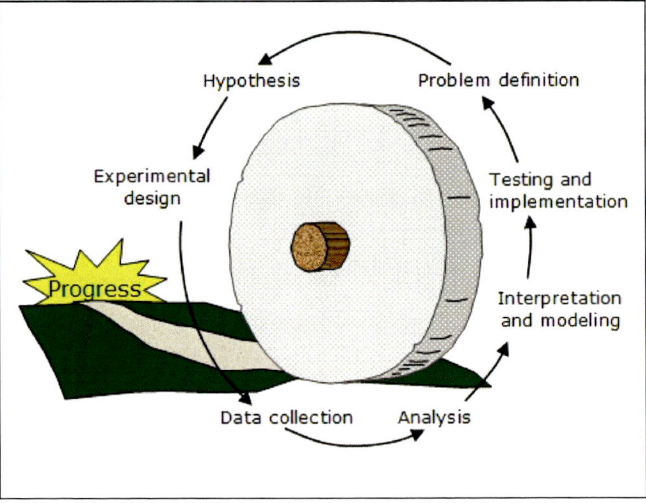


## Problem definition

In **problem definition,** the limitation of the management practice is identified. For example, a crop consultant believes that the N recommendations are too high for highly productive land and too low for less productive lands. Based on this problem, several hypotheses can be developed. One may be that N fertilizer efficiency is higher in highly productive soils and lower in less productive soils. Another hypothesis may be that there is not enough water in the low productive soil to increase yield and N could be reduced in this area. A third hypothesis may be that the crop in the low productive soils does not grow well because of pH problems. No matter which of these hypotheses is chosen, it can be seen that the problem and hypothesis must be clearly defined prior to starting a large experiment. Perhaps simple tests to gather preliminary data, such as soil pH, N level, or soil characterization, should be done prior to full scale experiments are implemented. Based on the final working hypothesis, an experiment is conducted, data are collected, analyzed, interpreted, and validated in a series of new experiments.

This scenario outlines how many practitioners determine whether or not a problem may exist. Among scientists themselves, this type of dialog is also engaged in. However, scientists will review published, peer-reviewed, scientific literature to identify factors that need to be considered when approaching the problem. Practitioners don't often have the time to perform such reviews themselves. Instead, they may rely on dialog with experts who are familiar with the scientific literature or have more experience. Regardless of how the information is obtained, awareness of a potential problem and identification of factors are the backbone of the scientific method.

## Hypothesis

A **hypothesis** is "a tentative assumption made in order to draw out and test its logical or empirical consequences" (Merriam-Webster Online Dictionary, 2007). Simply put, a hypothesis is a statement that can be tested using the scientific method. A hypothesis might be: "Nitrogen rates can be improved." To be functional, the hypothesis must not be grandiose or too general. It should be as specific as possible. Our hypothesis can be improved if we incorporate some other factors that affect the appropriateness of N rates in our area. We identify these factors when defining the problem. For instance, weather, soil, crop, and crop rotation are known to be important factors. So a better hypothesis might be, the N rates used for corn grown after soybean can be improved by considering soil water.

By limiting the extent of the hypothesis, we increase the chance that it can be tested. However, by limiting it, we also reduce our **inference space.** The **inference space** is the set of conditions to which we can extend our results. So controlling a lot of factors (like soil, crop, and crop rotation) helps test the hypothesis, but limits the inferences to other situations.

So is there any hope that the scientific method can address enough factors to be extended to a large geographical area? Two approaches are commonly used to address this question. The first approach is to conduct a series of controlled experiments across a wide range of conditions. The second approach is to develop a model that explains the results at each location.

## Experimental unit and treatment

An **experimental unit** is the unit of material to which one application of a treatment is applied. The unit may be a 10 by 50 ft plot, an individual animal in a feeding study, a pen with 10 animals, or a field. The **treatment** is the procedure whose effect is to be measured and compared to other treatments. A treatment may be a fertilizer rate, a feeding ration, or a level of shade. Treatment selection must be carefully considered so that treatments provides efficient answers related to the experiment's objectives.

## Experimental design

**Experimental design** is an approach to conducting a study that allows statistical analyses to be performed. Consequently, designing and conducting experiments must be done in a way that will allow statistical methods to test the hypothesis (see Knighton, 2000: Wittig and Wicks III 2000). This usually requires the hypothesis itself to be restated into a statistical hypothesis, a more rigorous statement of the problem, discussed in subsequent chapters. The type of statistical analysis to be conducted and treatment comparisons should be planned in advance. The number of **replications** (times a treatment appears in the study) also needs to be taken into account. Replications are needed because there is variation in experimental material. A large number of replicates are needed if high variation is expected. When choosing the number of replicates, it may also be important to consider sampling and analysis costs, labor, time, and available space.

Besides statistics, experimental design also relies on science. A practioner must consider the factors that influence the results, determine the importance of the factor, and if the factor can or cannot be controlled. To the extent possible, all factors other than the ones to be altered should be held constant. For natural resource managers, inexperienced in experimental design, dialogs with scientists and statisticians are a good course of action prior to putting treatments out in the field. Fortunately, standardized experimental designs can often be used.

## Data collection

**Data collection** is a systematic gathering of factual information. The scientific method relies on observations that can be measured. This keeps things unbiased and produces data that can be used mathematically. Finding measurable characteristics that reflect what you want to monitor is sometimes difficult and may be limited by available technologies.

To see how scientists approach data collection, consider your lawn. If you have a dog, you have probably noticed patches that are greener than others. Since you know that the dog is fertilizing those spots, you might like to know what effect fertilization has upon lawn greenness. But how will you measure the response? How green is green? You can't just say, that a fertilized spot is greener than an unfertilized spot, because that is too subjective, and another person wanting to do the same experiment may not agree with your idea of what is green and may come up with different conclusions. You need a way to measure greenness, using some standard scale. As a solution, you develop a color chart with five different shades of green. Each shade of green has a number, with one being the most yellow shade and five being the deepest shade of green. This scale allows you to do two important things. First, it provides you with a standard scale by which you can categorize the shades of green you encounter in your lawn. Second, it is something you can give someone else who may want to take the same measurements. Because you are both using the same scale, you can compare your results and see if you come up with the same conclusions.

Statistics can be used to decide how many samples should be collected. A **sample** is a subset of a population that represents the population from which it was derived (Freund and Wilson, 1997). The sample represents a subset of manageable size. The size of the sample depends on the variability of the system. If there is very little variability, a sample might consist of one or two measurements. However, if the variability is high, the sample might consist of many measurements. Scientific principles and common sense can be used to determine how and how often to collect measurements. For example, in cold northern climates, sampling flying insects in mid-winter would most likely result in no insects being collected. Collecting insects in mid-summer, however, would yield much different results and interpretation.

## Analysis

**Analysis** is the examination of data. In statistics, analysis refers to the mathematical examination of samples to gain information about the population from which they were collected and represent. In the laboratory, analysis refers to techniques used to determine something about the physical and chemical properties of samples.

Analysis requires expertise and resources. Consequently, if you are a practitioner, you will need to form an association with those who possess this expertise. When it comes to laboratories, be careful to select laboratories that have a high quality standard. Sometimes, analysis costs can be reduced or even waived if the experiment you are conducting is of interest to the laboratory. There may be opportunities for partnering that should be investigated during the planning phase. If experimental design was done carefully, the appropriate data analysis techniques should already be known by the time this step is performed. In this book, we provide guidance on how to analyze an experiment based on widely-used designs.

## Interpretation and modeling

The results of an experiment may be positive, negative, or inconclusive. In order to determine the result, data must be interpreted. **Interpretation** is the act of explaining the meaning of the information gathered from the analysis of data. After the analysis, it is important to determine how your results fit within the context of findings from other similar studies. Interpretation also is necessary to assess if the findings are consistent with others. To minimize biases an attempt should be make to emotionally detach yourself from the analyses. It is human nature to want results to come out a certain way. After all, you've invested a lot of time and energy into the scientific method. Regardless, nonbiased interpretation is an important component in the scientific method.

**Modeling** is describing the relationships among various parts of a system. Models can be conceptual and mathematical. **Conceptual models** do not quantify relationships, while **mathematical models** quantify relationships between factors.

To understand the difference between conceptual models and mathematical models, think about a bathtub filling up with water. We would like to know how long it will take to fill the bathtub. We want to start the water, leave to do something else, then come back, shut it off, and get in, but we don't want it to overflow while we're gone. The first step is to create a conceptual model that shows the relationships among the factors at work in this system. An example is shown in **Box 15-2**. The two factors that are of primary importance in determining the time needed to fill the tub are: 1) the rate of water coming out of the faucet and 2) the volume of the tub. Conceptual models are often drawings that show or demonstrate relationships. They are very useful for identifying key components of a system.

**Box 15-2.** Conceptual model of the factors needed to describe a tub filling with water, expressed as a relational diagram.

Volume of the tub

Time for the tub to fill

Rate of water coming out of the faucet

Instead of using pictures, we can also express our conceptual diagram as a relational diagram. A **relational diagram** shows the relationships among factors in a system, using graphical representations. An example is shown in **Box 15-3**. In this diagram, the rate at which water comes out of the faucet is denoted by an arrow connecting two geometric shapes representing the volume of the tub and the time needed to fill the tub.

**Box 15-3.** A relational diagram of the key factors describing the time needed to fill a tub with water.

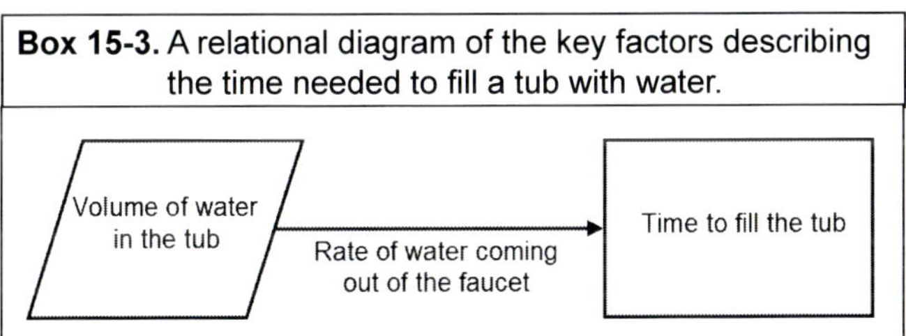


Now the mathematical model can be developed. This model will allow us to transform our conceptual model into an expression that provides a quantitative answer. We need to state some assumptions as we proceed. First, assume that when we start filling the tub, we plug the drain and it seals tightly enough not to leak. This means that the rate at which the tub fills is solely dependent on the rate of water flow. Second, let's assume that the rate at which the water comes out of the faucet doesn't vary over time, but remains constant. To make mathematical modeling easier, we will use variables. **Variables** are represented by symbols and are used in equations. Variables can take on a range of values or denote the level of a factor in a mathematical expression. The letter V will be used to represent the volume of the tub. The letter r will be used to represent the rate at which the water flows from the faucet. The letter t is used to represent the time needed to fill the tub.

We are now ready to express the relationships among the factors in a quantitative way. We may know this from previous training, or we may use other resources to help define the model. To define this model, the relationship among the three variables, *V, r*, and *t* must be considered. If units are added to the variables, this may help decide on a reasonable mathematical expression. For *V*, gallons or liters may be appropriate units, for *t*, minutes may be appropriate, and for *r*, gallons/min, or liters/min would be reasonable. The relationship can be represented as this expression:

$$V = r \cdot t.$$

When adding units the expression would be

Gal = (gal/min) • (min)

We want to know how long it will take to fill the tub, so the expression is solved for the variable *t*:

$$t = \frac{V}{r}.$$

This general equation gives an expression that can be solved for a range of values for *V* and *r*. So if our tub holds 45 liters of water ($V = 45$ L), and the water flows out of the faucet at 5 liters per minute ($r = 5$ L/min$^{-1}$), then the amount of time it takes to fill the tub is:

$$t = \frac{45\text{ L}}{5\text{ L/min}} = 9 \text{ min to fill the tub.}$$

However, if the $V = 1{,}000$ gal and the $r = 50$ gal/min then, it would take 20 min to fill the tub. By expressing our relationships in a mathematical equation, we were able find numerical solutions using a wide range of values for many variables rather than one specific solution.

The original equation was based on the assumptions associated with the calculation. These assumptions included that the drain plug fit tightly, the water flow was constant over time, and the tub was filled to its full capacity. If we did not want to make these assumptions, our conceptual model, relational diagram, and mathematical model would become more complex. For example, if we wanted the tub to fill to only $^2/_3$ capacity, then *V* would need to be multiplied by $^2/_3$, which would affect *t*. If *r* was not constant or if the tub had a slow leak, these assumptions could also be included in the equation. It is important to see that each change in the conditions results in more complex model systems. However, all can be related back to the original basic equation.

Models should attempt to explain the basic processes controlling the system. Many resource managers find that the process of building models is an important part of developing a detailed understanding of the problem. However, if the assumptions are not correct, then the resulting prediction will not be accurate.

## Testing and Implementation

**Testing** is the process of evaluation. Possible improvements resulting from the previous steps in the scientific method need to be examined to see if they work in production settings. There is uncertainty and risk in adopting new practices into commercial operations, since conditions are not as controlled as in experimental units. Often, testing is not done as rigorously as the original experiment. If only small changes were observed in the experiment, no changes may be seen in 'real world' applications. An area may be managed with the new techniques and monitored over time to evaluate performance. Trying a new practice in production settings can surface many unforeseen issues, particularly with logistics. Testing can also be done formally in these settings, starting the discovery cycle over again.

**Implementation** is the process of putting something into effect. For a new practice to be implemented, it must meet several criteria:

- It must be practical,
- The perceived risk of putting it into action must be less than or equal to current practices, and
- It must provide tangible benefits beyond what is currently being done.

Implementation may be slow or it can be very rapid. In industrial settings, implementation is usually done in a stepwise fashion. Small test areas are evaluated over time to see if the new practice is generally better than previous management practices. A select group of progressive people are usually selected for this testing. Then, if results are positive, adoption grows over time. Complete adoption, however, is not usually possible, since there will always be some people resistant to change.

As shown in **Box 15-1,** testing and implementation lead back to the start of the cycle of discovery. As new ideas are tried, new questions arise and the process starts over again. This is why science is ongoing and is never finished. The more we learn, the more we realize that there are new questions. While this may seem discouraging, our body of knowledge and our understanding of relationships increases, equipping us ever better to find real solutions to difficult problems.

## Chapter Summary

In summary, the scientific method is an approach that finds solutions to problems or reveals relationships that exist in nature. This approach relies on clearly defined problems and hypotheses. Experiments are designed to test the effects of only the factors and treatments that are of interest and can be controlled. Data are collected and subjected to analyses. Relationships among factors are then modeled so that we can understand better what we know and what we have yet to learn. Attempting to implement new approaches reveals new questions that start the process over again. The scientific method encourages creativity and tries to minimize bias. It provides a defendable and reproducible approach that leads to two primary outcomes: 1) the probability of getting a certain outcome, and 2) the magnitude of that outcome. Such information is the basis upon which informed decisions and recommendations are made.

### Exercise 15-1. What are the various parts of the scientific method?

### Exercise 15-2. In Box 15-1, why is the scientific method represented as a circle?

### Exercise 15-3. What is bias and how does it affect recommendations?

## References

Freund, R.J. and W.J. Wilson. 1997. Statistical methods (rev. ed.). Academic Press, New York, NY.

Knighton, R.E. 2000. Setting up on-farm experiments. SSMG-17. *In*: Clay, D.E. et al. (eds.) Site-Specific Management Guidelines. PPI, Norcross, GA.

Merriam-Webster Online Dictionary. 2007. Hypothesis [Online]. Available at http://www.m-w.com/ (accessed 4 May 2007; verified 4 May 2007).

Wikipedia contributors. 2007. Scientific method [Online]. Wikipedia, the free encyclopedia. Available at http://en.wikipedia.org/w/index.php?title=Scientific_method&oldid=128107779 (modified 4 May 2007; accessed 4 May 2007; verified 4 May 2007).

Wittig, T.A. and Z.W. Wicks III. 2000. Simple on-farm comparisons. SSMG-18. *In*: Clay, D.E. et al. (eds.) Site-Specific Management Guidelines. PPI, Norcross, GA.

# SCIENTIFIC EXPERIMENTATION, STATISTICAL ANALYSIS AND INTERPRETATION

**Chapter concepts:** Independent variable, dependent variable, correlation, observation, sampling, survey, case study, replication, factorial experiment, mean, median, standard deviation, Type I error, Type II error, population, hypothesis testing

**Chapter purpose:** A brief discussion and overview of experimentation, sampling theory, and data analysis are presented. To make informed, science-based decisions, understanding data collection, simple arithmetic calculations used for statistical analysis, and data interpretation are critical. In this chapter, the types of experiments that will be discussed are: experimental, correlation, observations, survey, and case studies. Techniques to determine sampling requirements and test hypothesis are discussed. For additional information, readers are encouraged to study an in-depth statistics book.

Information for scientific-based decisions can be derived from published work and experimentation. To improve management, new techniques are frequently used to retest old hypotheses. In statistical analysis results from experiments are evaluated. For example, you may intuitively think that football linemen weigh more than quarterbacks or that house cats weigh less than mountain lions. The differences or similarities in weight can be determined by weighing a random sample of individuals in each class, determining the mean and variation of weight around the mean, and using statistical analysis to compare their weights.

## Types of Research

### Experimental research

In experimental research, the influence of treatments (independent variables) on a measurable variable (dependent) is investigated. ***Independent variables*** are those that are manipulated, whereas ***dependent variables*** are only measured. For example, in an experiment where different N rates are applied to corn, the independent variable is N rate and dependent variable is corn yield. In an erosion study where the influence of cultivation on erosion, the independent variable might be the number of times that a field is cultivated while the dependent variable is the amount of erosion. Research experiments are designed to provide insight into cause and effect. Disadvantages with the experimental approaches are that planning is required, research can be expensive, training in statistical methods are needed, and quality control procedures must be used. The results from experimental approaches can be complemented with the development of mathematical models.

Experimental research is used when determining the optimum N rate or plant population level for a field. In experimental studies, interactions between independent variables can occur. An ***interaction*** means that the answer for one factor depends on the level of another factor. For example, the relationship between N rate and yield might be different for irrigated and non-irrigated fields. A factorial experiment (all combinations of 5 nitrogen rates and 3 population levels (i.e., 15 individual treatments) can be used to evaluate interactions.

In addition to treatments, experimental studies usually contain multiple replications of each treatment. Replications provide assurance that the response to a treatment is real and repeatable and not due to error, or random chance. For example, if you flip a coin once and get heads, with no replication, you may predict that heads will always occur. As more replications of the flip occur, you should see that heads comes up 50% of the time and tails comes up the other 50% (unless the coin is weighted or it lands on its edge). This actual outcome does not match the original predicted outcome of 100% heads. Knowing the actual expected outcome, we can now use the results to reevaluate the original problem.

One of the most common approaches for incorporating replication into field experiments is to conduct ***randomized block experiments*** (**Box 16-1**). In a randomized block experiment, the number of blocks represents the number of times a treatment is replicated in the experiment. Each block contains a plot where one of the treatments is placed. Each treatment combination is randomly located within a block. **Randomized block experiments are used when environmental variability is not constant across the experimental site.**

**Box 16-1.** Example of a randomized blocked experiment. In a randomized blocked experiment, each treatment is replicated in each block and the treatments are randomly located within a block.

| Block 1 | | Block 2 | | |
|---|---|---|---|---|
| A | F | B | I | |
| B | G | J | A | ← Treatment A in block 2 |
| C | H | F | D | |
| D | I | C | E | |
| E | J | H | G | |
| Block 3 | | Block 4 | | |
| J | C | H | A | ← Treatment A in block 4 |
| H | G | E | J | |
| I | B | F | I | |
| F | E | D | C | |
| A | D | B | G | |

## Correlation analysis

An ***observation*** is the act of making or recording a measurement. Correlation analysis can be used to evaluate relationships between observations. For correlation analysis, two sets of observations must be collected. Correlation analysis is designed to determine the strength (i.e., how much are the variables correlated, a lot or a little?) and direction (positive, i.e., as one increases the other increases, or negative, i.e., as one increases the other decreases, correlation) of the relationship among these variables. Strengths of this approach are that direct manipulation of observations is not required (e.g., irrigation is not needed as a treatment to examine the effect of different amounts of water on yield) and predictive models can be developed. The greatest limitation of correlation is that it does not show cause and effect. For example, if in a production field there is a strong negative correlation between weed density and crop yield, it does not prove that weeds reduce yields; it only shows that there is a relationship between these variables. The low crop yield may also have been caused by poor crop growth that was limited by some other factor (excess water, planter skip, or poor germination), thereby providing an opportunity for weed seed germination and growth.

## Observational approach

In the observational approach, behavior or phenomena are carefully observed and recorded. Experiments testing the impact of independent factors on the behavior or phenomena are not conducted. Individual observations may consist of yields or soil characteristics at specific points in the field or observation involving animals as they go about their daily activities. The strength of this approach is that researchers can observe behavior in the setting where it occurs. Disadvantages are that: 1) the findings are difficult to extend beyond the boundaries of the individual observations, 2) observing behavior is very time consuming, 3) behavior is often disrupted during the observation (e.g., would the elephants have continued grazing and not stampeded if the car with observers had not driven into their territory?), and 4) cause and effect are not investigated. Correlation analysis can be integrated into the observational approach.

## Surveys

Surveys collect data from targeted populations. This research approach is used extensively to track human preferences but other types of surveys can also be undertaken. In surveys, finite populations are randomly sampled. For example, Minnesota has 10,000 lakes but a researcher may choose 50 lakes randomly across the state to examine the fish species that are present. These data may then be used to extrapolate the species present in the nonsampled lakes. Another type of survey is to ask individuals to respond to specific questions. This method does not involve direct observation by an observer. Correlation analysis can be integrated into the survey approach.

## Case study

A case study is an in-depth study of a specific question or event. The advantage of the case study method is that it allows for intensive analyses of appropriate research results from quality sources using experimental, correlation, observational, and surveying approaches. By its nature, a case study is specific and is difficult to generalize to other situations. This approach is often used in problem solving by exploring the many varied options that are (were) possible. The case study approach can be used to investigate such diverse issues as potential solutions to declining fish populations in the Grand Banks off Newfoundland to the implications of starting a 1000 cow dairy on land use, labor requirements, storage facilities, and costs. Case studies can incorporate the results from many projects that have used experimental, correlation, observational, and survey approaches.

## Analysis of Experiments

### Populations

Most experiments produce data that must be analyzed to test the hypothesis. Data collected from different treatments is viewed as data collected from different populations. A ***population*** is a group of individuals with similar characteristics and consists of all possible values. For example, all deer within a given area or all walleyes within a given lake are populations.

A population may or may not be limited (finite). If the entire population is measured, then the results can be described without ambiguity. For example, if 5 students in a population consisting of 20 students have blue eyes, then 25% of this population has blue eyes. Elections are another example where the outcome is based on measuring the entire voting population. In natural resource management, payment for products delivered may be based on measurements of the entire population. For example, a farmer receives payment for the total weight of the grain from a given field delivered to the elevator or the total gallons of milk delivered to a dairy from a dairy herd. In these situations, we can say with 100% certainty that the amount of harvested corn delivered to the elevator was 105.7 bushels/acre, or that the herd produced 1,010 gal of milk in 3 days.

In most situations, measuring the desired trait(s) of an entire population is not possible, so a portion of the population is sampled. Based on the information from the sample, a statistical analysis is used to make inferences about the whole population. An excellent example of drawing inferences about populations is the tracking of voter opinions using public opinion polls. When reported, the results from the poll often include the number of people surveyed and the margin of error of the responses. In some situations, poll results do not match election results. These differences may result from the between poll and election results. One of the reasons could be that the poll randomly sampled *eligible voters,* while the election measured preferences of the voting population.

In natural resource management, small samples of soil, plant, fish, or water are often collected from which inferences and management recommendations are made about the whole population. For example, in the surface 6 inches of an acre, there are approximately 2,000,000 lb of soil. Many fertilizer recommendations are based on the amount of nutrients contained within this soil. Analyzing all of this soil is impossible; therefore, a sample is collected, analyzed, and recommendations are given based on results from the small sample. To develop a valid, meaningful recommendation, the small sample must represent the larger population. If the sample does not represent the population a faulty recommendation can result.

### Sampling

A key component of all experiments is collecting representative samples or observations. The three most utilized sampling approaches are random, systematic, and stratified sampling. In ***random sampling*** each portion of the population has equal chance of being selected. When soil sampling a field for fertilizer recommendations, a random sample is often collected from the entire field. The sampling is accomplished by collecting several subsamples that are combined into a larger composite sample (**Box 16-2**).

***Systematic sampling*** is the collection of information or observations from specific locations. Grid sampling is a type of systematic sampling. In grid soil sampling, a sample is collected at each specified point. Another type of systematic sampling is to stop every 5$^{th}$ car at check point. This approach provides excellent information as long as samples do not contain hidden order.

***Stratified sampling*** is an approach where a large population is separated into distinct categories based on similarity of characteristics from which random samples are collected. The stratified approach may have a lower sampling error than random sampling.

The management zone approach for site specific field applications of nutrients is a modification of the stratified approach. The management zone approach separates fields into different zones based on the similarity of characteristics within the zone (elevation, growth rates, yield, soil type). For nutrient recommendations, several soil samples from within each zone are pooled into a single sample.

**Box 16-2.** A schematic showing random, systematic, and stratified sampling approaches.

| Location of randomly selected sampling locations | Example of grid sampled field | Random sampling within a stratified sampling approach |
|---|---|---|
| x x<br>x<br>x<br>x x<br>x x | x x x<br>x x x<br>x x x | x x o zone 2<br>x x o o<br>x o o<br>zone 1 x |

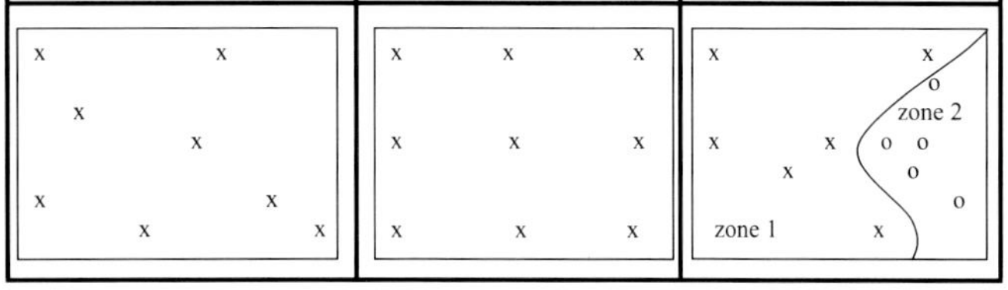

## Statistics

***Statistics*** is a tool used to help design experiments and objectively evaluate numerical data in order to make informed decisions. Statistics numerically evaluates the sample, or population from which it was derived, and provides a measure of the uncertainty of an inference made from the data. Statistics involves computation and arithmetic manipulation of data.

Measured variables can be either qualitative or quantitative. Qualitative variables have values that can not be defined numerically (e.g., sex, plant species, or marital status). Quantitative variables have values that can be defined numerically (e.g., yield, weed density, weight). Quantitative variables can either be discrete (whole values) (e.g., eggs per nest, petals per plant) or continuous (e.g., yield, weight).

## Means or averages

***Means*** and ***medians*** are used to describe the average value or central tendency of the data set. The arithmetic mean is determined by summing all the values within the data set and then dividing this sum by the number of observations. One of the most widely used operators in statistics is the summation symbol, Σ. The summation term is best described by an example. If 4 measurements (e.g., 1, 5, 6, and 3) are obtained from a population, then the sum of these values (1 + 5 + 6 + 3) is 15. Each individual value can also be written in terminology where $X_1$ is 1, $X_2$ is 5, $X_3$ is 6 and $X_4$ is 3. The subscripts 1, 2, 3, and 4 refer only to the physical location of the data in the data set and not the value.

Using the Σ terminology, it is possible to write universal mathematical expression that can be used to calculate a variety of factors. The general summation equation is $\sum_{i=1}^{n} Xi$ and means to sum all the Xi values where the i subscript is used to identify a specific sample and the subscript n represents the total number of observations in the data set. When summing the 4 values given above (1 + 5 + 6 + 3), the equation would be written as $\sum_{i=1}^{4} Xi$ and when i=1 the first sample in a list is substituted into the equation, when i=2 then the second sample in the list is added, when i=3, the third sample is added, and when i=4, the fourth sample is added.

The ***mean***, or average, of a sequence of numbers is the sum divided by the number of observations contained in the data set. A notation for the mean or average is $\overline{X}$. The mean for the example above is

$\overline{X} = \frac{15}{4} = 3.75$. Mathematically this is expressed by the equation, $\overline{X} = \frac{\sum_{i=1}^{n} Xi}{n} = \frac{15}{4} = 3.75$. This equation indicates that all values of X from the first to the last ($n^{th}$) will be summed and divided by the total number of observations ($n$).

The data provided in **Box 16-3** will be used as another example of calculations for the sum and the average values of a data set. In this example, 18 growers were contacted and asked to provide the amount of nitrogen (N) that was applied per acre of corn. The number of observations ($n$) is 18 and the sum of all the values (100 + 170 + 145 +...+ 180) is 2662 and the mathematical expression for the summation of this data set is $\sum_{i=1}^{n} Xi$.

The mean is then calculated by summing all the values and dividing by the total number of observations: $\overline{X} = \frac{\sum_{i=1}^{n} Xi}{n} = \frac{2662}{18} = 148$ lb N/acre. This result indicates that the average N rate applied per acre of corn was 148 lb. Note: some growers applied much less N and some much more N than the average. The ***varia-***

***tion*** of the individual values from the mean value is important information in data interpretation and will be discussed in detail later in this chapter.

**Box 16-3.** Hypothetical grower survey responses for the lb of N applied per acre to corn.

| | | | | | |
|---|---|---|---|---|---|
| 100 | 170 | 145 | 150 | 143 | 152 |
| 110 | 200 | 135 | 180 | 148 | 161 |
| 124 | 140 | 125 | 160 | 139 | 180 |

**Exercise 16-1. How would you write a summation statement for the first ten observations and last 8 observations contained in Box 16-3?**

Solution:

first 10, $\sum_{i=1}^{10} Xi$

last 8, $\sum_{i=11}^{18} Xi$

Note: Instead of manually calculating the data in **Box 16-3**, spread sheets can be developed (such as in Microsoft Excel program). In this exercise, the student should enter the data in **Box 16-3** into the 'B' column of a Microsoft Excel spread sheet. The average value is calculated by using the average command by typing in general terms [=average(start list:end list)] in the desired cell where you want the answer (note: do not enter this command in any of the cells that already contains a value) and pressing enter. In specific terms, if the data is in column B and starts in row 1, then type [=average(B1:B18)] and press enter.

**Exercise 16-2. Calculate the average of the first 10 observations using the data in Box 16-3.**

Solution:
In the Excel program type, =average(B1:B10), in the desired cell where you want the answer. Mean of first ten numbers is 139.9.

**Exercise 16-3. Calculate the average of the last 8 observations using the data in Box 16-3.**

## Median

The ***median*** is a measure of central tendency and is the value that 50% of the values are above and 50% are below. The median can be determined by placing the values in sequential order, and then determining by manual inspection. As shown in **Box 16-4**, the median for the data shown in **Box 16-3** is between 145 and 148 lb N/a.

**Box 16-4**. The data from **Box 16-3** arranged in sequential order from smallest to largest.

| | | | | | |
|---|---|---|---|---|---|
| 100 | 110 | 124 | 125 | 135 | 139 |
| 140 | 143 | 145 | 148 | 150 | 152 |
| 160 | 161 | 170 | 180 | 180 | 200 |

**Exercise 16-4. Calculate the median value for the data presented in Box 16-4 using Microsoft Excel.**

Solution:

If the data is in a Microsoft Excel spread sheet, this value can be determined by using the median command by typing in general terms [=median(start list:end list)] in the desired cell. If the data is in the B column then type[=median(B1:B18)] in the desired cell. The resulting calculated value of 146.5 is the median of this data set.

Note that the average value (calculated above) (147.9) and the median (146.5) are similar. When the mean and median values are not similar the population may be skewed and may not follow a normal distribution (bell shaped curve).

## Precision vs Accuracy

Even though ***precision*** and ***accuracy*** are strived for simultaneously, they have different meanings. ***Precision*** is the ability of a measurement to be consistently produced. For example, a thermometer in a temperature controlled room may have the same reading each time the temperature is checked because the temperature does not vary. The thermometer is said to be precise. ***Accuracy***, on the other hand, is the ability to match the reading with the actual value of the quantity being measured. If the temperature in the room is 33 °C and the reading on the thermometer is 33 °C, then the thermometer is both accurate and precise. If the thermometer reads 29 °C and the reading does not change because the temperature does not change, then the thermometer reading **is precise** but does **not accurately** measure the temperature. If the readings on the thermometer fluctuate but the room temperature does not, then the thermometer is neither precise and nor accurate.

## Variance and standard deviations

Unless each observation in a sample is the same number, there is variation around the average value. The ***variance*** ($s^2$) provides a measure of this variation. The variance is a measure of precision, not accuracy. The variance ($s^2$) is calculated using the equation,

$$s^2 = \frac{\sum_{i=1}^{n}(X_i - \overline{X})^2}{n-1}$$

where $\overline{X}$, is the mean of the observations, $X_i$ are the individual measurements, and $n$ is the number of observations in the data set. Using the data set where $X_1$=1, $X_2$=5, $X_3$=6, and $X_4$=3, the variance is calculated as follows:

$$\overline{X} = \frac{1+5+6+3}{4} = \frac{15}{4} = 3.75$$

$$s^2 = \frac{\sum_{i=1}^{n}(X_i - \overline{X})^2}{n-1} =$$

$$\frac{(1-3.75)^2+(5-3.75)^2+(6-3.75)^2+(3-3.75)^2}{4-1} = 4.922$$

The ***standard deviation*** ($s$) is defined as the square root of the variance ($s^2$). In this example the standard deviation is 2.22 ($4.922^{0.5}$).

The ***coefficient of variation (CV)*** provides an indication on the degree of precision relative to the mean. The *CV* is a percent and is calculated using the equation, $CV = \frac{s}{\overline{X}} \cdot 100\%$, where $s$ is the standard deviation (square root of the variance, $s^2$) and $\overline{X}$ is the mean. The CV provides a relative measure of the variation contained within the data set. The CV has been used to assess the relative precision of the measurements.

**Exercise 16-5. Using Microsoft Excel, calculate the variance ($s^2$) and standard deviation (s) for the data in Box 16-3.**

Solution:

In Microsoft Excel, variance ($s^2$) is determined using the command [=var(startlist:endlist)] and standard deviation ($s$) is determined by use the command [=stdev(startlist:endlist)] in the respective cells where the answer is desired.

The variance for the data in **Box 16-3** is 633.5 and the standard deviation is 25.17.

**Exercise 16-6. Calculate the coefficient of variation (CV) for the data in Box 16-3.**

**Exercise 16-7. Using Microsoft Excel, calculate the variance ($s^2$) and standard deviation of the following data.**

| | | |
|---|---|---|
| 10 | 3 | 34 |
| 15 | 7 | 8 |
| 12 | 8 | 22 |
| 6 | 20 | 10 |

## Estimating sampling requirements

Recommendations are only as good as the samples they are based on. The following discussion provides guidance on how many subsamples should be collected from a population to ensure that the recommendation has the desired precision and accuracy.

To estimate the number of subsamples required, three values are needed: an estimate of the variance ($s^2$), a value for the desired precision ($D$), and a t-value (that is obtained from a t-table commonly found in statistics books). The equation used to calculate the variance of a population is discussed above. However, the variance can be estimated from data collected from other populations with similar characteristics. The $D$ value represents the desired precision of the decision maker. For example, one manager might like to estimate the amount of feed for a dairy cow within 10 lb/day of the true value, whereas another manager may want this same information but want the estimate within 5 lb/day. It makes intuitive sense that the sampling requirement is directly related to the desired precision. The subsampling requirement for a given precision level can be calculated using the following equation:

$$\textit{Subsampling Requirement} = \frac{t^2 \cdot s^2}{D^2}$$

The equation has the terms $t^2$, $s^2$, and $D^2$. The ***t*** refers to a ***t*-value** that is obtained from a ***t*-table** (found in most statistics books) rather than a value that is calculated **(Box 16-5)**. The $s^2$ is the variance either estimated from another similar experiment or calculated from a relevant data set, and $D$ is the desired precision of the population estimate. For example, if the desired precision is ± 10 lb/day, then $D$ is 10.

The *t value* in the subsampling requirement equation is based on the degrees of freedom associated with the variance and the desired confidence level. A sample of the **t-table** is shown in **Box 16-5**. A more complete discussion of what the t-table represents and specific values for different probability levels and degrees of freedom are available at http://www.statsoft.com/textbook/sttable.html. Most t-tables have α level across the top of the table (columns) and degrees of freedom going in the vertical direction (rows). For our purposes, in calculating sampling requirements, 2 times the α value is the type I error associated with the estimate (type I error = 2 · α). ***Type I errors*** are often called the **p-value** (probability) and represent the error associated with concluding that there is an effect (or difference) when there is none (e.g., concluding that adding N increased yields, when it did not). If the p-value or Type I error is 0.05 then there is a 5% chance that you will conclude that two populations are different when they are not.

For the following examples, a ***two-sided test*** is desired, and therefore the p-value is twice the α value (α = p/2). A two-sided test is used when the hypothesis states that two treatments are equal and the alternative hypothesis is that they are not equal. For this test, there is a chance that the second treatment could be either greater than or less than the first treatment. The Type 1 error is the summation of both these chances (5%) and the α value is the probability associated with Treatment 1 being greater than (2.5%) or less than (2.5%) Treatment 2. For this reason, the α value for a two sided test is ½ the Type I error or p-value.

The row values of most t-tables are the ***degrees of freedom (df)***. The number of degrees of freedom is determined by subtracting one from the number of total number of observations (n-1). For example in a two-sided t-test with $\alpha = 0.10$ (probability level of 20%) and 21 observations, the *t value* for 20 degrees of freedom (21-1) is 1.325. For every degree of freedom and $\alpha$ value there is a unique *t*-value. The *t*-value decreases with increasing number of observations (e.g., the degrees of freedom increase), and increases with increasing desired precision (e.g., $\alpha = 0.10 \rightarrow \alpha = 0.025$).

**Box 16-5.** A simplified sample t distribution table. The $\alpha$ value represents the degree of significance desired in the calculations. For a two sided test the type I error is two times the $\alpha$ value.

| degrees of freedom | $\alpha$=0.10 | $\alpha$=0.05 | $\alpha$=0.025 |
|---|---|---|---|
| 1 | 3.078 | 6.314 | 12.706 |
| 2 | 1.886 | 2.920 | 4.303 |
| 3 | 1.638 | 2.353 | 3.182 |
| 4 | 1.533 | 2.132 | 2.776 |
| 5 | 1.476 | 2.015 | 2.571 |
| 6 | 1.440 | 1.943 | 2.447 |
| 7 | 1.415 | 1.895 | 2.365 |
| 8 | 1.397 | 1.860 | 2.306 |
| 9 | 1.383 | 1.833 | 2.262 |
| 10 | 1.372 | 1.812 | 2.228 |
| 20 | 1.325 | 1.725 | 2.086 |
| ∞ | 1.282 | 1.645 | 1.960 |

**Exercise 16-8. What is the α-value for a two sided test that has a Type I error of 0.10?**

Solution:

0.10/2= 0.05

**Exercise 16-9. What is the one sided *t* value for 9 degrees of freedom and an α of 0.05?**

**Exercise 16-10. What is the *t* value of 20 degrees of freedom and an α of 0.10?**

The number of subsamples needed (sampling requirement) is determined by inserting *s*, *t*, and *D* into the subsampling requirement equation (*Subsampling Requirement* = $\frac{t^2 \cdot s^2}{D^2}$). For example, if a natural resource manager desires to determine the soil test phosphorus (P) value from a field within ± 3 ppm of the true value 80% of the time. The manager knows that that in a previous field he collected 21 samples that were analyzed separately. The variance ($s^2$) of these samples was 14. Because 21 samples were collected the degree of freedom (df) associated with this experiment was 20 (21-1) and *t*-value with an $\alpha$ of 0.10 [(1-0.80)/2] is 1.325 (**Box 16-5**). Once the $t$, $s^2$, and $D$ values are determined, they are substituted into the subsampling requirement equation,

$$Subsamp.\ Requirement = \frac{t^2 \cdot s^2}{D^2} = \frac{1.325^2 \cdot 14}{3^2}$$

$$Subsamp.\ Requirement = \frac{1.76 \cdot 14}{9}$$

$$Subsamp.\ Requirement = \frac{24.6}{9} = 2.73$$

This solution means that 3 individual subsamples should be combined into a single combined sample and this combined sample will provide results within 3 ppm of the true value 80% of the time.

In this next example, the sampling requirement will use $\alpha$ 0.10 Type 1 error [(1-0.8)/2] and 20 degrees of freedom (e.g., the *t* value is 1.325) but the variance ($s^2$) will increase to 50 (from 14) and the desired precision (*D*) will increase to 2 ppm. Once again, the sampling requirement will be calculated using equation 2. When these values are then substituted into subsampling requirement equation, the results are,

$$n = \frac{t^2 \cdot s^2}{D^2} = \frac{1.325^2 \cdot 50}{2^2} = 22$$

This calculation suggests that for a soil test result to be within 2 ppm of the true value 80% of the time a composite soil sample should contain at least 22 randomly collected individual cores. Note that the sampling requirement increased with the variance, (e.g. high variance = high sampling requirement and high desired precision = high sampling requirement).

**Exercise 16-11. How many composite cores should be combined into a single soil sample if the variance for 10 samples collected in an adjacent field is 20 ppm and the desired confidence interval is ±4 ppm 90% of the time?**

Solution:

The 90% of the time is referring to the desired Type I error. The Type I error is determined by converting the percent to decimal form, subtracting the resulting 0.90 from 1, and dividing this difference by 2 ($\alpha$ = 0.05 = [(1.0-0.90)/2])

The *t* value for 9 degrees of freedom with an $\alpha$ value of 0.05 is 1.833.

The exercise states that the *D* value is 4 ppm and the variance ($s^2$) is 20 ppm. These values are then substituted into the subsample requirement equation.

$$n = \frac{t^2 \cdot s^2}{D^2} = \frac{1.833^2 \cdot 20}{4^2} \approx 4 \text{ cores}$$

**Exercise 16-12. How many composite cores should be combined into a single soil sample if the variance for 10 samples collected in an adjacent field is 600 ppm and the desired confidence interval is ±4 ppm 90% of the time?**

## Frequency distributions and histograms

To assist in data interpretation, data can be reorganized in order from smallest to largest, separated into several different groups, and the number of observations contained within a group plotted against a value representing the class. This type of graph is called a ***histogram***, or frequency distribution. The data in **Box 16-3** was separated into several classes (**Box 16-6**). The first class contained data points between 95 and 115, while the second class included all data between 116 and 135. Classes do not need to be of equal intervals. The number of points contained within a class is then plotted against the class value to develop a histogram (**Box 16-7**). Histograms are useful for revealing skewing in the data set and the range of potential values.

**Box 16-6.** Interval spacing and number of observations of data collected from a hypothetical survey where the data is shown in **Box 16-3**.

| Interval spacing | # of Observations |
|---|---|
| 95-115 | 2 |
| 116-135 | 3 |
| 136-155 | 7 |
| 156-175 | 3 |
| 176-195 | 2 |
| >195 | 1 |

**Box 16-7.** Histogram of the data collected summarized in **Box 16-6**.

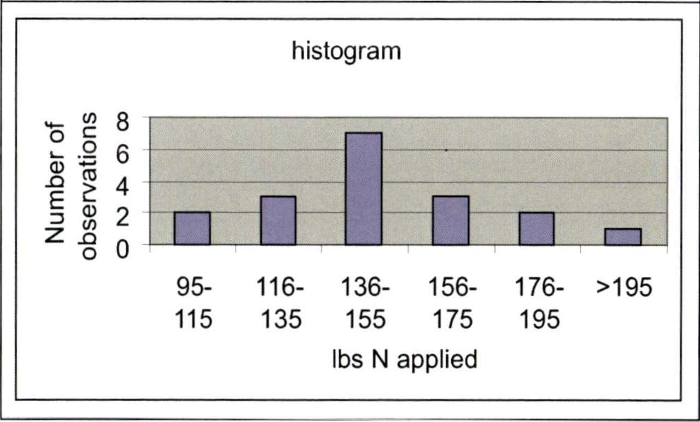


Charts of histograms can be created in Excel. This is accomplished by:

1. Enter the data shown in **Box 16-3** in column A;
2. In column B, enter the classes into which the data will be distributed;
3. In column C, enter the lower limit of each class;

4. On the Data menu, click Data Analysis. If you not find this function, it should be enabled;
5. Select Histogram and click OK.
6. In the dialog box that opens, the Input Range field, select the data in column A. In Block Interval field, select the data in column C; finally, mark the Result option Chart and click OK.

## Hypothesis testing

***Hypothesis testing*** is the process used to accept or reject hypothesis statements. In hypothesis testing, two statements are identified: a hypothesis and an alternative hypothesis. The hypothesis is identified as $H_o$ and the alternative hypothesis is identified as $H_a$. A hypothesis might be that the average N rate in a county is equal to 110 lbs N/a, with the alternative hypothesis being the average N rate is not equal to 110 lbs N/a. Mathematically these would be written as,

$H_o : \mu = 110$ lb N/acre,

$H_a : \mu \neq 110$ lb N/acre

where μ is the entire population mean. Since most of the time entire populations are not sampled, the $\overline{X}$ term, which indicates a sample mean, is used in place of μ. In hypothesis testing, two types of error can occur. ***Type I error*** occurs when the hypothesis is accepted when in fact it is false. ***Type II error*** occurs when the hypothesis is false and it is accepted. Many different types of analysis approaches can be used in hypothesis testing. The following discussion will use two different approaches.

## The t-test

A *t*-test is used to compare two treatments. Random samples are collected from these two treatments and the mean and variance of the desired trait(s) of each population are determined. If the assumption is made that the two treatments have a common variance and are normally distributed (bell shaped curve), then a *t*-test can be used to compare the means and determine if the mean of each population has the same average value ($H_o$ accepted) or different values ($H_a$ accepted). The *t*-test is conducted by,

1. Collecting random samples from each population (treatment).
2. Determining the standard deviation and variance of each population (treatment) mean.
3. Determining the weighted average of the calculated standard deviations (also called the pooled standard deviation or $s_p$) of the two population means. The weighting is based on the number of samples used to estimate each variance. The weighted average, or pooled standard deviation, is calculated with the following equation:

$$s_p = \left[\frac{(n_1 - 1)s^2_1 + (n_2 - 1)s^2_2}{(n_1 - 1) + (n_2 - 1)}\right]^{0.5}$$

where $s^2_1$ and $s^2_2$ values are the variances of treatments 1 and 2, respectively, and the values of $n_1$ and $n_2$ are the number of samples collected from treatments 1 and 2, respectively.

4. Calculating the test statistic with the equation,

$$Test\ Statistic = \frac{(\overline{X}_1 - \overline{X}_2)}{s_p\sqrt{\left(\frac{1}{n_1} + \frac{1}{n_2}\right)}}$$

where $\overline{X}_1$ *and* $\overline{X}_2$ are the means of the population 1 and population 2, respectively. The test statistic is then compared with the t value associated with the appropriate level of degrees of freedom and error level. The degrees of freedom (df) associated with the *t* values in the *t* table (**Box 16-8**) is $[(n_1 - 1) + (n_2 - 1)]$ or $[n_1 + n_2 - 2]$.

**Box 16-8.** Hypothetical data collected from a study evaluating the influence of two corn hybrids on corn yields. A random sample of farmers growing these two hybrids was obtained.

| | Hybrid 1 | Hybrid 2 |
|---|---|---|
| Yield Mean (bu/acre) | 100 | 130 |
| Yield Variance ($s^2$) | 800 | 900 |
| Yield Standard deviation (s) | 28.28 | 30.00 |
| Number of samples | 6 | 6 |

For example, six producers grew and harvested 2 corn hybrids. The hypothesis was that both hybrids had similar yields. To solve this exercise, the values provided in Box 16-8 are substituted into pooled standard deviation ($s_p$) and test statistics equations. The substitutions of these values into the appropriate equation are shown below.

$$s_p = \left[\frac{((6-1)(800)) + (6-1)(900)}{(6-1) + (6-1)}\right]^{0.5} = 29.15$$

$$Test\ Statistic = \frac{(130 - 100)}{29.15\sqrt{\left(\frac{1}{6}\right) + \left(\frac{1}{6}\right)}} = 1.78$$

The test statistic is then compared with a $t$ value with the appropriate degrees of freedom and level of significance in **Box 16-5**. The $t$ value at the 95% probability (Type I error = 0.05, two sided test and therefore $\alpha = 0.025$) and 10 degrees of freedom is 2.228. Because the test statistic 1.78 is less than 2.228, the hypothesis is not rejected (two hybrids have similar yield). If the test statistic would have been greater than 2.228, then the hypothesis would be rejected in favor of the alternative hypothesis that the two populations are not equal. Note, if the probability of the $t$ value was 80% (Type I error = 0.20, two sided test and therefore $\alpha = 0.1$), rather than 95%, then the alternative hypothesis (that one hybrid population had a different yield than the other) would have been accepted. Determining the level of probability that you are willing to accept is important and must be considered when interpreting results! In hypothesis testing, the ability to detect differences improves with replications ($n$).

**Exercise 16-13. Use Microsoft Excel to determine if the yields from two varieties are the same or similar. Yield data was randomly collected from different fields in a county.**

| samples | variety 1 | variety 2 |
|---|---|---|
| 1 | 110 | 132 |
| 2 | 100 | 133 |
| 3 | 99 | 154 |
| 4 | 75 | 99 |
| 5 | 47 | 88 |
| 6 | 112 | 120 |
| 7 | | 243 |

Solution:

1. Type data into Excel
2. Select tools
3. Select Data Analysis (If Data Analysis does not show up, go to Add-Ins, and check Data Analysis)
4. Select t-test two-sample assuming equal variances
5. OK
6. Highlight data for variable 1 and 2, select OK
7. Results are shown in the following chart.

| t-Test: Two-Sample Assuming Equal Variances | | |
|---|---|---|
| | Variable 1 | Variable 2 |
| Mean | 90.5 | 138.4286 |
| Variance | 627.5 | 2614.286 |
| Observations | 6 | 7 |
| Pooled Variance | 1711.201 | |
| Hypothesized Mean Difference | 0 | |
| df | 11 | |
| t Stat | -2.08256 | |
| P(T<=t) one-tail | 0.030713 | |
| t Critical one-tail | 1.795885 | |
| P(T<=t) two-tail | 0.061426 | |
| t Critical two-tail | 2.200985 | |

For this analysis a two tailed test will be used. Based on a P(T<=t) two-tail value of 0.061 (1-0.061 = .939 = 93.9%) indicates that there is a 93.9% probability that the treatmants are different.

## Paired t-test

A different type of test that can be used is the paired $t$ test. In this test, two measurements are linked. For example, before and after measurements are linked observations. The following example shows how the paired $t$ test can be used to test the hypothesis that fully inflating tires does not change gas mileage. In this experiment, gas mileage of each car is measured before and after inflating tires. This data is shown on the following page.

| **Box 16-9.** Data collected from a hypothetical experiment testing the effect of inflating tires on gas mileage. | | | |
|---|---|---|---|
| | Before | After | |
| Car | GPM | GPM | difference |
| 1 | 12 | 13 | 1 |
| 2 | 16 | 15 | -1 |
| 3 | 22 | 25 | 3 |
| 4 | 40 | 43 | 3 |
| 5 | 13 | 15 | 2 |
| 6 | 12 | 12 | 0 |
| 7 | 14 | 15 | 1 |
| | | Mean difference | 1.285714 |
| | | Standard deviation | 1.496026 |

The test statistic for this analysis is,

$$TestStatistic = \frac{(\overline{X}_{difference})}{s_{difference}\sqrt{\left(\frac{1}{n}\right)}} = \frac{1.28}{1.50\sqrt{\frac{1}{7}}} = 2.258$$

After substituting the values for mean of the differences ($\overline{X}$ difference = 1.28) the standard deviation of the difference ($s$ difference = 1.50) and the number of cars in the study ($n$ = 7) into the equation, the test statistic value was determined to be 2.258. This value is compared with the $t$ value obtained from the $t$ value table in **Box 16-5**. A critical $t$ value (2.365) is found for 6 degrees of freedom (7-1) and an $\alpha$ value of 0.025 (type I error = 2* 0.025= 0.05) (i.e., 95% probability). The hypothesis that inflating the tires influences gas mileage is rejected, at the 5% level, because the calculated test statistic, 2.258, is less than 2.365 (the tabled $t$ value). If we change our probability criteria from 95% to 90%, the interpretation of the results would change since the test statistic of 2.258 is greater than the $t$ table value of 1.943 (6 df and $\alpha = 0.05$).

**Exercise 16-14. What would the test statistic be if the pooled standard deviation for the data in Box 16-8 was 12.25 instead of 29.15 (calculated earlier in this chapter)? Would two treatments be significantly different at the 95% probability level?**

Solution:

Substitute 29.15 in the test statistic equation with 12.25 and calculate the test statistic. The calculation is shown below.

$$Test\ Statistic = \frac{(\overline{X_1 - X_2})}{sp\sqrt{\left(\frac{1}{n1}+\frac{1}{n2}\right)}} = \frac{(130 - 100)}{12.25\sqrt{\frac{1}{6}+\frac{1}{6}}} = 4.24$$

The critical t value in Table 1.3 for $\alpha$ = 0.025 and 10 degrees of freedom is 2.228. The critical t value of 2.228 is less than the test statistic of 4.24 and therefore the populations have different means.

**Exercise 16-15. Calculate the test statistic for the data in Box 16-8 if each sample was replicated 10 times (n = 10 for each mean) instead of 6. Would the means have been considered different or similar at the 95% probability level?**

## Chapter Summary

This chapter provides an introduction to several different types of experiments that can be conducted, how to sample, and preliminary statistical analysis approaches that can be used on data. Each of the approaches has unique strengths and weaknesses. Understanding these strengths and weaknesses is important for designing efficient experiments.

## References


Davis, B. Introduction to Agricultural Statistics. 2000. Delmar Thomson Learning.

Steel, R.D.G., J.H. Torrie, and D.A. Dickey. 1997. Principles and Procedures of Statistics a Biometrical Approach, Third Edition, McGraw-Hill Publisher.

# UNDERSTANDING AND QUANTIFYING CHANGE WITH MODELS

**Chapter concepts:** Empirical modeling, mechanistic modeling, relational diagram, boundary conditions, linear models, exponential decay functions, half-life, exponential growth, logistic models.

**Chapter purpose:** To provide an introduction to building and selecting appropriate models that can be use for predictive purposes.

Mathematical models can be used to improve decisions. Mathematical models can be either empirically or mechanistically based. In ***empirical modeling*** a numeric technique is used to find the coefficients that maximize the amount of variability explained by the model. In these models, a detailed understanding of the system is not needed. In ***mechanistic modeling,*** the fundamental knowledge of the chemistry, biology, and physics is used to define relationships between factors. Most models are constrained by the data sets used to develop the models. These constraints are called the ***boundary conditions***.

## Conceptual and Relational Diagrams

A useful step in modeling is to first draw a diagram that conceptualizes the problem. This drawing is called a ***conceptual model***. A diagram helps identify the important components of the problem and provides insights into how to solve the problem. Conceptual models are used to identify assumptions, known values, and focus the question. For example, two trains full of grain were going in different directions after starting at the same point, and we want to know how far apart the trains are now. To solve this problem, we need to know the direction of travel. If the trains are going in opposite directions (i.e. one headed east and one headed west), then the solution will be derived differently than if the trains are on a perpendicular (or some other) course. In this example, we will assume that the trains are moving perpendicular to one another, Train 1 headed east and Train 2 headed north. A conceptual model showing the problem is presented in **Box 17-1**.

## Converting Relational Diagrams to Mathematics

The mathematics used to solve a problem depend on how the problem was conceptualized. One solution to the train problem above is to measure the distance between the trains at the specified time. In this solution, the speeds of the trains are not known and not needed. However, the only time this solution would be correct would be for the time when the measurement is taken.

**Box 17-1.** A diagram showing how a conceptual model can be used to identify assumptions, components of a problem, and relationships between what is known. Conceptual models are graphical representations of problems.

Question: How long is the line between the trains if the angle between the two trains is 90 degrees?

Train 2 traveling north at 60 miles/hour

Train 1 traveling east at 50 mile/hour

A more general solution to this problem would involve developing equations with different levels of complexity. One scenario might be that Train 1 is traveling east at 50 mile/hour and Train 2 is traveling north at 60 mile/hour. How far apart are the trains after 2 hours? The distance between the two trains after 2 hours could be determined by solving the equation,

$$\text{distance} = (x^2 + y^2)^{0.5} = \left[\left(\frac{50 \text{ mile}}{\text{hour}} \cdot t\right)^2 + \left(\frac{60 \text{ mile}}{\text{hour}} \cdot t\right)^2\right]^{0.5},$$

where x is the distance of the Train 1 traveling east and y is the distance of Train 2 traveling north, and $t$ is time. When $t$ is 2, the solution to this equation is 156 miles. A different solution to the problem might take the form,

$$\text{Distance}_t = \left[\left(t\frac{dy}{dt}\right)^2 + \left(t\frac{dx}{dt}\right)^2\right]^{0.5} = \left[\left(t\frac{50 \text{ miles}}{\text{hour}}\right)^2 + \left(t\frac{60 \text{ miles}}{\text{hour}}\right)^2\right]^{0.5}$$

where $t$ is time, $\frac{dy}{dt}$ and $\frac{dx}{dt}$ are the speeds of the train traveling north (60 mile/hour) and east (50 miles/hour), respectively.

Using the more general equations, the distances between the trains at 1, 4, and 10 hours can be calculated by solving for distance after substituting 1, 4, and 10 for $t$. The distances are 78, 312, and 781 miles. These data then can be graphed and a linear model developed (**Box 17-2**).

**Box 17-2.** Distance between the trains as measured with tape at specified time intervals.

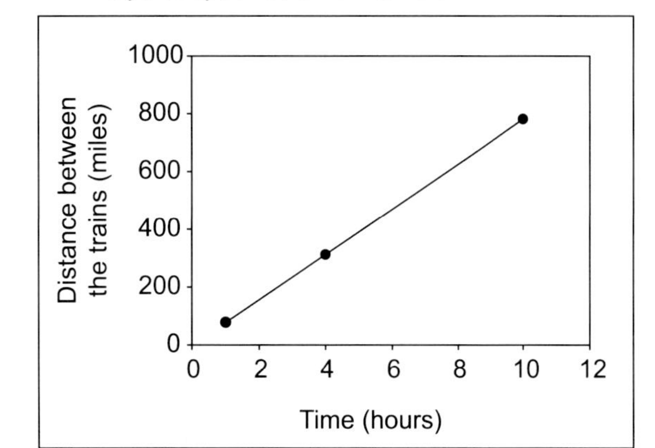


All three approaches produce reliable results. The differences between the approaches result from different ways to conceptualize the problem.

In summary, models are used to:

- Gain a better understanding of the physical behavior of the system;
- Discover special relationships within the system;
- Predict behavior under different conditions; and
- Predict the effectiveness of different approaches.

All models are constrained by the ***boundary conditions*** used to define the problem. Boundary conditions are the range of environments observed in the experiment.

## Boundary Conditions

All models are constrained by the conditions under which the data were collected and the assumptions associated with the mathematics. Extrapolating the predictions beyond these constraints contains risk. These constrains are the boundary conditions. Considering the constraints of the model is extremely important when considering the consequences of management decisions. For example, the application of 1 oz of herbicide per acre may kill weeds and not injure the crop, whereas a 4 oz application of the same herbicide may not only kill the weeds, but also kill the crop and have a long-term residual to injure the next year's crop.

To use models for predictive purposes, the model needs to provide an accurate representation of the system. Models need to be based on "good" data, and should be tested under a range of conditions. The model complexity depends on the problem and the acceptable error. For example, a model that only considers yield potential and soil nitrate-N may provide the precision and accuracy required for nitrogen fertilizer recommendations. However, for a more accurate recommendation, a more complex model may be needed. Both simple and complex models have their role in understanding and managing ecosystems. The ability to explain the complexity of natural resources with mathematics is critical for developing improved recommendations.

Explaining complexity requires a good understanding on how change is measured and integrated into decision tools. Change has the units associated with a rate. The rate explains how quickly something changes. Rates can be used to explain the speed of a car or the growth of plants, microbes, or livestock. A rate can be anything as long as it describes numerically how something has changed with respect to time. The first derivative of an equation represents the rate of change. In the figure shown in **Box 17-2**, the rate is the slope of the equation.

## Linear Models

A ***linear model*** can be used to explain the relationship between two values **(Box 17-2)**. A linear equation has the form, $y = mx + b$, where $m$ is the slope and $b$ is the $y$-intercept. Linear equations have also been called zero order rate equations. A characteristic of zero order equations is that the rate of change is equal to a constant $m$.

A ***relational diagram*** relating gas contained in a car and distance traveled is shown in **Box 17-3**. A relational diagram shows the different components of the problem and the relationships between the compartments. In this problem, the distance the car is traveling ($m$) is independent of the amount of gas in the tank or the distance it has traveled. Note, $m$ is not influenced by either the amount of gas in the car or the distance traveled.

**Box 17-3.** A relational diagram showing a system that might follow a linear or zero order model. In this example, the rate constant ($m$) relating gas in the car and the distance traveled is not influenced by or linked to either the amount of gas in the car or the distance traveled.

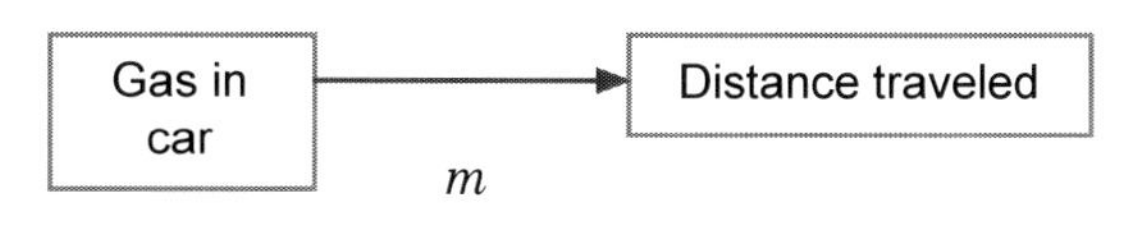


The model for **Box 17-3** can be developed by plotting gas remaining in the tank on the Y axis and time on the X axis. The $b$ value (or intercept) in the resulting linear equation represents the amount of gas at the start of the trip and the $m$ value (slope or $\frac{dy}{dt}$) represents gas use per unit of time. In this situation, the amount of gas remaining in the tank decreases with time.

The relationships outlined above are simplistic and are based on the assumption that distance and gas used are independent of each other.

A zero order equation can also be used to predict how much cat food a cat will eat in 7 days. If a cat eats 1 lb of food a day, then in 7 days it will eat 7 lb. This model will have the form,

$Food\ Eaten = mt + b = \frac{dy}{dt}x + b = \left(\frac{1\ lb\ food}{day}\right) \bullet days + b$ where $b$ is zero because the cat did not eat anything at time zero.

In summary, zero order models can be used to describe the gain or loss of a substance over time when the rate of change is constant. The form of this equation is,

$$y = mt + b$$

If $m$ is negative, then y decreases with increasing time. For zero order kinetics the half life, or length of time required for one-half to be decomposed, is calculated with the equation,

Half life = So/(2 • $m$)

where So is the amount of a substance at time 0, and $m$ is slope of the linear equation.

## Exponential Decay Function (First Order Models) and Half Lives

In many situations, the rate of change is related to the amount of substrate. This system can be modeled with an exponential model. Imagine the situation where herbicides are applied to soil. Over time these herbicides are decomposed. The rate that they are decomposed is dependent on the amount of herbicides in the soil. A relational diagram can show the feedback loop on the rate constant, $k$ (**Box 17-4**).

**Box 17-4.** A relational diagram showing how herbicides are decomposed to $CO_2$. The model shows this relationship by drawing an arrow between the amount of gas in the car and the rate constant.

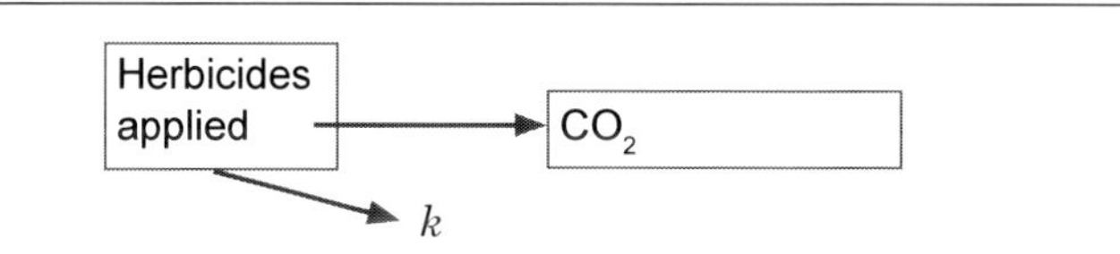


***Exponential decay functions or first order models*** can be used to model systems where the rate of change is related to the concentration of either the substrate or the product. If the reaction rate is dependent on the amount of substrate, then the rate of change in y can be expressed as:

$$\frac{dy}{dt} = -ky$$

This equation means that the rate of change in y is dependent on the rate constant $k$ and the herbicide concentration. In this example, the herbicide remaining in the soil decreases with increasing time but instead of a linear relationship, the relationship is explained by an exponential decay function (**Box 17-5**).

**Box 17-5.** A hypothetical relationship between time and the amount of herbicide remaining in the soil.

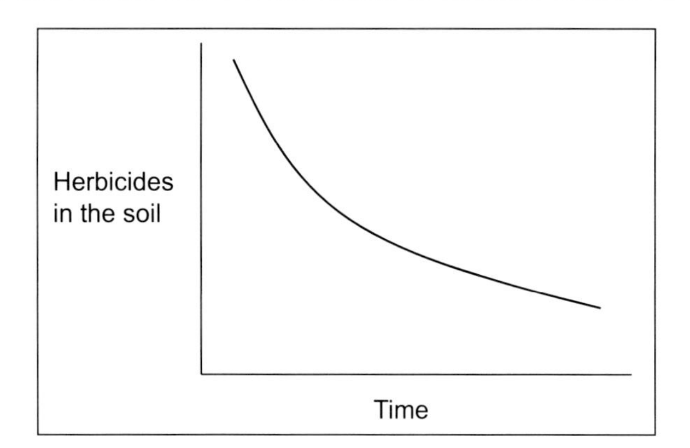


The exponential decay derives its name from the fact that the rate constant is negative (indicating decay) and the data set fits an exponential equation. The generalized form of the exponential decay function is,

$$(y_t) = y_0 e^{-kt}$$

where $y_t$ is the concentration at some time $t$. This equation can be rewritten into the form,

$$\ln(y_t) = \ln\,(y_0 e^{-kt})$$

where ln is the natural log and $e$ is the exponential function. Because

$$\ln(y_0 e^{-kt}) = \ln y_0 + \ln e^{-kt}, \text{ and } \ln e^{-kt} = -kt$$

can be defined by the equation,

$$\ln(y_t) = -kt + \ln y_0$$

where $y_t$ is the concentration at time $t$ and $y_0$ is the starting concentration at time zero. This equation will produce a straight line when the natural log of $y_t$ is plotted against time. The $k$ value in this equation is called the first order rate constant. The amount that has been lost is calculated with the equation,

$$y_{lost} = y_o(1-e^{-kt})$$

Exponential decay functions, or first order equations, have been used to describe N and C mineralization in soil and radioactive decay. Exponential decay functions can be used to determine the ***half-life,*** length of time required for half of the reactants to react. The half-life and residence times are calculated with the equations,

$$t_{\text{half-life}} = \frac{\ln 2}{k}$$

$$t_{\text{residence}} = \frac{1}{k}$$

For example if the rate constant is 0.0379 day$^{-1}$, then the half life is 18.4 and residence time is 914 hours.

$$t_{\text{half-life}} = \frac{\ln 2}{k} = \frac{0.693}{0.0379} = 18.4\,hours$$

$$t_{residence} = \frac{1}{k} = \frac{1}{0.0389} = 914\,hours$$

## Radioactive Decay

Many radioactive isotopes can be described using first order rate kinetics. Radioactivity is measured in the Curie (Ci). Two terms, disintegrations per minute and counts per minute, are used to describe decay rates. Counts per minute are generally lower than disintegrations per minute because counting efficiency is not 100%. Half lives are determined using the same calculations discussed above. For example, what is the half life if the specific decay constant is 0.012/day.

$$t_{\text{half-life}} = \frac{\ln 2}{k} = \frac{0.693}{0.012} = 58\text{ days}$$

Selected decay constants

| Isotopes | Decay constants ($d^{-1}$) | Half life |
|---|---|---|
| $^{14}C$ | $3.3 \cdot 10^{-7}$ | 5760 years |
| $^{3}H$ | $1.5 \cdot 10^{-4}$ | 12.3 years |
| $^{13}N$ | 78.3 | 13 min |
| $^{32}P$ | 0.048 | 14.3 days |
| $^{33}P$ | 0.028 | 25.2 days |
| $^{40}K$ | $1.5 \cdot 10^{-12}$ | $1.3 \cdot 10^{9}$ years |
| $^{35}S$ | 0.008 | 87.2 days |

## Exponential Growth

The *exponential growth model* is used to predict the outcome of interest on money in the bank or unconstrained birthrates on population levels.

**Box 17-6.** A relational diagram showing a system that might follow an exponential increase model. In this example, the actual rate that the population is growing is influenced by the number of deer and is not constrained by food.

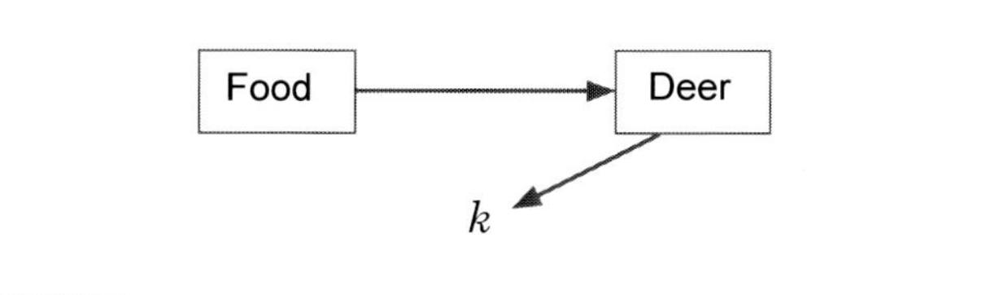

An exponential growth model predicts that if growth is not constrained, then it will increase exponentially. The exponential model takes the form,

$$y_t = y_0 e^{kt}$$

where $k$ is the growth rate and $t$ is time and is illustrated in **Box 17-7**. If payments are not made on cars, houses, credit cards, and motorcycles loans, the total amount of the loan increases in this manner.

**Box 17-7.** An example of an exponential model where deer are increasing exponentially.

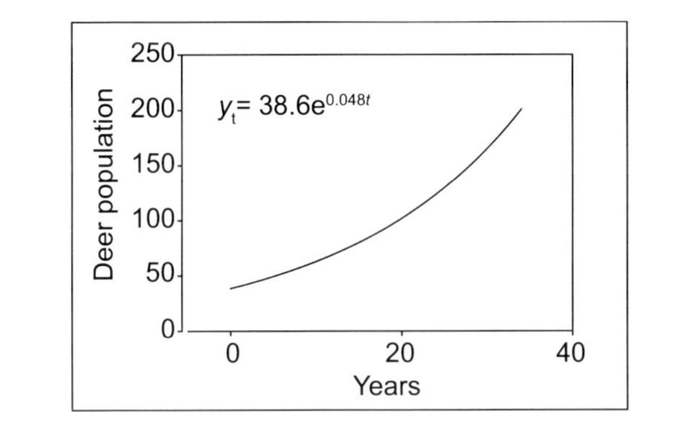


Exponential models can be used to describe uncontrolled population growth. In the example shown in **Box 17-7**, the rate constant can be defined as the difference between the birth and death rates. Similar to the exponential decay function, exponential growth model can be expressed by the equation,

$$\ln (y_t) = kt + \ln(y_o)$$

This model is derived by taking the natural log (ln) of both sides of $y_t = y_o e^{kt}$. The equation $\ln(y_t) = kt + \ln(y_o)$ suggests that if $\ln y$ is plotted against time, a linear line will result with the slope representing the rate constant $k$. The assumptions of this model are that

1) Continuous growth is occurring and

2) Resources are unlimited.

This model predicts that a population will increase exponentially without bounds at some rate $k$.

## Logistic Model

In many situations, the rate of growth is limited by the resources and population density. This situation can be conceptualized by the relational diagram in **Box 17-8.**

**Box 17-8.** A relational diagram showing a system where the growth rate is influenced by the deer population level and the carrying capacity (amount of food) of the system. The arrows linking both the food and deer components to the rate constant show this relationship.

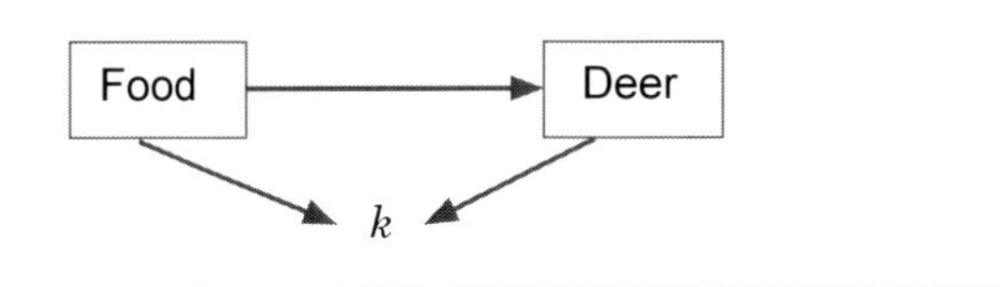


Theoretically, it makes sense that it should be possible to predict deer population based on the deer growth rate ($r_o$) and the carrying capacity ($K$). Mathematically these terms are used to predict the population level using this equation,

$$Deer_t = \frac{Deer_o \cdot K}{[Deer_o + (K - Deer_o)e^{-r_o t}]}$$

where $Deer_o$ is the population of the deer at time zero. This equation is referred to as the ***logistic model***. If $Deer_o$ is less than the carrying capacity ($K$), then the population will increase until it reaches a plateau. If $Deer_o = K$, then the population level will not change and if $Deer_o$ is greater than $K$, then the population will decline until $Deer \leq K$. This model combines two processes: growth and competition. The graphic representation of this equation when $Deer_o < K$ results in an S curve (**Box 17-9**).

**Box 17-9.** A hypothetical relationship between deer population and time. This is the type of curve that can be fit to a logistic model.

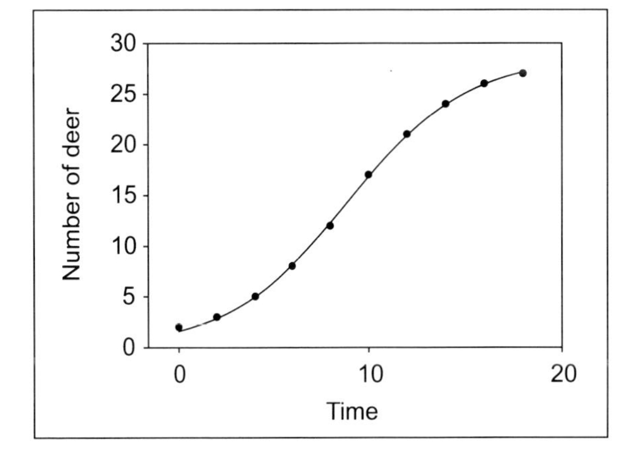

## Model Selection Influences the Recommendation

In natural resource management, managers may be more interested in relationships between inputs and outputs than the rate of change functions described above. The power of mathematics is that the solutions for the rate equations described above can be easily modified to consider relationships between inputs and outputs. The simplest approach is to replace time on the X-axis with a specific input and replace a product on the Y-axis with an output such as yield. This type of expression can be defined by the function,

Yield = f(nutrients, genetics, pests, environment, other)

To solve this problem, most scientists simplify it. For example, rather than looking at the influence of N and water on yields, an experiment may consider only the effect of N or water on yields. Numerous natural resource management experiments have used this approach. The following example demonstrates the influence of the model used to define the system on the resulting recommendations.

## Importance of Selecting the Appropriate Models

A hypothetical experiment was conducted to determine the influence of aphids on soybean yields. The experiment has two treatments (the control area had all aphids controlled, and an area where aphids were not controlled). The yield in the area with zero aphids was 45.0 bu/acre, and the yield in the area with 15,000 aphids/plant was 8.1 bu/acre.

Regression analysis programs can be used to develop many mathematical models. In solving problems, it is important to remember that two independent equations are required to solve problems with two unknowns and that three independent equations are required to solve problems with three unknowns. In a linear equation ($y = mx + b$), the $x$ and $y$ terms are measured and the m and b terms must be determined. This equation contains two unknowns ($m$ and $b$). Each set of paired data that is collected from a site can be arranged into an independent equation. For example, by substituting the soybean yields and aphid populations for $y$ and $x$, respectively, two independent equations were developed. These equations are,

8.1 bu/acre = $m$(15,000 aphids/plant) + $b$

45 bu/acre = $m$(0 aphid/plant) + $b$

This linear equation for these two points is determined by solving one equation for one of the unknowns and then substituting this solution into the other equation. For example, based on the given data,

$b$ = 45 bu/acre.

When this value is substituted into, 8.1 bu/acre = $m$(15,000 aphids/plant) + $b$, the equation,

8.1 bu/acre = $m$ (15,000 aphids/plant) + 45 bu/acre

was derived. The value for $m$ is then determined by solving the equation for $m$. The resulting equation is

$m$ = [(8.1-45)/15,000] = -0.00246.

**Box 17-10.** Hypothetical relationship between soybean yields and aphid population.

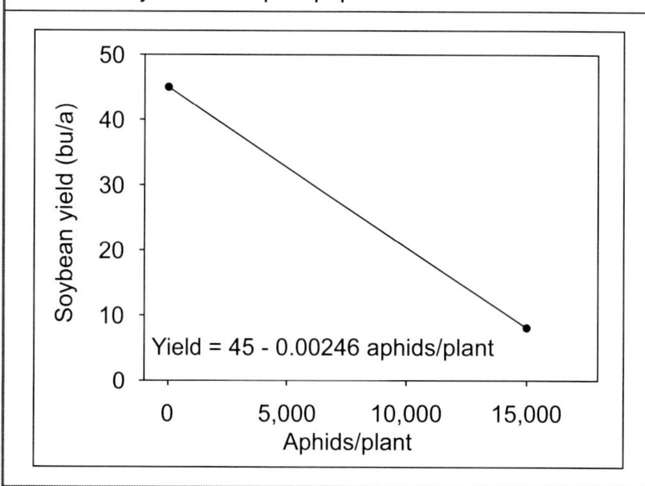


The resulting equation and linear relationship is shown in **Box 17-10** and indicates that for each aphid, 0.00246 bu/acre of soybean is lost. The economic threshold level for a treatment can now be determined based on this equation. This economic threshold level is the point when the cost of aphid control and the value of soybean saved by treatment are equal. If aphid control costs \$10/acre and soybean value is \$5.00/bu, the economic threshold level is determined by calculating the yield loss required to equal the cost of the chemical (\$10/acre/\$5.00/bu = 2 bu/acre). This solution (2 bu/acre), is then used to calculate the number of aphids required to produce this yield loss [2 bu/acre = (0.00246 aphids/plant • aphid population)] or [(2/0.00246) = aphid population/plant]. In this example, the number of aphids required to produce a 2 bu yield loss at the V5 growth stage is 813 aphids/plant. This solution assumes that each aphid produces an identical yield loss.

An exponential decay function ($y_t = y_0 e^{-kt}$) could also be used to describe the relationship between yields and aphids. In this solution, the negative exponential decay function is converted to its natural log form by taking the natural log of both sides of the equation. Using the approach described above, the equations,

$\ln(45 \text{ bu/a}) = m \text{ (0 aphid/plant)} + \ln y_0(\text{bu/a})$
$\ln(8.1 \text{ bu/a}) = m \text{ (15,000 aphid/plant)} + \ln y_0 \text{ (bu/a)}$
are derived.. The equation, $\ln(45 \text{ bu/acre}) = m(0 \text{ aphid/plant}) + \ln y_0(\text{bu/a})$, is simplified to $\ln y_0 = \ln 45$ bu/a, which is substituted into $\ln(8.1 \text{ bu/a}) = m \text{ (15,000 aphid/plant)} + \ln y_0(\text{bu/a})$. The resulting equation is,

$$\ln(8.1 \text{ bu/a}) = m(15{,}000 \text{ aphid/plant}) + \ln 45 \text{ (bu/a)},$$

which is rearranged into the equation,

$$\ln(8.1 \text{ bu/a}) - \ln 45 \text{ (bu/a)} = m \text{ (15,000 aphid/plant)}.$$

This equation is simplified to,

$$m = \{[\ln (8.1/45)]/15{,}000\} = -0.0001143.$$

The rate constant (m) for this equation can be used directly to calculate the economic threshold level. This is accomplished by determining the acceptable amount of yield loss (2 bushels as calculated above), and then solving the following equation,

$$\text{Economic threshold} = \frac{\ln(45 - 2) - 3.8}{-1.113 \cdot 10^{-4}} = 343 \text{ aphid/plant}$$

This equation suggests that each plant should be infested with 343 aphids/plant before treatment, less than 50% of the aphid number calculated by the linear equation.

The linear and exponential decay functions had different solutions to this problem. The economic optimum control rate was much lower for the exponential decay function than the linear model solution. The different optimum rates are the direct result of different assumptions of the models. The linear model assumed that each aphid has an equal impact on yield, while the exponential decay model assumed that the first aphid has the largest impact on yield with each succeeding aphid having less of an impact.

How would you design an experiment to determine which model should be used?

## Chapter Summary

This chapter provides a framework for evaluating resource allocation questions. In many systems, various factors interact to influence reactions. Both empirical and mechanistic models can be used for predictive purposes. Examples on the use of several of these models and techniques to determine the coefficients were provided. All models have assumptions and boundary conditions associated with them and should not be used outside of their constraints. This is why it is important that models be tested or validated prior to widescale use.

**Exercise 17-1 Mathematical models can be either __________ or __________ based.**

Solution:
Empirical or mechanistic

**Exercise 17-2. Which type of modeling utilizes the knowledge of chemistry, biology and physics to define relationships between factors?**

Solution:
Mechanistic

**Exercise 17-3. Conceptual models and __________ __________ are often used to identify the assumptions, known values and focus on the question through visualization.**

Solution:
Relational diagrams

**Exercise 17-4. The ______ ______ is the time required for half of the reactant to react.**

Solution:
Half life

**Exercise 17-5. What is the difference between the linear and exponential decay models?**

Solution:
They have different assumptions. In the linear model, it is assumed that the rate of change is constant, whereas in the exponential decay model, the rate changes with time.

**Exercise 17-6. For a zero order problem, the rate constant is 5 g/d. If 50 g are added, how long will it take before half and all of the material is gone?**

Solution:
$t_{1/2}$ = So/2•$k$
$t_{1/2}$ = 50/2•5 = 5 days
Mean residence time for zero order equations = So/$k$
$t_{mrt}$ = 50/5 = 10 days

**Exercise 17-7. If a zero order model (linear) is used to describe herbicide degradation, how long will it take for half of the herbicide to degrade if 100 g are added and the rate constant is 10 g/day?**

**Exercise 17-8. If a first order model is used to describe herbicide degradation, how long will it take for one-half and all of the herbicide to degrade if 100 g are added and the rate constant is 0.02/day?**

## Simulation Models

A free programmable modeling environment for simulation of natural and social phenomena is available at http://ccl.northwestern.edu/netlogo/. Netlogo is designed to model complex systems. The website contains programs that allow students to explore the impact of different conditions on system behavior.

## References

Coyne, M.S., and J.A. Thompson. 2006. Math for Soil Scientists. Thomson Delmar Learning.

Limpkin, L., and D. Smith. 2001, Logistic growth model. J. Online Mathematics and its Application. Available at http://mathdl.maa.org/mathDL/4/?pa=content&sa=viewDocument&nodeId=484&pf=1

Hale, B.M., and M.L. McCarthy. 2005. An introduction for plant ecology. J. Online Mathematics and its Application. Available at http://mathdl.maa.org/mathDL/4/?pa=content&sa=viewDocument&nodeId=634&pf=1.

# EVALUATING COSTS AND RETURNS OF MANAGEMENT DECISIONS

| | |
|---|---|
| KEY CONCEPTS: | Economic optimum rates, best management practices, integrated pest management, partial budgeting, enterprise budget, cash flow budget, whole farm analysis. |
| MATHEMATICAL SKILLS: | Use of exponents and review of addition and multiplication. |
| CHAPTER PURPOSE: | To provide an introduction to the importance of considering economics in management decisions. |

A background discussion on economics is included because an economic analysis is a critical part of most management decisions. In an economic analysis, the costs and returns of an operation are considered. The different types of analysis approaches are whole farm analysis, cash-flow budgets, enterprise budget, and partial budgets. Different analysis approaches are used for different problems. ***Cost of capital analysis*** is used to quantify the resources invested in capital items. An ***enterprise budget*** is a financial plan for a specific enterprise. Enterprise budgets can be developed for pieces of equipment (combine), individual fields, or a feed-lot operation. ***Partial budgets*** are used to evaluate the economic consequences of minor management adjustments. These budgets can be used to determine the economic consequences of individual management decisions. ***Cash flow budgets*** ensure that adequate cash is available to meet the cash obligations of the farm and ranch. Cash flow budgets consider loan repayment plans. ***Whole farm analysis*** is a detailed listing of the entire farm or ranch. These budgets are useful for setting long-term priorities and planning.

Many of the examples below are canned programs available in Microsoft Excel. For example, the command =ispmt (rate,per,nper,pv) calculates interest paid during a specific time. Other commands can be identified by using help. Numerous other tools are available on the internet.

## Cost of Capital Analysis

The rules associated with capital costs for tax purposes are beyond the scope of this chapter, and it is recommended that the reader contact an accountant for assistance in tax calculations. The following discussion is intended to provide assistance for on-farm decisions.

The annualized capital expense for equipment is the sum of the interest (annual cost of debt) and depreciation. All capital items, even those owned, have annualized capital expenses because the equipment can be sold and proceeds reinvested into something else. Annual interest can be calculated using the equation,

$$I = C\,(1 + {}^{r}/_{n})^{n} - C$$

where

I = Interest C = initial value

r = interest rate (expressed as a fraction: eg. 0.06)

n = number of times per year interest is compounded

**Exercise 18.1 What is the annual interest on a \$10,000 loan at a 5% rate if compounded daily and annually?**

Solution:

Annual

$I = C\,(1 + r/n)^{n} - C$

$I = 10{,}000\,(1 + 0.05) - 10{,}000$

$I = \$500$

Daily

$I = 10{,}000\,(1 + 0.05/365)^{365} - 10{,}000$

$I = 10{,}000 \cdot 1.051267 - 10{,}000$

$I = \$512.67$

**Exercise 18-2 What is the annual interest on a $10,000 loan at a 6% rate if compounded semi-annually?**

**Depreciation** is the amount of value that a capital item such as machinery will lose annually as a result of use and/or age. There are many different methods to calculate depreciation. Techniques to calculate depreciation include the: straight line, declining balance, and units of production methods. The annual depreciation using the **straight line method** is calculated with the equation,

$$Annual\ depreciation = \frac{(cost\ of\ asset - scrap\ value)}{life\ span}$$

**Exercise 18-3. A combine is purchased for $100,000, that has a scrap value of $10,000, and a life span of 15 years. What is its annual depreciation amount using the straight line method?**

Solution:

$$\frac{(\text{cost of asset - scrap value})}{\text{life span}}$$

$$\frac{(100,000 - 10,000)}{15 \text{ years}} = \$6,000$$

In the **declining balance method**, the annual percentage of decrease remains constant over the time frame. The annual depreciation is equal to the value of the asset times the depreciation (decimal). For example, if a tractor is purchased for $50,000 and the depreciation is 10% per year, then the depreciation in year 1 is $5,000 (50,000 • 0.10), and the depreciation in year 2 is $4,500 (45,000 • 0.10). Using this method, the depreciation decreases with increasing time. The equation for determining the value of equipment (initial value – total depreciation) is,

Value after n years = $(1\text{-dr})^n \times init_{val}$

where dr is the annual decimal depreciation rate, n is the number of years, and $init_{val}$ is the initial value of the equipment. The table in Exercise 18-4 shows the value of a machine (starting out being worth $10,000) at the end of 5 years, 10 years, and 15 years with constant depreciation rates from -4% (appreciation of 0.04/year) to 8% (0.08/year).

**Exercise 18-4. Using the declining balance method determine the machine value that was purchased for $10,000 after 5, 10, and 15 years for depreciation rates of -0.04, 0, 0.04, and 0.08**

Solution:

Value after n years = $(1\text{-dr})^n \times init_{val}$

Value = $(1-0.04)^5 \cdot 10,000 = \$8,154$

| Depreciation rate, per year | years after purchase | | |
|---|---|---|---|
| | 5 years | 10 years | 15 years |
| -0.04 | $12,167 | $14,802 | $18,009 |
| 0 | $10,000 | $10,000 | $10,000 |
| 0.04 | $8,154 | $6,648 | $5,421 |
| 0.08 | $6,591 | $4,344 | $2,863 |

**Exercise 18-5. Using the declining balance method, determine the machine value that was purchased for $10,000 after 5, 10, and 15 years for depreciation rates of -0.02, 0, 0.02, and 0.06.**

**Exercise 18-6. If a piece of machinery costs $10,000 and depreciates on the average at a rate of about 10% per year, what is it worth at the end of 10 years? What is the total depreciated value over the 10-year period? What is the depreciation between years 5 and 6?**

Solution:
$(1\text{-}dr)^n \times init_{val} = (1\text{-}0.1)^{10} \cdot 10{,}000 = \$3{,}487$
What is the total depreciated value over the 10-year period?
$10,000-3,487 = $6,513
What is the depreciation between years 5 and 6?
$= (1\text{-}dr)^n \times init_{val} = (1\text{-}0.1)^5 \times 10{,}000 - (1\text{-}0.1)^6 \times 10{,}000 =$
$590

**Exercise 18-7. Using the declining balance method, estimate the annual depreciation cost of a $10,000 piece of machinery in years 1 and 14. Assume a 9% depreciation rate.**

**Exercise 18-8. What are the capital costs/acre (machinery) for a producer who farms 1,000 acres, has $300,000 in machinery, would receive 5% on the money if invested, and the equipment depreciates at a rate of 7% per year? In this exercise, use the declining balance method for depreciation and assume interest is compounded daily.**

Solution:
Interest: $I = C\,(1 + {}^{r}/_{n})^n - C$
$I = 300{,}000(1+0.05/365)^{365} - 300{,}000$
$I = 300{,}000(1.051267) - 300{,}000 = \$15{,}380$
Depreciation
$\text{Depreciation} = init_{val} - (1\text{-}dr)^n \times init_{val}$
$\text{Depreciation} = 300{,}000 - (1\text{-}0.07)^1 \cdot 300{,}000$
Depreciation = $21,000
Capital costs are $15,380 + 21,000 = $36,380 or $36.38/acre

**Exercise 18-9. What are the yearly capital costs/acre for a combine harvesting 2,000 acres that is valued at $143,000, has an annual interest rate of 6%, and has equipment that depreciates 10% per year? In this problem, use the declining balance method for depreciation and the interest is compounded annually.**

## Cash Flow (Loan Balance and Interest Paid)

For many small businesses, it is important to be able to calculate the loan balance after $t$ number of years. Loan balances (B) are calculated with the equation,

$$B = A\left(1 + {}^{r}/_{n}\right)^{nt} - P\left[\frac{\left(1 + {}^{r}/_{n}\right)^{nt} - 1}{\left(1 + {}^{r}/_{n}\right) - 1}\right]$$

where

B = balance after $t$ years
A = amount borrowed
n = number of payments per year
P = amount paid per payment
r = annual percentage rate (APR)
nt = number of years

**Exercise 18-10. If you buy a $200,000 combine, make annual payments of 25,000, and have an annual interest rate of 7.5% interest, what is your balance in year 7?**

Solution:

$$B = A\left(1 + {}^{r}/_{n}\right)^{nt} - P\left[\frac{\left(1 + {}^{r}/_{n}\right)^{nt} - 1}{\left(1 + {}^{r}/_{n}\right) - 1}\right]$$

$$= 200{,}000\left(1 + {}^{0.075}/_{1}\right)^{7} - 25{,}000\left[\frac{\left(1 + {}^{0.075}/_{1}\right)^{7} - 1}{\left(1 + {}^{0.075}/_{1}\right) - 1}\right]$$

$B = 331.810 - 219.683 = \$112.127$

...continued on next page

Total payments = 7•\$25,000 = \$175,000

Toward principal = 200,000 – 112,127 = \$87,873

Interest = \$175,000 - \$87,873 = \$87,127

**Exercise 18-11. If you buy a \$150,000 tractor, make annual payments of 31,000, and have an annual interest rate of 6.0% interest, what is your balance at the end of year 5?**

## Different Types of Economic Analysis

An ***enterprise budget*** is a financial plan for a specific enterprise. A farm or ranch can be separated into several different enterprises. Enterprise budgets can be developed for pieces of equipment (combine), individual fields, or a feed-lot operation. Enterprise budgets are useful for developing financial plans and assessing the profit potential of specific activities. For example, if crop harvesting is considered as an enterprise, then the enterprise analysis would determine the harvesting costs. The net result of this analysis could be that a new combine will be purchased or that costs can be reduced by contracting a commercial harvesting company.

***Partial budgets*** are used to evaluate the economic consequences of minor management adjustments. For example, should herbicide be applied or not applied, should N be applied at 50 or 25 lb/acre? In partial budgets, only the economic factors influenced by the change in management are considered. For example, if reducing the N rate only influences the cost of the inputs (N applied) and the value of the crop produced, then the partial budget would consist of the reduced cost of fertilizer and reduced value of the crop.

The focus of this manual is to provide a framework on how to incorporate input costs, environmental costs, and product values into individual decisions and therefore will rely on the partial budgeting and enterprise analysis approaches. These approaches can be incorporated into daily and short-term planning purposes. Examples of using the partial budgeting approach are provided in chapters 19 and 20.

Most agricultural management decisions are decisions about resource allocation. The resources that are most often allocated are equipment, labor, or supplies. Enterprise and partial budgeting approaches can be used to answer many production questions including,

- Do I spray a field for weeds?
- Do I purchase a new combine or grain dryer?
- Do I harvest my hay today or next week?, or
- Do I move the cattle from one pasture to another?

Many universities have developed guidelines to assist in these decisions. For weed and insect problems the guidelines may be reported as economic threshold levels, i.e. where the cost of control will be less than the loss of revenue due to the pest. Many economic threshold values have been integrated into ***integrated pest management*** (IPM) decisions. Integrated pest management is a management approach that considers production requirements, economic returns, threshold levels of competition, and environmental consequences. Soil fertility recommendation guidelines are based on the assumption that the response from a given fertilizer application is indirectly related to the concentration or amount of nutrients contained in the soil. Many soil and fertilizer-based guidelines are reported as "best management practices", or BMPs.

All chemicals applied to control pests or provide nutrients for crops have some risk at being transported from the target to the non-target area. Some chemicals have higher risk potentials than others. For example, phosphate ($PO_4^{3-}$) is primarily moved through erosion while nitrate ($NO_3^-$) is lost with percolating water. There are numerous examples where individual decisions may have a negative impact on profitability and the environment. Examples of these activities might include:

- Controlling weeds when the weed population is below the economic threshold level;
- Applying pesticides to control aphids when the field does not contain aphids; and
- Applying fertilizer when the soil test results are greater than the recommended soil test level.

In partial budgeting, only the product costs and values changed by the modified management are considered. This analysis approach typically has not considered environmental consequences. In the future it is likely that environmental consequences will be integrated into these analyses. If these costs are known, it is relatively easy to integrate them into the analysis by including them as input costs. For example,

Total input costs = input costs + enviro costs

If the expected costs outweigh expected returns, then implementing an individual decision will have a negative impact on the financial returns. The economic optimum value is the value where cost of the inputs equals the value of the products produced. The economic optimum value is a function of the price of the inputs and the value of the products. Mathematically this is expressed by the equation,

d(total costs of inputs) = d (total value of products)

The *d* value in the equation means the change in total cost of inputs or change in total value of products. An implication of this equation is that if the price of N fertilizer doubles, then less N will be applied and the amount of product produced will decrease.

Including environmental costs in these calculations will have the same impact as increasing the cost of the inputs. For these reasons, increasing the costs of inputs or including environmental cost in economic calculations tends to reduce production, i.e. less inputs = less production. Evaluating individual decisions requires forecasting the net effect of a treatment on the expected returns. Analysis must be conducted in a systematic and organized manner. Economics and production requirements can be linked through mathematics.

**Exercise 18-12. Why should economic analysis be integrated into daily planning and decision-making?**

Solution:
By integrationg costs into decisions, profitability will be improved.

**Exercise 18-13. Most natural resource management decisions are really decisions about ____ and _____.**

Solution:
Costs and returns

**Exercise 18-14. Name three economic assessment approaches.**

## Estimating Cost of Production

To make good decisions, it is necessary to estimate the cost of production and the net cost per unit produced. The cost of production is the summation of all costs used to produce a product. Production costs can be separated into fixed and variable costs. Fixed costs can not be changed within the time frame of the production cycle. Land and machinery costs are examples of fixed costs. Variable costs can be changed within the time frame of a production cycle. Fertilizer, herbicide, and seeding costs are examples of variable costs. Examples on determining fertilizer and herbicide costs are provided in chapters 5, 6, and 9. Cost of production should consider all variable and fixed costs.

Costs of production are extremely variable. For example, reducing the N application rate from 150 lb N/acre to 100 lb N/acre can reduce the cost of production \$25.00/acre. Information and guidance for determining cost of production are available in Klein et al. (2011) and Duffy (2011). Understanding the cost of production is one of the most important calculations that can be made. Estimated cost of production for several crops grown in Iowa, South Dakota, and Nebraska are shown on the next page.

| **Box. 18-1.** Cost of production for 2006 and 2007 for selected crops in Iowa, South Dakota (SD), and Nebraska. | | | | | | | |
|---|---|---|---|---|---|---|---|
| | Iowa estimated cost of production (Duffy, 2011) | | | | | | |
| | Corn following corn | | Corn following soybean | | Soybean following corn | | SD estimated costs No-tillage, (Carlson) corn following soybean |
| | 2006 | 2007 | 2006 | 2007 | 2006 | 2007 | 2006 |
| Machinery | 100.07 | 102.94 | 97.39 | 100.12 | 45.9 | 46.76 | 30 |
| Seed, chemical, etc | 201.62 | 222.22 | 169.26 | 189.33 | 106.79 | 107.58 | 149 |
| Labor | 29.93 | 31.35 | 27.3 | 28.6 | 25.73 | 26.95 | 29 |
| Land | 145 | 155 | 145 | 155 | 145 | 155 | 80 |
| Total cost/acre $/acre | 476.61 | 511.51 | 438.95 | 473.05 | 323.41 | 336.29 | 288 |
| Assumed yield (bu/a) | 140 | 145 | 155 | 160 | 45 | 50 | 120 |
| Cost/bu )$/bu) | 3.4 | 3.53 | 2.83 | 2.96 | 7.19 | 6.73 | 2.4 |
| | Iowa estimated | | | | | | |
| Alfalfa hay, annual costs | 2006 | 2007 | | | | | |
| 1/3 planting costs ($/acre) | 36.83 | 37.27 | | | | | |
| Annual fertilizer | 103.36 | 103.46 | | | | | |
| Harvest machinery | 107.1 | 90.4 | | | | | |
| Labor | 56 | 58.67 | | | | | |
| Land | 95 | 100 | | | | | |
| Total cost/acre | 398.29 | 389.79 | | | | | |
| Assumed yield, tons | 6 tons | 6 tons | | | | | |
| Cost/ton, ($/ton) | 66.38 | 64.97 | | | | | |
| From Duffy, 2011 | | | | | | | |

| Nebraska cost of production 2006 (Seeley and Klein, 2006) | | | | | | |
|---|---|---|---|---|---|---|
| Crop | | Previous crop | | | Yield goal/acre | Cost, $/acre |
| Corn | Dryland | Soybean | No-tilled | Bt | 110 bu | 271.45 |
| Corn | Irrigated | Corn | No-tilled | Bt | 195 bu | 538.21 |
| Soybean | Dryland | Corn | No-tilled | | 40 bu | 166.96 |
| Wheat | Dryland | Row crop | No-tilled | | 40 bu | 176.37 |
| Grass hay | Dryland | | | | 2 tons | 72.72 |

## Additional Information


Dalsted, N.L. and P.H. Gutierrex, 2006. Partial budgeting. Publication #3.760. Colorado State University Cooperative Extension-Agriculture. Available at http://www.ext.colostate.edu/PUBS/FARMMGT/03760.html.

Duffy, M. 2011. Estimated cost of crop production in Iowa. 2011. Iowa State University. Available at http://www.extension.iastate.edu/agdm/crops/pdf/a1-20.pdf

Greaser, G.L. 1994. Agricultural alternatives, Enterprise budget. The Pennsylvania State University. Available at http://agalternatives.aers.psu.edu/farmmanagement/enterprise/enterprise_budget_analysis.pdf

Manura, D. 1995-2005. Math reference tables. Available at http://math2.org/math/general/interest.htm (accessed 6/8/08).

Kein, R.N., R.K. Wilson, H.D. Jose, P.A. Burgener, and T.N. Barrett. 2011. Crop Budgets - Nebraska 2011. EC 872. University of Nebraska Extension, Lincoln NE. Available at http://www.ianrpubs.unl.edu/epublic/live/ec872/build/ec872.pdf

Seeley, R.A. and R.N. Klein. 2006. Nebraska Crop Budgets - 2006 (EC872). University of Nebraska-Lincoln.

Swinton, S., C.M. Lansen, and W. Zhang. 2005. Economic Analysis of Sustainable Agriculture and Food Systems. Michigan State University. Available at http://safs.msu.edu/econ/index.htm.

# USING LEAST SQUARES PREDICTION MODELS TO ESTIMATE CORN YIELD LOSSES

**Chapter concepts:** To use linear models and standard deviation to estimate corn yield losses from unevenly planted fields.

**Mathematical skills:** Use regression analysis in Microsoft Excel.

**Chapter purpose:** Prediction models can be developed using a variety of approaches. The least squares regression-based approach for developing polynomial models ($y = b + mx + cx^2 + dx^3$) is a commonly used technique. One of the most commonly used polynomial models is the linear model ($y = mx + b$). Linear models have been used to determine fertilizer recommendations and assess the impact of unevenly-spaced corn plants on yields.

This exercise will demonstrate how Microsoft Excel can be used to determine polynomial equations and calculate standard deviations. In this example, a linear model will be used to determine yield losses from unevenly spaced corn.

## Impact of Crop Planting Uniformity on Corn Yield

Does an evenly spaced plant stand have greater yield potential than an unevenly spaced stand? The distances between adjacent corn plants in many fields can range from identical (uniform spacing) to highly variable distances. Doerge and Hall (2000) showed that on average, calibrating planters to reduce planting distance variation on average improved yields 4.2 bu/acre. At some locations, the advantage for calibration exceeded 20 bu/acre. Others have had different results. Nielsen (1997) reported that in uneven stands, yield losses range from 7 to 15 bu/acre. Erbach et al. (1972) reported that stand uniformity had inconsistent impacts on yield, while Krall et al. (1977) reported that yields increased with planter spacing uniformity. Lessons learned from Dorge and Hall (2000) were:

1. Planters with a plant spacing standard deviation that is > 3.0 inches should be calibrated.
2. Yields were increased 83% of the time by planter calibration.
3. Standard deviations of planter spacing distances could be minimized by planting at a reasonable speed.
4. A standard deviation of the distances between adjacent corn plants of 2.0 in. was the best spacing uniformity that can be expected.
5. Uniform plant distributions were more important in wide than narrow row spacing.

Findings from these studies show that there is a high probability that yields will be increased by increasing stand uniformity. The approach to determining the potential impact of planter spacing uniformity relies on the ability to calculate plant density standard deviations and determine linear equations. The goal of this problem is to provide an example on how to design experiments to answer problems. The specific question addressed in this chapter is how to predict the impact of corn distribution on yield.

## Approach

Solving this problem can be separated into three steps. First, measure plant spacing uniformity and associated grain yields in production fields. Second, develop a predictive equation relating the measured values. Third, calculate potential grain yield losses.

## Measuring Plant Spacing Uniformity and Impact on Yield

Plant spacing uniformity is determined by measuring the distances between adjacent plants. The variation of these estimates is then determined by calculating the standard deviation of the distances between adjacent plants. Spacing differences between adjacent plants can be determined by placing a 20 ft or longer measuring tape on the ground next to the row of plants (**Box 19-1**). The location of each plant along the tape is measured and recorded.

**Box 19-1.** Tape measure placed next to a corn row.

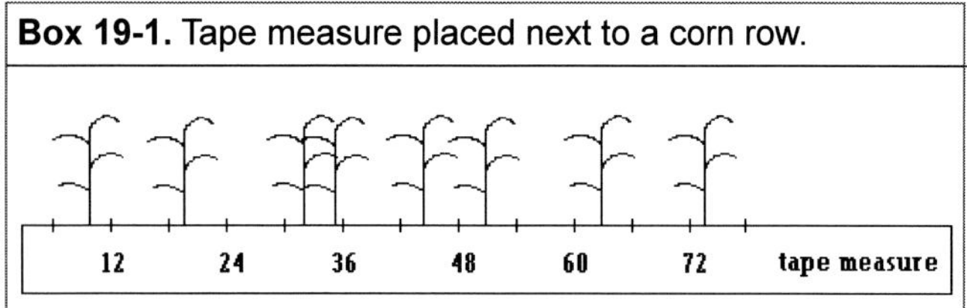


In **Box 19-1**, plants are located at 11, 20, 32, 35, 44, 50, 63, and 73 inches. To obtain an accurate measure of uniformity these measurements should be repeated at a number of locations. Collecting plant spacing data from more areas or for longer lengths of row improve the reliability of data. These locations should also be on different rows and/or different units from the same planter. The locations of the plants along the tape should be typed into a spreadsheet (**Box 19-2**). Based on these locations, the differences between adjacent plants are calculated by inserting the appropriate equation into the spread sheet.

**Box 19-2.** Hypothetical measured locations and distances between adjacent corn plants along a 5.75 ft transect. The data is types into an Excel spread sheet.

| | A | B | C |
|---|---|---|---|
| | Measured locations | Spacing between plants | Row Width |
| 1 | 0 | | 30 |
| 2 | 3 | 3 | |
| 3 | 10 | 7 | |
| 4 | 15 | 5 | |
| 5 | 25 | 10 | |
| 6 | 39 | 14 | |
| 7 | 41 | 2 | |
| 8 | 49 | 7 | |
| 9 | 55 | 6 | |
| 10 | 63 | 8 | |
| 11 | 69 | 6 | |
| 12 | Average | 6.9 | |
| 13 | Standard deviation | 3.45 | |
| 14 | Estimated yield loss bu/acre | | |
| 15 | Plants/acre | | |

Enter distance between plants in a row

=A2-A1
=A3-A2
=A4-A3

=average(B2:B12
=stdev(B2:B12)
=(B13-2)*4.6-0.27
=(1/(C1*B12))*144*43560

For example, = A2-A1 should be placed in cell B2. This equation will determine the difference between the plants locations in cells A2 and A1 and this difference is then stored in cell B2. The rest of the cells in the B column must be filled in, i..e. =A3-A2 is typed into cell B3 and =A4-A3 is typed into cell B4.

After all the differences are calculated, the standard deviation is determined using the equation,

$$s = \left[\frac{\sum_{i=1}^{n}(\chi_i - \bar{\chi})^2}{n-1}\right]^{0.5}$$

where: $\bar{\chi}$ is the average value, $X_i$ is the i[th] observation, and n is the number of observations. In this example, the observations are the distances between two adjacent plants. In a perfectly planted corn field where all the plants are 7.0 in. apart, the average spacing is 7.0 in. and the standard deviation would be zero. If half of the spacings are 6.0 in. and half are 8.0 in., then the average spacing is 7.0 in. but the standard deviation is greater than zero. The standard deviation increases with increasing variability in distances. In Excel, the standard deviation of the differences is determined by entering the equation =stdev(B2:B11) in cell B13 and the average of the differences is determined by entering =average(B2:B11) in cell B12.

**Exercise 19-1. Determine the standard deviation of the following data.**

| | | |
|---|---|---|
| 5 | 5 | 9 |
| 6 | 10 | 5 |
| 3 | 1 | 6 |
| 7 | 9 | 2 |
| 9 | 7 | 4 |
| 4 | 2 | 1 |

Solution:
Type the data into Excel and use the =stdev(start, end) command.
2.845131

## Plant Population

The Excel spread sheet can be used to calculate the population. This is accomplished by solving the expression,

$$= \frac{1\text{ Plant}}{6.9\text{ in.} \cdot 30\text{ in.}} \cdot \frac{144\text{ in.}^2}{\text{ft}^2} \cdot \frac{43,560\text{ ft}^2}{\text{acre}} = \frac{30,300\text{ plants}}{\text{acre}},$$

where 6.9 in. represent the average distance between plants. This mathematics is conducted in Excel by typing,
=(1/C1*B12))*144*43560), into cell B15. Yield should also be measured at the locations where plant uniformity is measured. Yield can be measured by hand harvesting, drying, and weighing the corn from this area.

**Exercise 19-2. Determine the plant population in plants/acre if 30 plants are counted on a row that is 20 feet long. Each row is 30 in. apart.**

Solution:

$$= \frac{30\text{ plants}}{20\text{ft} \cdot 30\text{ in.}} \cdot \frac{12\text{ in.}}{\text{ft}} \cdot \frac{43,560\text{ ft}^2}{\text{acre}} = \frac{26,136\text{ plants}}{\text{acre}}$$

## Estimating Yield Improvement from Planter Calibration

This experiment contains two treatments: area where the field was planted with a calibrated planter and area where corn was planted with an un-calibrated planter. Yields and standard deviation of the distances between the plants in both areas are measured. Data for this exercise was derived from Doerge and Hall (2000). A portion of the data from this study is provided in **Box 19-3**. Linear regression will be used to determine the linear relationship between yield and planting spacing uniformity improvements. In this experiment yields and standard deviation of planter calibrated and un-calibrated areas are measured. Data in **Box 19-3**, represents the differences between these two treatments. The equation is determined by,

1. Typing data shown in **Box 19-3** into an Excel spread sheet. These should be the 50 rows of data.
2. After the data is entered into Excel, select Tools, Data analysis, and Regression.
3. In the regression menu enter A1:A52 into the y-range and B1:B52 into the x range. After entering this information select OK.
4. **For this exercise if the data analysis tools are not available, this program must be down loaded in Excel 2003. This download is accomplished by selecting Tools, add-ins and checking analysis tool box, and then following directions.**
5. The results from the statistical analysis are shown in **Box 19-4**. A plot of the data can be obtained by highlighting the data, selecting Chart

**Box 19-3.** Data derived from Doerge and Hall (2000). Data under column A represents improvement in bu/a resulting from planter spacing distance calibration. Data under column B represents the improvement in standard deviation (in) of the distances between adjacent plants resulting from planter calibration. When entering this information it should be entered in two columns of number running vertically down the page.

| # | A | B | # | A | B | # | A | B | # | A | B | # | A | B | # | A | B |
|---|---|---|---|---|---|---|---|---|---|---|---|---|---|---|---|---|---|
| 1 | 20 | 4.5 | 11 | 3 | 1.9 | 21 | -4.7 | 0.1 | 31 | 6 | 1.5 | 41 | 4.9 | 0.5 | 51 | 0 | 0.4 |
| 2 | 20 | 3.5 | 12 | 8 | 1.1 | 22 | -3 | 0.2 | 32 | 5 | 1.4 | 42 | 4.8 | 0.6 | 52 | 0 | 0.4 |
| 3 | 23 | 2 | 13 | 9 | 0.3 | 23 | -0.5 | 0 | 33 | 4 | 1.4 | 43 | 3 | 0.6 | | | |
| 4 | 20 | 2 | 14 | 7 | 0 | 24 | -3 | 0.8 | 34 | 3 | 0.9 | 44 | 2.5 | 0.7 | | | |
| 5 | 16 | 1 .3 | 15 | -5 | -0.2 | 25 | -3 | 0 | 35 | 3 | 1 | 45 | 2 | 0.7 | | | |
| 6 | 11 | 2 | 16 | 4 | -0.6 | 26 | -0.5 | 1.5 | 36 | 3 | 1.1 | 46 | 1.5 | 0.8 | | | |
| 7 | 9 | 1.8 | 17 | 0 | -0.5 | 27 | 3 | 1.5 | 37 | 3 | 1.2 | 47 | 1 | 0.8 | | | |
| 8 | 5 | 2.1 | 18 | -5 | -0.9 | 28 | 3 | 1.5 | 38 | 7 | 0.8 | 48 | 3 | 0.2 | | | |
| 9 | 5 | 1.9 | 19 | -5 | 0 | 29 | 3 | 1.5 | 39 | 5 | 0.4 | 49 | 2 | 0.3 | | | |
| 10 | 3 | 2 | 20 | -4.8 | -0.1 | 30 | 5 | 1.5 | 40 | 5 | 0.45 | 50 | 1 | 0.3 | | | |

Wizard; selecting XY scatter; next, next, next, and finished.

## Determining Yield Loss from Non-calibrated Planter

The intercept and slope of the line are under the heading coefficients. This approach can be used to determine the coefficients associated with more complicated polynomial models. For example to determine the coefficients for the $y = b + mx + cx^2$ model, the $x$ values in cells B1 through B52 would be squared and placed in cell C2 through C52. The $y$-values in the regression menu would then be changed from B1:B52 to C1:C52.

The value for the y-intercept is adjacent to Intercept and the value for the slope is adjacent to X Variable 1. Inserting these coefficients results in the model,

(Yield calibrated -Yield not calibrated) = 4.9 * $(s_{non\text{-}calibrated} - s_{calibrated}) - 0.28$ where $s_{calibrated}$ is the plant distance uniformity standard deviation in calibrated areas and $s_{non\text{-}calibrated}$ is plant distance uniformity standard deviation in non-calibrated areas. Replacing $s_{calibrated}$ with the experimentally measured value of 2 and by replacing (Yield calibrated -Yield not calibrated) with yield improvement the following model was developed.

$$Yield_{improvement} = 4.91\ ((s_{non\text{-}calibrated} - 2) - 0.28.$$

To calculate yield losses for the data in **Box 19-3**, this model (= 4.91*(B13-2) – 0.28) should be incorporated into cell B14. Based on this model, the estimated yield loss resulting from a non-calibrated planter was 6.8 bu/acre.

In summary this chapter showed how data collected from a field can be summarized and analyzed using Data Analysis tools available in Excel. Similar analysis can be conducted for a number of situations including: 1) the importance of tree spacing in forest plantations; 2) the importance of Christmas tree spacing in Christmas tree farms; and 3) the influence of planter spacing on tomato harvest, and 4) the effect of crowding of animals in a pen.

**Box 19-4.** Results from a linear regression analysis of the data in **Box 19-3**.

| | | | | | | | | |
|---|---|---|---|---|---|---|---|---|
| SUMMARY OUTPUT | | | | | | | | |
| Regression Statistics | | | | | | | | |
| Multiple R | 0.646575 | | | | | | | |
| R Square | 0.418059 | | | | | | | |
| Adjusted R Square | 0.40642 | | | | | | | |
| Standard Error | 4.937054 | | | | | | | |
| Observations | 52 | | | | | | | |
| ANOVA | | | | | | | | |
| | df | SS | MS | F | Significance F | | | |
| Regression | 1 | 875.5157 | 875.5157 | 35.91933 | 2.24E-07 | | | |
| Residual | 50 | 1218.725 | 24.3745 | | | | | |
| Total | 51 | 2094.241 | | | | | | |
| | Coefficients | Standard Error | t Stat | P-value | Lower 95% | Upper 95% | Lower 95.0% | Upper 95.0% |
| Intercept | -0.28298 | 0.999441 | -0.28314 | 0.778239 | -2.29042 | 1.724455 | -2.29042 | 1.724455 |
| X Variable 1 | 4.916901 | 0.820403 | 5.993274 | 2.24E-07 | 3.269073 | 6.564729 | 3.269073 | 6.564729 |

The R square value tells you that 42% of the variability is explained by the linear model.

The multiple R value is the square root of the r squared value.

Intercept is the y-intercept in the linear model.

Variable 1 is the slope of the linear model

The t Stat and P-values provide information about the significance of the model parameters.

The higher the t stat and higher the P-value the more the significant the values are. See the Basic Statistics chapter for more information. In this example, the intercept is not different from zero (p>0.05) and the slope (X Variable 1) is different from 0 (p<0.05).

**Exercise 19-3. Using the model developed above, what is the expected yield loss if the standard deviation is 5.12 in.?**

Solution:
Expected yield loss =

$$\left(\frac{\frac{4.91\,\text{bu loss}}{\text{acre}}}{\text{in. Standard Deviation}}\right)\cdot(5.12 - 2\,\text{in. Standard Deviation}) - 0.28 = \frac{15.0\,\text{bu Yield Loss}}{\text{acre}}$$

**Exercise 19-4. What is the plant population if the row spacing is 24 in. and the average distance between the plants is 8 in.**

**Exercise 19-5. What is the standard deviation, estimated yield loss based on the equation derived above and the plant population for the data below.**

| | A | B | C |
|---|---|---|---|
| Plants | Locations (in.) | Differences | Row width |
| 1 | 0 | | 30 |
| 2 | 3 | | |
| 3 | 10 | | |
| 4 | 15 | | |
| 5 | 25 | | |
| 6 | 39 | | |
| 7 | 41 | | |
| 8 | 49 | | |
| 9 | 55 | | |
| 10 | 63 | | |
| 11 | 69 | | |
| 12 | Average | | |
| 13 | Standard deviation | | |
| 14 | Estimated yield loss bu/acre | | |
| 15 | Plants/acre | | |

## Chapter Summary

Excel can be used to solve data that you have collected or have obtained from published reports. Understanding how to analyze this data can help improve your management. This section discussed how to use standard deviations, which is a measure of variability, and linear models.

## References


Doerge, T. and T. Hall. 2000. The value of planter calibration using the MeterMax System. Crop Insights 10(23):1-4, Pioneer Hi-Bred International.

Erbach, D.C., D.E. Wilkins, and W.G. Lovely. 1972. Relationships between furrow opener, corn plant spacing, and yield. Agron. J. 64:702-704.

Krall, J.M., H.A. Esechie, R.J. Raney, S. Clark, G. TenEyck, M. Lundquest, N.E. Humburg, L.S. Axthelm, A.D. Dayton, and R.L. Vanderlip. 1977. Influence of within-row variability in plant spacing on corn grain yield. Agron. J. 69:797-799.

Nielsen, R.L. 1997 Stand Establishment Variabiity in Corn, AGRY-91-01, Purdue University.


*MeterMax is a trademark of Precision Planting, Inc.

# USING ITERATION TO DEVELOP PREDICTIVE EQUATIONS FOR POLYNOMIAL, MITSCHERLICH, HYPERBOLIC, AND LOGISTIC MODELS

**Chapter concepts:** Iterative approach, linear model, second order polynomial, constraints, hyperbolic model, Mitscherlich model, logistic model.

**Mathematical skills:** Iteration for solving non-linear equations.

**Chapter purpose:** This chapter will show how an iterative approach can be used to develop coefficients associated with a variety of equations that are routinely used in natural resource management. The Mitscherlich equation is used in soil fertility work, the hyperbolic model is used to calculate yield losses due to weed stress, and the logistic model is used to calculate carrying capacity. This iterative approach complements the Least Squares approach that is used to develop polynomial equations. The program Solver, available in Microsoft Excel, will be explained and demonstrated.

## Importance of Non-linear Equations

In natural resource management, non-linear equations are used to describe relationships between several factors. Non-linear equations can not be solved using standard regression approaches. Specific equations are used because their coefficients have biological meanings. Expensive software can be purchased that will develop solutions to these equations. However, it is not necessary because Microsoft Excel has routines for solving these problems. No matter which software is used, all solutions must be inspected.

The coefficients associated with many equations can only be determined by using an **iterative approach**. In the iterative approach, the equation's coefficients are determined by following a process of systematic trial and error. The prediction equation's coefficients are those that minimize the error or maximize the amount of variation explained by the equation. The basic principle of this approach will be demonstrated on the data provided in **Box 20-1**.

**Box 20-1.** Hypothetical data from a N rate experiment.

| N rate, lb/A | Yield, bu/acre |
|---|---|
| 50 | 90 |
| 80 | 120 |
| 100 | 140 |
| 120 | 170 |

## Solving Polynomial Equations

Polynomial equations have the form, $y = b + mx + cx^2 + dx^3$ where the $y$ and $x$ are the measured factors. Coefficients $b$, $m$, $c$, and $d$ are unknown and must be determined. Polynomial equations can be solved using the Least Squares approach available in most statistical packages. This exercise will demonstrate how an iterative approach can be used to solve these problems. The following example will fit data from **Box 20-1** to the model, $yield = N_{rate} \bullet m + b$, where $m$ is the slope and $b$ is the y-intercept.

When using the iterative approach, a selection criteria must be selected. Most scientists chose to maximize the amount of variation explained by the equation ($R^2$) or minimize the sum of squares for error (SSE). The equations for $R^2$ and SSE values are

$$SSE = \sum(PredictedValue - ObservedValue)^2$$

$$R^2 = \frac{\sum_{i=1}^{n}(y_i - \bar{y})^2 - SSE}{\sum_{i=1}^{n}(y_i - \bar{y})^2}$$

In the $R^2$ equation, $\bar{y}$ is the average value of all the y-values and $y_i$ are the individual y values. Regardless which criteria are selected they should produce similar results.

The technique used by the iterative approach is best demonstrated by an example that uses the data shown in **Box 20-1**.

1. Estimate the $m$ and $b$ values for the linear equation. For this example lets assume that $m$ is 1.1

and $b$ is 50. The resulting equation is $yield = 1.1 \cdot N_{rate} + 50$.

2. Determine the predicted $y$ value for each $x$ value based on these guesses. In this example, each $y$ value is determined by adding 50 to the $N_{rate}$ multiplied by 1.1. These values are shown in **Box 20-2**.
3. Determine the sum of squares for error (SSE) by squaring the difference between the measured and predicted values.
4. Based on these results, select two new coefficients. In the second iteration $m$ is reduced from 1.1 to 1.0, and $b$ stays constant (**Box 20-3**). Because the new SSE of 700 is less than the old value of 1973, the second model fits the data better than the first. By repeating this process many times the equation that minimizes SSE can be obtained.

**Box 20- 2.** Measured yields, predicted yields, and squared differences using in the iterative approach. The calculations are conducted in Microsoft Excel. The equation used to predict yields was, yield = 1.1 · $N_{rate}$ + 50.

| | A | B | C | D |
|---|---|---|---|---|
| 1 | | Measured yield | | |
| 2 | $N_{rate}$ | Yield, bu/acre | Predicted Yield | Squared Difference |
| 3 | 50 | 90 | 105 | 225 |
| 4 | 80 | 120 | 138 | 324 |
| 5 | 100 | 140 | 160 | 400 |
| 6 | 120 | 150 | 182 | 1024 |
| 7 | | | SSE = | 1973 |

=A3*1.1+50
=A4*1.1+50
=A5*1.1+50
=A6*1.1+50

=sum(D3:D6)

=(C3-B3)^2
=(C4-B4)^2
=(D5-B5)^2
=(D5-B6)^2

**Box 20-3.** Measured yields, predicted yields, and squared differences using in the iterative approach. The equation used to predict yields was, yield = 1.0 • $N_{rate}$ + 50.

| | Measured yield | Predicted | Squared |
|---|---|---|---|
| $N_{rate}$ | Yield, bu/acre | Yield | Difference |
| 50 | 90 | 100 | 100 |
| 80 | 120 | 130 | 100 |
| 100 | 140 | 150 | 100 |
| 120 | 150 | 170 | 400 |
| | | SSE = | 700 |

**Box 20-4a.** Screen views when "Solver" is installed in Microsoft Excel

File Edit View Insert Format Tools Data Window
Solver...
Add-Ins...
Options...
Data Analysis...
L21
B C D

Add-Ins
Add-Ins available:
Analysis ToolPak
Analysis ToolPak - VBA
Conditional Sum Wizard
Euro Currency Tools
Internet Assistant VBA
Lookup Wizard
Solver Add-in
OK
Cancel
Browse...
Automation...

The iterative approach is very systematic and computer programs can be written to automate this process. The program "Solver", within Microsoft Excel, is one such program. To use "Solver" it must be installed in Microsoft Excel 2003 (**Box 20-4a**). If "Solver" has not been installed (i.e. the option is not available when the "Tools" menu is clicked) under the "Tools" menu, do so now. Select "Office" button upper left corner • select "Office" options • select "Solver Add-Ins" • click "Go" • select "Analysis Tool Pak" and "Solver Add-in" • select "OK". In Microsoft Excel 2007, follow the direction in **Box 20-4b.**

**Box 20-4b.** Directions for installing solver in Excel 2007.

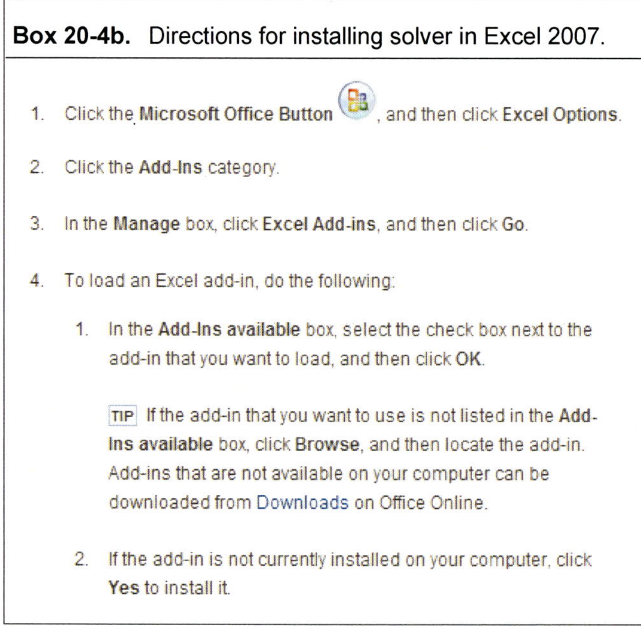
1. Click the **Microsoft Office Button**, and then click **Excel Options**.
2. Click the **Add-Ins** category.
3. In the **Manage** box, click **Excel Add-ins**, and then click **Go**.
4. To load an Excel add-in, do the following:
    1. In the **Add-Ins available** box, select the check box next to the add-in that you want to load, and then click **OK**.

       TIP If the add-in that you want to use is not listed in the **Add-Ins available** box, click **Browse**, and then locate the add-in. Add-ins that are not available on your computer can be downloaded from Downloads on Office Online.
    2. If the add-in is not currently installed on your computer, click **Yes** to install it.

**Box 20-5.** Equations and data that must be entered into Excel

| | A | B | C | D |
|---|---|---|---|---|
| 1 | | | | |
| 2 | | m estimate | b estimate | |
| 3 | | 1.1 | 50 | |
| 4 | | | | |
| 5 | | | | |
| 6 | | Measured yield | Predicted | Squared |
| 7 | $N_{rate}$ | Yield, bu/acre | Yield | Difference |
| 8 | 50 | 90 | 105 | 225 |
| 9 | 80 | 120 | 138 | 324 |
| 10 | 100 | 140 | 160 | 400 |
| 11 | 120 | 150 | 182 | 1024 |
| 12 | | | SSE = | 1973 |

=$B$3*A8+$C$3
=$B$3*A9+$C$3
=$B$3*A10+$C$3
=$B$3*A11+$C$3

= (C8-B8)^2
= (C9-B9)^2
= (C10-B10)^2
= (C11-B11)^2
=Sum(D8:D11)

Now you are ready to determine a linear equation ($y = mx + b$) for the N rate study described above using the "Solver" program. The first step is to type in the data (cells A8 through B11) as shown in **Box 20-5**. When entering this data make sure that you enter the equations, not the values in cells C8-C11 and D8-D12. For the predicted yields in cells C8 – rather than typing in the linear model ($y = mx + b$) – enter =$B$3*A8+$C$3 into cell C8. The $B$3 identifies that value in cell B3 as a constant (slope or m). When the $B$3 is used, this value will not change when you copy down. After the first value is entered in C9, the values for C10 and C11 can be entered by copying down. This can be accomplished by:

1) Centering the cursor on C8;
2) Left click on edit;
3) Left click copy;
4) Left click and highlight C9-C11;
5) Left click on edit; and
6) Left click on paste.

When the $ is used, the copy routine treats the value as a constant and will change the cell value when you copy. The $B$3*A8 +$C$3, represents linear equation with the slope ($m$) being 1.1 and the y-intercept ($b$) being 50. The same process is used to enter the squared error differences [=(C8-B8)^2] in cell D8. After the data is entered, copy down the equation to cells D9, D10, and D11. In cell D12 the equation, =[sum(D8:D11)] should be entered. This equation sums all the squared errors. This value represents the SSE.

To operate "Solver", the first step is to enter the location of SSE ($D$13) into set target cell (**Box 20-6**). This value identifies where the selection criteria are located. Because you want to minimize the error, select "min". Note, if $r^2$ is used, then "max" should be selected. The values $B$3: $C$3 should be typed into **By Changing Cells**. For this problem, **constraints** were not required. Using the original data and estimates for the slopes and intercepts, calculate the square difference. **Constraints** are used when specific values for the coefficients are known. For example, if the slope can only be a positive number, the constraints $B$3 ≥ 0 should be used.

**Box 20-6.** Screen from "Solver" after the data in the program is entered correctly.

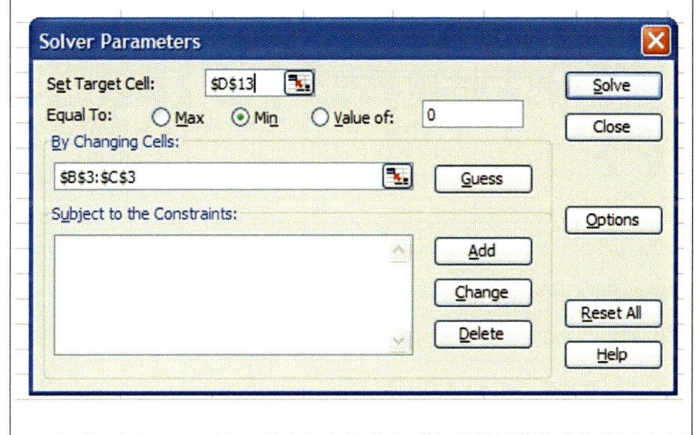


Using the original data and estimates for the slope as shown in **Box 20-2**, calculate the $R^2$ value as shown in cell D12 (**Box 20-7**). Using this original data, the $R^2$ value is 0.06, which indicates that only 6% of the variability is explained by the original linear prediction model. Now using "Solver", click Solve to determine the coefficients using the iterative approach. The coefficients for the linear equation are located in cells B3 ($m$) and C3 (B) (**Box 20-7**). Based on this solution the resulting linear equation is, Yield = 48.1 + 0.87*$N_{rate}$. This equation has an $R^2$ value of 0.983 which means that it explains 98.3% of the yield variability. Clearly this is an improvement over the initial guess ($m$ = 1.1 and $b$ = 50).

**Box 20-7.** Excel spread sheet after "Solver" solved the problem.

| | A | B | C | D | E |
|---|---|---|---|---|---|
| 1 | | | | | |
| 2 | | m estimate | b estimate | | |
| 3 | | 0.878504 | 48.13092 | | |
| 4 | | | | | |
| 5 | | | | | |
| 6 | | Measured | Predicted | Squared | Squared Dif |
| 7 | $N_{rate}$ | bu/acre | Yield | Difference | Meas-aver |
| 8 | 50 | 90 | 92.05611 | 4.227599 | 1225 |
| 9 | 80 | 120 | 118.4112 | 2.524198 | 25 |
| 10 | 100 | 140 | 135.9813 | 16.14991 | 225 |
| 11 | 120 | 150 | 153.5514 | 12.61231 | 625 |
| 12 | Average | 125 | R2 | 0.983089 | 2100 |
| 13 | | | SSE | 35.514 | |

Coefficients for the equation. Cell values placed in By Changing Cells

=average(B8:B11)

$R^2$=(E12-D13)/E12
SSE=sum(D8:D11)

=($B$12-B8)^2
=($B$12-B9)^2
=($B$12-B10)^2
=($B$12-B11)^2
=sum(E8:E11)

Now use $R^2$ instead of SSE in the target cell. Once again go to Tools, Solver, and in target cell replace D13 with D12 and select Max. Highlight solve, and compare the results for the max "$R^2$" and min "SSE" solutions. In both cases, the solved *m* estimate, *b* estimate, SSE, and $R^2$ values are identical. The equation yield = 0.87 N applied + 48.1 is interpreted as if no N is applied – then 48.1 bu would be expected. For each 1 lb of N added, yields are increased 0.87 bu.

In this example, the iterative approach was used to develop the polynomial equations. As previously shown, polynomial predictive models can also be determined using Regression Package available under tool/data analysis. The regression package used the method of Least Squares to determine this model. The Least Squares approach can not be used to solve non-linear equations. Non-linear equations are those that do not follow the polynomial format ($y = b + mx + cx^2 + d x^3$).

**Exercise 20-1. Determine the linear ($y = b + mx$) yield and N using solver for the following data.**

| $N_{rate}$, lb/acre | Yield, bu/A |
|---|---|
| 0 | 100 |
| 30 | 140 |
| 60 | 150 |
| 90 | 160 |
| 120 | 180 |
| 150 | 185 |

Solution:
The steps as discussed above should be used to solve this problem. First, set up the Excel spread sheet as shown below. Select "Solver", and press "solve" as discussed above. The resulting equation is yield = 112.857 + 0.529x. The $R^2$ value of 0.919 means the 91.9% of the yield variability was explained by the model.

## Using Solver to Solve the Mitscherlich Equation

The Mitscherlich equation is a non-linear model that can be fit to any data source that has the general shape shown in **Box 20-8**. This equation is routinely used to describe the relationships between fertilizers and yield. This equation is very similar to the first order models discussed in chapter 17. The Mitscherlich equation has the form,

$$Y = A \cdot (1 - e^{-BX})$$

where A and B are regression constants and Y and X values are the measured parameters. The A value represents the maximum value while the B value provides an indication of curvature. The curvature value represents how quickly the yield response curve approaches the maximum. In the following example, a hypothetical field experiment was conducted where 60, 90, 120, and 150 lb N/acre were applied to different plots. The yields for these plots were 119, 140, 149, and 152, respectively. To use "Solver", the criteria for model selection must be identified. In this example, the criterium to minimize the sum of squares of error (SSE) value was selected. The equation for SSE is $SSE = \sum_{i=1}^{n}(Y - Y_{est})^2$, where $y$ and $y_{est}$ are the measured and estimated values, respectively. As shown in **Box 20-9**, the data and equations must be entered into the Excel spreadsheet. The model selection criteria is located in cell E12. As stated above, the goal is to minimize this value (error) and therefore "Min" is checked. The initial estimates for A and I are found in cells B3:C3. When the data and equations are entered, highlight "Solve" as shown in **Box 20-10**.

**Box 20-8.** Data that is suitable for fitting to a Mitscherlich model

**Box 20-10.** Solver and spreadsheet prior to using Solver.

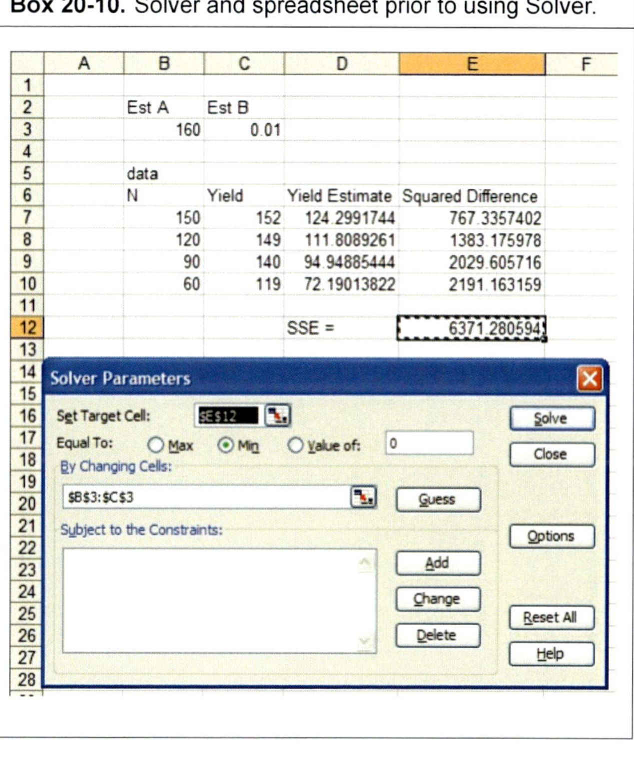


**Box 20-9.** Equations and data entered into Microsoft Excel prior to using "Solver".

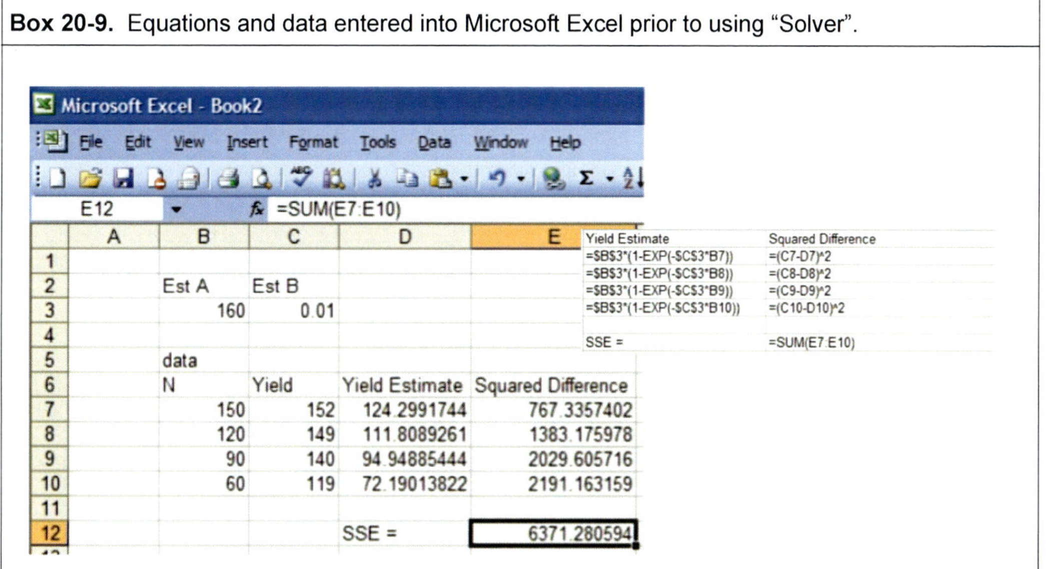


After the solve button is pushed, you will have the option of "Keep Solver Solution" or "Restore Original Values". If "Keep Solver Solution" is selected and "OK" is clicked then the solution coefficients for the problem will be shown in cells B3 and C3 (**Box 20-11**). These coefficients are inserted into the Mitscherlich equation to produce the function,

$$Y_{est} = 157 \cdot (1 - e^{-0.0239X})$$

**Exercise 20-2. Determine the Mitscherlich Equation for an experiment that studied the impact of P rate on strawberry yields.**

| P rate, lb $P_2O_5$/ acre | Strawberry yield, lb/acre (fresh) |
|---|---|
| 0 | 4,000 |
| 30 | 6,000 |
| 60 | 8,000 |
| 90 | 9,000 |
| 120 | 9,200 |

**Box 20-11.** Excel spreadsheet showing analysis results for Mitscherlich Equation.

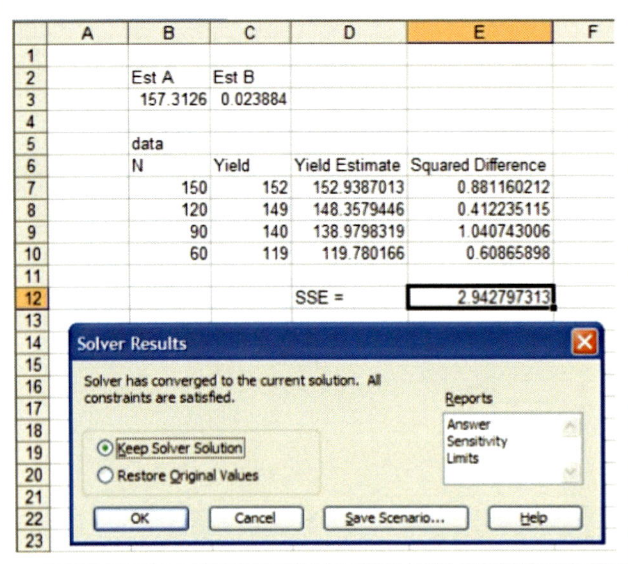

| | A | B | C | D | E | F |
|---|---|---|---|---|---|---|
| 1 | | | | | | |
| 2 | | Est A | Est B | | | |
| 3 | | 157.3126 | 0.023884 | | | |
| 4 | | | | | | |
| 5 | | data | | | | |
| 6 | | N | Yield | Yield Estimate | Squared Difference | |
| 7 | | 150 | 152 | 152.9387013 | 0.881160212 | |
| 8 | | 120 | 149 | 148.3579446 | 0.412235115 | |
| 9 | | 90 | 140 | 138.9798319 | 1.040743006 | |
| 10 | | 60 | 119 | 119.780166 | 0.60865898 | |
| 11 | | | | | | |
| 12 | | | | SSE = | 2.942797313 | |



## Using the Iterative Approach to Solve the Hyperbolic Model

The hyperbolic model has been used to define the impact of weeds on crop growth. This model has the form,

$$yield\ loss = \frac{I*D}{1+\frac{I*D}{A}}$$

where D is the density of weeds/unit area, A is the maximum amount of yield loss/unit area at high weed pressures, I is the incremental yield loss per unit area $\left(\left(\frac{Yield\ Loss}{Weed\ Density \cdot Unit\ Area}\right)\right)$ at low weed densities, and Yield loss (YL) is the difference between the optimum yield and the measured yield (YL = optimum yield – measured yield). Data fit to a hyperbolic model typically has a curved shape showing increasing yield loss with increasing weed population (**Box 20-12**).

**Box 20-12.** Weed vs crop yield data suitable for fitting to a Hyperbolic model.

% Yield loss (0–10) vs Weed population (0–25)

When using "Solver", the selection criteria must be carefully identified. As before, the selection criteria will be to minimize the summed differences between the measured and calculated values (SSE). To solve this problem, the hyperbolic equation must be rewritten in terms used by Microsoft Excel. The rewritten equation now has the form,

YL = (I*D)/(1+(I*D)/A).

In this problem, YL (yield loss) and D (weed densities) were measured and therefore appropriate cell locations for these values must be provided. Initial guesses for I (incremental yield loss) and A (maximum yield loss) must also be provided. When all the data points and equations are entered, the spreadsheet should resemble the data shown in **Box 20-13**.

**Box 20-13.** Data and equations entered into Excel for fitting the Hyperbolic model.

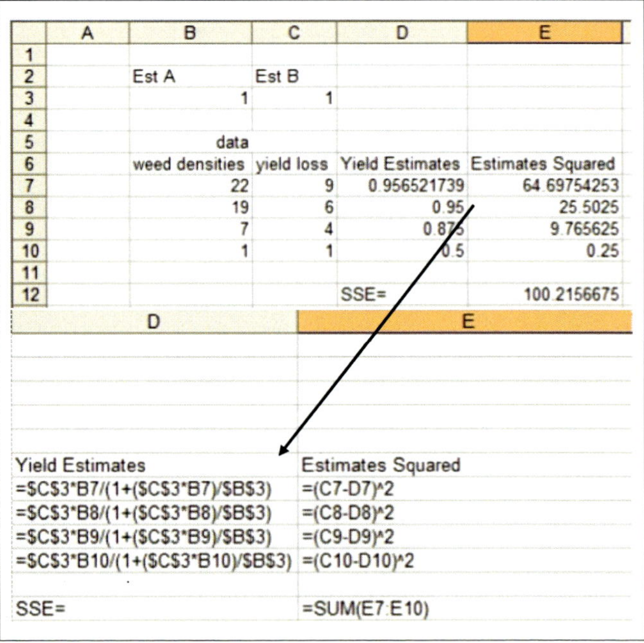

| | A | B | C | D | E |
|---|---|---|---|---|---|
| 1 | | | | | |
| 2 | | Est A | Est B | | |
| 3 | | 1 | 1 | | |
| 4 | | | | | |
| 5 | | data | | | |
| 6 | | weed densities | yield loss | Yield Estimates | Estimates Squared |
| 7 | | 22 | 9 | 0.956521739 | 64.69754253 |
| 8 | | 19 | 6 | 0.95 | 25.5025 |
| 9 | | 7 | 4 | 0.875 | 9.765625 |
| 10 | | 1 | 1 | 0.5 | 0.25 |
| 11 | | | | | |
| 12 | | | | SSE= | 100.2156675 |

| D | E |
|---|---|
| Yield Estimates | Estimates Squared |
| =$C$3*B7/(1+($C$3*B7)/$B$3) | =(C7-D7)^2 |
| =$C$3*B8/(1+($C$3*B8)/$B$3) | =(C8-D8)^2 |
| =$C$3*B9/(1+($C$3*B9)/$B$3) | =(C9-D9)^2 |
| =$C$3*B10/(1+($C$3*B10)/$B$3) | =(C10-D10)^2 |
| SSE= | =SUM(E7:E10) |

When compared to the previous exercise, the individual data points must be changed as well as the equations in cells D7 through D10. In this example, the **hyperbolic model** must be placed in cells D7 through D10. The code that Excel understands is provided in **Box 20-13**. The error (SSE) is the same as the linear model. In this example, the initial guesses for I and A are in cells B3 and C3. You must make an estimate, even though you know it is wrong. Because "Solver" performs an iterative process to obtain a solution, the values in these cells are changed during the solution process. The final values for these coefficients will be found in cells B3 and C3.

When "Solver" is selected, the "Solver" menus will appear (**Box 20-14**). The target cell for the selection criteria should be identified. In this example, the target cell is located in E12; to identify this cell, type $E$12 into the target cell. "Min" should be checked because the goal is to minimize the error term (SSE). The cells (B3 and C3) where the initial guesses for the A and I values should be placed in, "By Changing Cells." To minimize mistakes, the guesses should be realistic. In this example, constraints (values are not negative) will be placed on the A and I values.

Box 20-14. Screens within Excel "Solver".

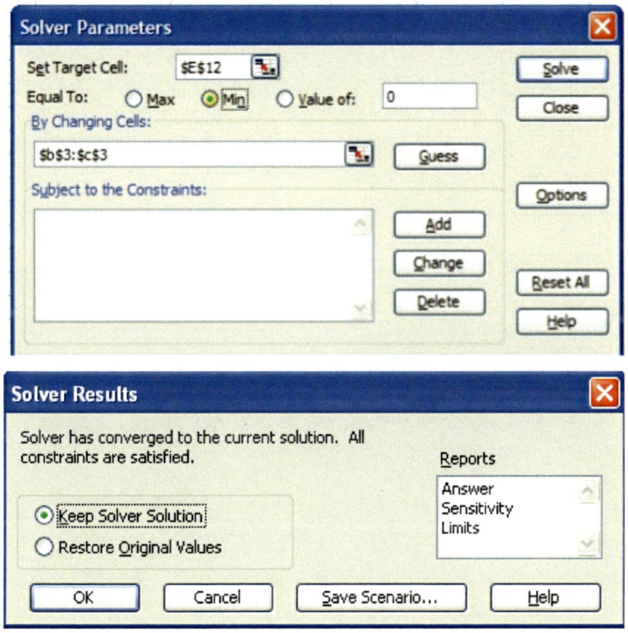

**Solver Parameters**

Set Target Cell: $E$12

Equal To: ○ Max ⊙ Min ○ Value of: 0

By Changing Cells: $b$3:$c$3

Subject to the Constraints:

Solve | Close | Guess | Options | Add | Change | Delete | Reset All | Help

**Solver Results**

Solver has converged to the current solution. All constraints are satisfied.

⊙ Keep Solver Solution

○ Restore Original Values

Reports: Answer, Sensitivity, Limits

OK | Cancel | Save Scenario... | Help

| | A | B | C | D | E |
|---|---|---|---|---|---|
| 1 | | | | | |
| 2 | | Est A | Est B | | |
| 3 | | 16.7564265 | 0.684871 | | |
| 4 | | | | | |
| 5 | | data | | | |
| 6 | | weed densities | yield loss | Yield Estimates | Estimates Squared |
| 7 | | 22 | 9 | 7.933480132 | 1.137464629 |
| 8 | | 19 | 6 | 7.324532692 | 1.754386853 |
| 9 | | 7 | 4 | 3.727609718 | 0.074196466 |
| 10 | | 1 | 1 | 0.657978099 | 0.116978981 |
| 11 | | | | | |
| 12 | | | | SSE= | 3.083026928 |

When all the equations and data are entered, click "Solve". As above, cells B3, C3, and E13 have changed appreciably. The values as shown below in B3 and C3 are the coefficients to the hyperbolic equation. The equation that minimizes SSE is $\textit{yield loss} = \dfrac{0.685 \cdot D}{\left(1 + \dfrac{0.685 \cdot D}{16.756}\right)}$.

The interpretation of this equation was that for each additional weed at low densities, 0.68% yield loss would occur and that the maximum yield loss at high densities was 16.7%.

**Exercise 20-3. Fit the data in Exercise 20-1 to a second order equation ($y = b + mx + cx^2$).**

**Exercise 20-4. Does the linear ($y = b + mx$) or the 2nd order ($y = b + mx + cx^2$) equation fit the data better?**

Solution:
The SSE value for the linear equation (SSE = 387) was higher than the SSE value for the 2nd order (SSE = 170) and therefore the 2nd order equation fit the data better.

**Exercise 20-5. Determine the amount of variation explained by the hyperbolic equation for the data shown in Box 20-13.**

Solution:
To solve this problem, the equation for $R^2$ (amount of variation explained by the equation) must be solved. The equation for $R^2$ is: $R^2 = \dfrac{\sum_{i=1}^{n}(y_i - \bar{y})^2 - SSE}{\sum_{i=1}^{n}(y_i - \bar{y})^2}$.

The SSE value in this equation was previously discussed and calculated. This equation contains a new value that has not been discussed. The value represents the mean or average. This is determined by calculating the average yield loss values. This solution builds on the Findings in Box 20-14. The resulting spreadsheet is shown on the next page.

| | A | B | C | D | E | F |
|---|---|---|---|---|---|---|
| 1 | | | | | | |
| 2 | | Est I | Est B | | | |
| 3 | | 0.684872 | 16.75642 | | | |
| 4 | | | | | | |
| 5 | Weed Density | measured yield (Bu/acre) | Estimated Yield | Squared Differ-ence (M-pred) | Square Dif (meas-aver) | |
| 6 | | | | | | |
| 7 | 22 | 9 | 7.933484 | 1.137456 | 16 | |
| 8 | 19 | 6 | 7.324537 | 1.754398 | 1 | |
| 9 | 7 | 4 | 3.727613 | 0.074195 | 1 | |
| 10 | 1 | 1 | 0.657979 | 0.116978 | 16 | |
| 11 | | | | | | |
| 12 | Average | 5 | SSE = | 3.083027 | 34 | |
| 13 | | | | | | |
| 14 | | | R2 | 0.909323 | | |

| Estimated Yield | Squared Difference (meas-pred.) | Squared Difference |
|---|---|---|
| =($B$3*A7)/(1+($B$3*A7)/$C$3) | =(C7-B7)^2 | =($B12-B7)^2 |
| =($B$3*A8)/(1+($B$3*A8)/$C$3) | =(C8-B8)^2 | =($B12-B8)^2 |
| =($B$3*A9)/(1+($B$3*A9)/$C$3) | =(C9-B9)^2 | =($B12-B9)^2 |
| =($B$3*A10)/(1+($B$3*A10)/$C$3) | =(C10-B10)^2 | =($B12-B10)^2 |
| | | |
| SSE = | =SUM(D8:D11) | =SUM(E7:E10) |
| | | |
| R2 | =(E12-D12)/E12 | |

Based on this analysis the hyperbolic model explains 90.9% of the yield loss variability.

## Using Solver to Solve the Logistic Model

A logistic model is used to determine the carrying capacity in a system where growth is limited by the resources and population density. The logistic model has a S shaped curve (**Box 20-15**). The equation for the logistic model is

$$y_t = \frac{y_0 \cdot K}{[y_0 + (K - y_0)e^{-r_0 t}]}$$

where $y_t$ is the population at time t, $y_0$ is the population at time zero, $K$ is the carrying capacity, and $r_0$ is the growth rate at time zero. This model predicts that if the population ($y$) at time t is less than $K$ (carrying capacity), the population will increase until it reaches $K$.

**Box 20-15.** S-shaped curve typical for data that can be fit to a logistic model.

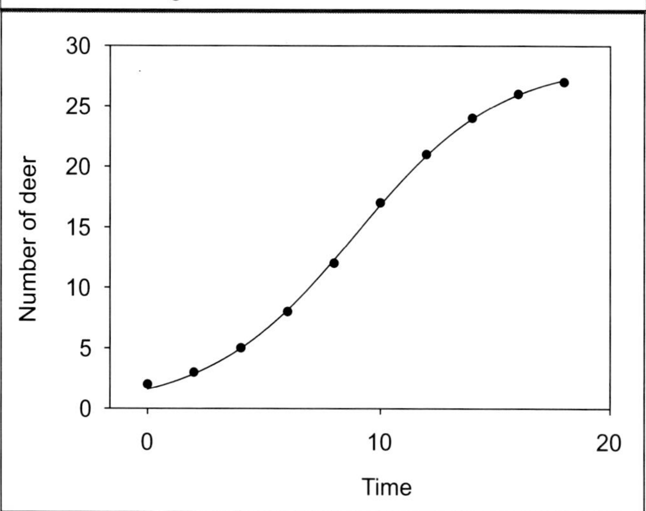

In the following example, "Solver" will be used to determine the carrying capacity of pheasants in a field. In this example, a game farm determined that a field had 10, 12, 15, 25, 30, 32, and 33 pheasants in the spring seasons of year 1, 2, 3, 4, 5, 6, and 7, respectively. As in the previous examples, the model selection criterion will be to minimize the SSE $\left(SSE = \sum_{i=1}^{n}(Y - Y_{est})^2\right)$. As before when using "Solver", initial guesses must be provided for $y_0$, $r_0$, and K. The guess for $y_0$ and $r_0$ should be around 1 and the guess for K should be around 36. After entering the data, the equations need to be entered and coefficients determined for solving the problem, select "Solver", insert the appropriate data, and solve the problem. The resulting spreadsheets and screens are shown in Box 20-16. For this problem, the logistic model must be converted into code that can be interpreted by Excel. This is accomplished by converting $y_t = \frac{y_0 \cdot K}{[y_0 + (K - y_0)e^{-r_0 t}]}$ to=($B$3*$D$3)/[$B$3+($D$3-$B$3)*exp(-$C$3*B7)]. The $ in the equation result in the values in those cells being treated as a constant, and therefore do not change when copying down.

After the data and equations are entered, select "Solver". The resulting spreadsheets and screens are shown in **Box 20-17**.

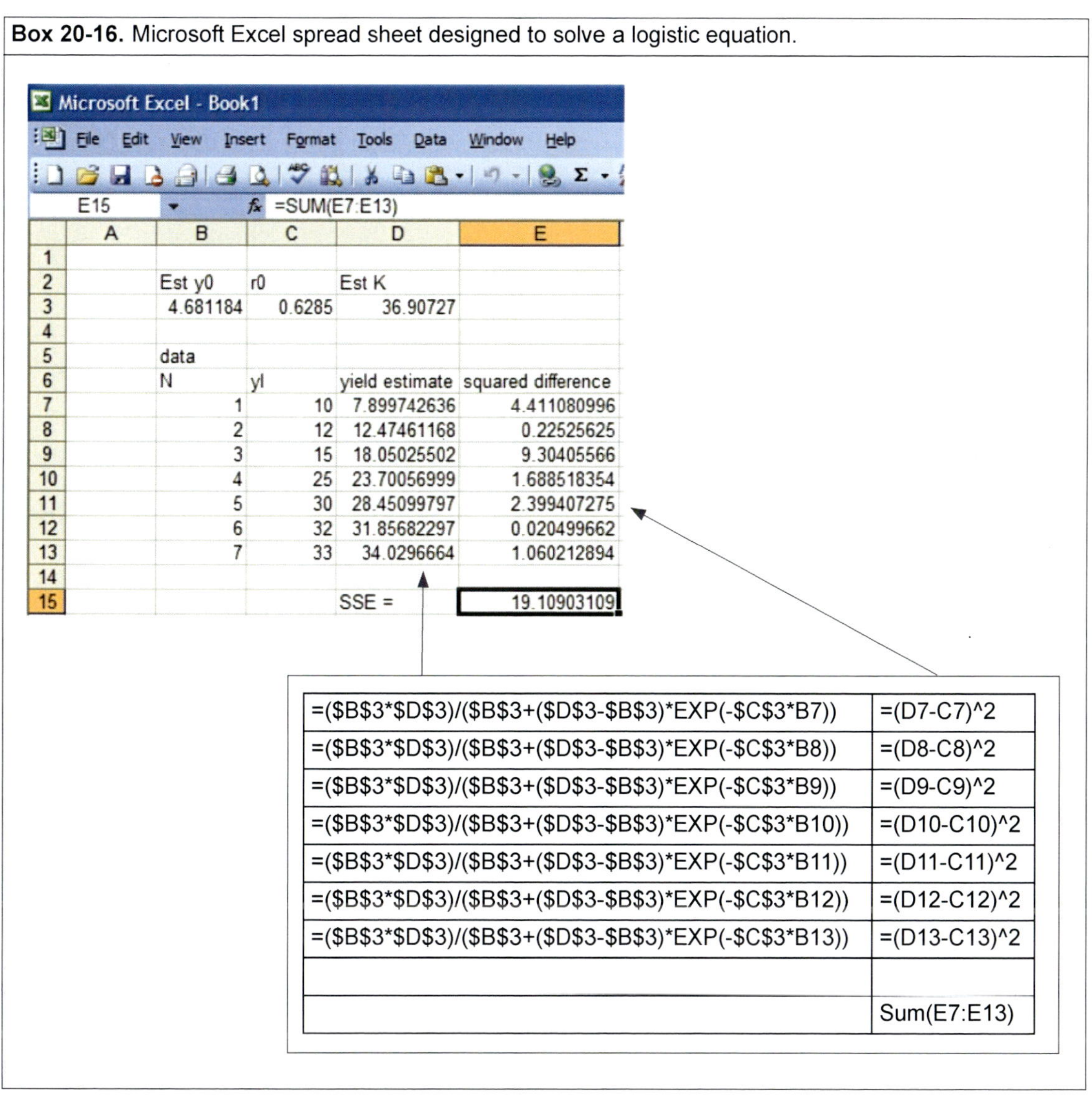

**Box 20-16.** Microsoft Excel spread sheet designed to solve a logistic equation.



| | A | B | C | D | E |
|---|---|---|---|---|---|
| 1 | | | | | |
| 2 | | Est y0 | r0 | Est K | |
| 3 | | 4.681184 | 0.6285 | 36.90727 | |
| 4 | | | | | |
| 5 | | data | | | |
| 6 | | N | yl | yield estimate | squared difference |
| 7 | | 1 | 10 | 7.899742636 | 4.411080996 |
| 8 | | 2 | 12 | 12.47461168 | 0.22525625 |
| 9 | | 3 | 15 | 18.05025502 | 9.30405566 |
| 10 | | 4 | 25 | 23.70056999 | 1.688518354 |
| 11 | | 5 | 30 | 28.45099797 | 2.399407275 |
| 12 | | 6 | 32 | 31.85682297 | 0.020499662 |
| 13 | | 7 | 33 | 34.0296664 | 1.060212894 |
| 14 | | | | | |
| 15 | | | | SSE = | 19.10903109 |

| | |
|---|---|
| =($B$3*$D$3)/($B$3+($D$3-$B$3)*EXP(-$C$3*B7)) | =(D7-C7)^2 |
| =($B$3*$D$3)/($B$3+($D$3-$B$3)*EXP(-$C$3*B8)) | =(D8-C8)^2 |
| =($B$3*$D$3)/($B$3+($D$3-$B$3)*EXP(-$C$3*B9)) | =(D9-C9)^2 |
| =($B$3*$D$3)/($B$3+($D$3-$B$3)*EXP(-$C$3*B10)) | =(D10-C10)^2 |
| =($B$3*$D$3)/($B$3+($D$3-$B$3)*EXP(-$C$3*B11)) | =(D11-C11)^2 |
| =($B$3*$D$3)/($B$3+($D$3-$B$3)*EXP(-$C$3*B12)) | =(D12-C12)^2 |
| =($B$3*$D$3)/($B$3+($D$3-$B$3)*EXP(-$C$3*B13)) | =(D13-C13)^2 |
| | |
| | Sum(E7:E13) |

**Box 20-17.** The solve parameters prior to selecting solve.

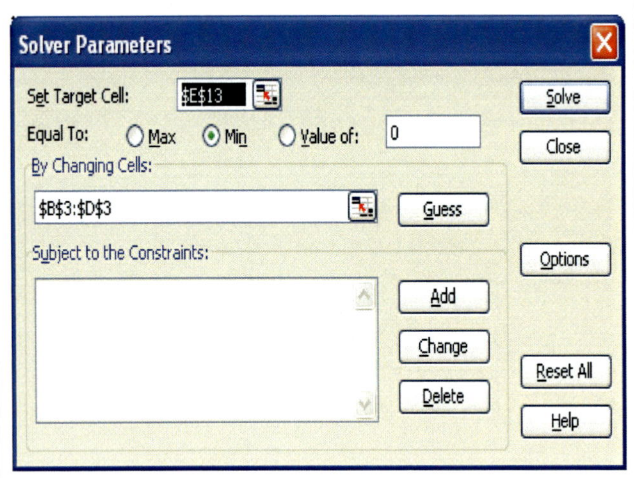


Based on this analysis, the logistic equation is,

$$y_t = \frac{4.68 \cdot 36.9}{[4.68 + (36.9 - 4.68)e^{-0.628t}]}$$

This analysis suggests that the carrying capacity is 36.9 birds.

**Exercise 20-6. What is the logistic equation if the times in the above example were doubled to 2, 4, 6, 8, 10, 12, and 14 years?**

**Exercise 20-7. Determine the Mitscherlich equation for the following data.**

| $P_2O_5$ rate, lb/acre | Yield, bu/a |
|---|---|
| 10 | 110 |
| 30 | 130 |
| 60 | 135 |
| 90 | 136 |

**Exercise 20-8. What is the N mineralization rate constant based on the equation $y_t$= N(1-$e^{-kt}$) for the following data?**

| Days | N mineralization ug/g soil |
|---|---|
| 0 | 20 |
| 30 | 50 |
| 60 | 70 |
| 90 | 75 |

## Chapter Summary

This chapter demonstrates an iterative approach for solving non-linear equations. Key components in using solver are to identify the target cell, what will be changed by the process, and any constraints of the model (for example values >0). In these examples, the sum of squares of error (SSE) was used to evaluate the models. The best results are observed when the initial guesses are close to the final parameter values. All models must be checked for viable results.

## Additional Information

Black, C.A. 1993. Soil Fertility Evaluation and Control, Lewis Publishers, Boca Raton, Florida 33431.

Lipkin, L. and D. Smith. 2004. The logistic growth model. Online J. Mathematics and its application. Available at http://mathdl.maa.org/mathDL/4/?pa=content&sa=viewDocument&nodeId=484&pf=1.

Park, S.E., L. Benjamin, and A.R. Watkinsen. 2003. The theory and application of plant competition models: An agronomic perspective. Annal Bot. 92:741-748.

# USING THE HYPERBOLIC MODEL AS A TOOL TO PREDICT YIELD LOSSES DUE TO WEEDS

**Chapter concepts:** Hyperbolic model, weed management decisions.

**Chapter purpose:** To show how coefficients for the hyperbolic model can be used to estimate yield losses from weed competition.

## Hyperbolic Model

Not controlling weeds when they should be controlled can cause severe yield losses, whereas weed control when not needed is a waste of chemical and money. The hyperbolic model has been used to convert weed density information into estimated yield losses (Cousens, 1985; Chikoye and Swanton 1995; Bensch et al., 2003; Moechnig et al., 2003). The mathematical form of the hyperbolic model is,

$$yield\ loss = \frac{I \cdot D}{\left[1 + \left(\frac{I \cdot D}{A}\right)\right]}$$

where D is the weed density, I is called the incremental yield loss, and A is the maximum yield loss for a given weed. The hyperbolic model, defines the curve shown in Box 20-12. To solve for I and A, a crop is grown with many different densities of the same weed, and % yield loss for each density is calculated, using the weed-free control. The I value is the incremental yield loss per weed. The A value is the maximum yield loss that occurs at high weed densities. Note that A cannot exceed 100% (i.e. no crop yield). A is often less than 100% as competition within the weed species limits additional crop loss.

Once the I and A values are defined (as in Box 21-1), yield loss can be calculated. At low weed densities, yield loss can be approximated as I x D, where D is weed density. At high densities, yield loss = A. To calculate with greater accuracy, the values for I, D, and A should be entered into the equation.

For example: if I = 0.11

D = 20

A = 48

then

$$\%\ yield\ loss = \frac{0.11 \cdot 20}{1 + \frac{(0.11 \cdot 20)}{48}} = \frac{2.2}{1 + 0.045} = \frac{2.2}{1.045} = 2.10\%$$

Research has shown that the incremental yield loss values (I) are not constant, but vary from year to year, site to site, crop to crop, and weed to weed. Incremental yield loss values have been measured for a variety of crops and weeds. Many of these values have been published in journals (Chikoye and Swanton, 1995; Lindquist et al., 1999; Banken 2000; Weaver 2001; Bensch et al. 2003; Conley et al. 2003; Cowbrough et al. 2003; Moechnig et al. 2003; OMAF 2003). These published values are important because they can be used to estimate boundaries of maximum and minimum yield losses from specific weeds in specific crops (**Box 21-1**). The incremental yield loss values are determined in experiments where the population of the crop and the weed species of interest are varied from pure stands of both species to mixtures of both species (Cousens, 1985; Rejmanek et al., 1989).

When using I values it is important to consider that I values are unit specific. The units in **Box 21-1** are estimated from several sources and were converted to a common unit [% yield loss/(weeds/10 ft$^2$)]. To convert these units to %yield loss/(weed/m$^2$), divide the values by 1.076. The D is the estimated density of the weed, and A is the maximum percentage of yield loss at high weed densities. The density of the weeds (D) is determined by making a 1 ft sampling square, i.e. a box with dimensions of one foot on a side. This sampling square is then randomly placed in a number of locations within the sampling zone. The number of weeds within the square is counted. If data are collected from 10 different locations, then these values are summed to determine the number of weeds contained with a 10 ft$^2$ area.

## Estimating Yield Losses from Incremental Yield Loss Coefficients (I)

The resource manager uses a 1 ft square to estimate yellow foxtail population levels at 10 locations within a sampling location. If the yellow foxtail populations are 0, 1, 0, 0, 0, 0, 1, 0, 1, and 2, what is the expected corn yield losses per acre and should the weeds be controlled? In this field the expected yield is 180 bu/acre.

This problem can be solved using a number of different approaches. The easiest solution is to estimate loss for the sampled area. For this problem look up the I (incremental yield loss) values in **Box 21-1.** For yellow foxtail, I values in corn range from 0.11 to 2.9 % yield loss/(weed/10 ft$^2$). When solving these problems, the units of measure must be maintained. The I units must be weeds/10 ft$^2$. If the units are in weeds/ft$^2$, then the data must be converted to weeds/10 ft$^2$ by multiplying the numerator by 10. The mathematics for this conversion is as follows, $\frac{\#\,weeds}{ft^2} \cdot \frac{10}{10} = \frac{10 \cdot \#\,weeds}{10\,ft^2} =$. This conversion means that to convert weeds/ft$^2$ to weeds/10 ft$^2$, multiply the original values by 10. In our example, the average weed density is 5 weeds/10 ft$^2$. This value is calculated as shown below,

$$\text{Average} = \frac{\frac{10\text{ weeds}}{10\text{ ft}^2} + \frac{0\text{ weeds}}{10\text{ ft}^2} + \frac{10\text{ weeds}}{10\text{ ft}^2} + \frac{0\text{ weeds}}{10\text{ ft}^2} + \frac{0\text{ weeds}}{10\text{ ft}^2} + \frac{0\text{ weeds}}{10\text{ ft}^2} + \frac{0\text{ weeds}}{10\text{ ft}^2} + \frac{10\text{ weeds}}{10\text{ ft}^2} + \frac{10\text{ weeds}}{10\text{ ft}^2} + \frac{20\text{ weeds}}{10\text{ ft}^2}}{10} = \frac{5\text{ weeds}}{10\text{ ft}^2}$$

High and low estimates for the yield loss values can be calculated based on the extreme I values. The low estimate is, $\%\text{ yield loss} = \frac{\frac{0.0011\text{ bu yield loss}}{\text{bu corn}}}{\frac{\text{weed}}{10\text{ ft}^2}} \cdot \frac{5\text{ weeds}}{10\text{ ft}^2} = 0.0055\text{ bu yield loss/bu for 5 weeds/10 ft}^2$

The estimated total loss is determined by multiplying the loss per bu times the expected yield

$$\text{Lost Corn} = \frac{0.0055\text{ bu yield loss}}{\text{bu corn}} \cdot \frac{180\text{ bu}}{\text{acre}} = \frac{0.99\text{ bu yield loss}}{\text{acre}}$$

The high yield loss was $\%\text{ yield loss} = \frac{\frac{0.029\text{ bu yield loss}}{\text{bu corn}}}{\frac{\text{weed}}{10\text{ ft}^2}} \cdot \frac{5\text{ weeds}}{10\text{ ft}^2} = 0.145\text{ bu yield loss/bu.}$

Total yield loss was determined by multiplying the loss per bushel times the expected yield,

$$\text{Lost Corn} = \frac{0.145\text{ bu yield loss}}{\text{bu corn}} \cdot \frac{180\text{ bu}}{1\text{ acre}} = \frac{26.1\text{ bu yield loss}}{1\text{ acre}}$$

Based on this information, a resource manager can then decide if the field should be sprayed. The range in yield losses (1 to 26 bu/acre) represents a large range of potential losses. This range of values can be narrowed down by considering the following.

- The date of weed emergence. This value is critical for estimating potential losses. The earlier the weeds emerge, the higher the I value (Bensch et al., 2003; Evans et al., 2003; Clay et al., 2005).
- Row spacing and planting density. Narrow rows and higher densities will canopy faster and have lower I values (Anderson, 2000; Forcella et al., 1992).
- Fertility and water availability constraints, I values generally decrease with increasing water and nutrient supplies (Banken, 2000; Clay et al., 2006; Evans et al., 2003).

There is risk associated with deciding to control or not control weeds based on weed density information. The error of these decisions can be reduced by understanding the biology of both the pest and the crop. Much of this information can be obtained by conducting literature searches on the internet or contacting the appropriate state extension pest management specialist.

**Exercise 21-1. Yellow foxtail is measured at 10 locations using a 1 ft$^2$ box in a production field. These populations are 2, 0, 2, 0, 0, 0, 0, 2, 2, and 4 plants/ft$^2$. What is the expected corn yield loss in this field?**

Solution:
First convert all data from weeds/ft$^2$ to weeds/10 ft$^2$. This is conducted in this example by summing all the values, (12 weeds/10 ft $^2$).

**Box 21-1.** Incremental yield loss (I=%YL/(weed/10 ft$^2$)) and maximum (A) yield loss values for green foxtail, yellow foxtail, giant foxtail, barnyardgrass, and common ragweed for corn and soybean.

| | Corn | | | | Soybean | | | |
|---|---|---|---|---|---|---|---|---|
| Weed species | I value[1] | A value[2] | Comments | Source[3] | I value[1] | A value[2] | Comments | Source[3] |
| Green foxtail | 0.33 | 31 | 1990, Harrow, Ontario | A | 0.75 | 78 | 1990, Harrow, Ontario | A |
| | 2.04 | 49 | 1991, Harrow, Ontario | A | 0.75 | 86 | 1991, Harrow, Ontario | A |
| | 0.65 | 28 | 1993, 1994, Fort Collins, CO | B | 2.15 | 29 | I yield loss at 1 plant m$^2$; A at 25 plants m$^2$ | C |
| | 2.15 | 25 | I yield loss at 1 plant m$^2$; A at 25 plants m$^2$ | C | | | | |
| Yellow foxtail | 2.9 | 48 | 1995, Brookings, SD | B | 1.08 | 20 | I yield loss at 1 plant m$^2$; A at 25 plants m$^2$ | C |
| | 0.11 | 48 | 1996, Brookings, SD | B | | | | |
| | 0.11 | 42 | 1995, Morris, MN, high fertility | D | | | | |
| | 0.65 | 46 | 1995, Morris, MN, low fertility | D | | | | |
| | 0.21 | 75 | 1996, Morris, MN, high fertility | D | | | | |
| | 0.43 | 75 | 1996, Morris, MN, low fertility | D | | | | |
| | 1.08 | 18 | I yield loss at 1 plant m$^2$; A at 25 plants m$^2$ | C | | | | |
| Giant foxtail | 1.94 | 55 | 1994, West Lafayette, IN | B | 11.19 | 82 | 1998, Arlington, WI | E, F |
| | 1.29 | 33 | 1994, Madison, WI | B | 2.04 | 82 | 1999, Arlington, WI | E, F |
| | 2.69 | 37 | 1995, Madison, WI | B | | | | |
| | 1.51 | 24 | 1957-1959, Urbana, IL | B | | | | |
| | 1.29 | 60 | 1994, 1995, East Lansing, MI | B | | | | |
| Barnyardgrass | 2.15 | 31 | I yield loss at 1 plant m$^2$; A at 25 plants m$^2$ | C | 1.83 | 64 | 1995, Woodstock, Ontario | G |
| | | | | | 5.16 | 88 | 1995, Harrow, Ontario | G |
| | | | | | 1.61 | 99 | 1996, Harrow, Ontario | G |
| | | | | | 3.23 | 36 | I yield loss at 1 plant m$^2$; A at 25 plants m$^2$ | C |
| Common ragweed | 4.84 | 80 | 1992, Harrow, Ontario | A | 15.06 | 65 | 1991, Harrow, Ontario | A |
| | 4.95 | 20 | 1993, Harrow, Ontario | A | 5.38 | 71 | 1993, Harrow, Ontario | A |
| | 5.38 | 45 | I yield loss at 1 plant m$^2$; A at 25 plants m$^2$ | C | 17.22 | 85 | 1999, aggregated distribution, Woodstock, Ontario | H |
| | | | | | 10.76 | 62 | 2000, aggregated distribution, Woodstock, Ontario | H |
| | | | | | 10.76 | 59 | I yield loss at 1 plant m$^2$; A at 25 plants m$^2$ | C |

[1]I is % yield loss to the crop per weed plant when weed densities are low. [2]A is the maximum % yield loss when weed densities are high.[3]Source: [A]Weaver, 2001; [B]Lindquist et al., 1999; [C]OMAF, 2002; [D]Banken 2000; [E]Conley et al., 2003 (I value); [F] Conley, 2001 (A value); [G]Cowan et al. 1998; [H]Cowbrough et al. 2003.

Second, estimate a low yield loss estimate

$$\% \text{ yield loss} = \frac{\frac{0.11\% \text{ yield loss}}{\text{bu corn}}}{\text{weeds}/10 \text{ ft}^2} \cdot \frac{12 \text{ weeds}}{\text{sample area}} =$$

$$\frac{1.32\% \text{ loss}}{\text{bu corn}} = \frac{0.0132 \text{ loss}}{1 \text{ bu}}$$

$$\frac{0.0132 \text{ bu yield loss}}{\text{bu corn}} \cdot \frac{180 \text{ bu}}{1 \text{ acre}} = \frac{2.376 \text{ bu corn loss}}{1 \text{ acre}}$$

Third, estimate a high yield loss estimate

$$\% \text{ yield loss} = \frac{\frac{2.9\% \text{ yield loss}}{\text{bu corn}}}{\text{weeds}/\text{sample area}} \cdot \frac{12 \text{ weeds}}{\text{sample area}} =$$

$$\frac{34.8\% \text{ loss}}{1 \text{ bu corn}} = \frac{0.345 \text{ loss}}{1 \text{ bu}}$$

$$\frac{0.345 \text{ loss}}{\text{bu corn}} \cdot \frac{180 \text{ bu}}{1 \text{ acre}} = \frac{62.1 \text{ bu corn loss}}{1 \text{ acre}}$$

Based on this information, corn yield losses will range from 2 to 62 bu/acre. The answer to the problem will depend on when the weeds are counted and their growth stage. If the weeds are counted early in the growing season, then yield losses will be more than if they are counted later in the season.

**Exercise 21-2. What is the expected loss of corn if the realistic yield goal is 100 bu/acre and the field contained five yellow foxtail/10 ft$^2$? Would this change your proposed management?**

Solution: (see Exercise 21-2)
First calculate yield losses as above

$$\% \text{ yield loss} = \frac{\frac{0.11\% \text{ yield loss}}{\text{bu corn}}}{\text{weeds}/\text{sample area}} \cdot \frac{5 \text{ weeds}}{\text{sample area}} =$$

$$\frac{0.55\% \text{ loss}}{1 \text{ bu corn}} = \frac{0.0055 \text{ loss}}{1 \text{ bu}}$$

$$\frac{0.0055 \text{ bu yield loss}}{\text{bu corn}} \cdot \frac{100 \text{ bu}}{1 \text{ acre}} = \frac{0.55 \text{ bu corn loss}}{1 \text{ acre}}$$

Second calculate yield losses as above

$$\% \text{ yield loss} = \frac{\frac{2.9\% \text{ yield loss}}{\text{bu corn}}}{\text{weeds}/\text{sample area}} \cdot \frac{5 \text{ weeds}}{\text{sample area}} =$$

$$\frac{14.5\% \text{ loss}}{1 \text{ bu corn}} = \frac{0.145 \text{ loss}}{1 \text{ bu}}$$

$$\text{Lost Corn} = \frac{0.145 \text{ bu yield loss}}{1 \text{ bu corn}} \cdot \frac{100 \text{ bu}}{1 \text{ acre}} = \frac{14.5 \text{ bu corn loss}}{1 \text{ acre}}$$

If corn is selling for $1.50/bu, then the maximum potential gain from controlling the weeds is $21.75. If corn is selling for $3.00/bu, then the gain is $43.50. For these calculations, the yield potential impacts the recommendation.

**Exercise 21-3. What is the low yield loss estimate if five common ragweeds per 10 ft$^2$ were found in a corn field with a yield potential of 180 bu/acre?**

## References


Anderson, R.L. 2000. Cultural systems to aid weed management in semi-arid corn (*Zea mays*). Weed Tech. 14:630-634.

Banken, K. 2000. Influence of yellow foxtail (*Setaria glauca* (L.) Beauv.) on corn (*Zea mays* L.) growth and western corn rootworm (*Diabrotica virgifera virgifera*, LeConte) development. M.S. thesis, South Dakota State University. 128 pg.

Bensch, C.N., M.J. Horak, and D. Peterson. 2003. Interference of redroot pigweed (*Amaranthus retroflexus*), Palmer amaranth (*A. palmeri*), and common waterhemp (*A. rudis*) in soybean. Weed Sci. 51:37-43.

Chikoye, D. and C.J. Swanton. 1995. Evaluation of three empirical models depicting *Ambrosia artemisiifolia* competition in white bean. Weed Res. 35:421-428.

Clay S.A., J. Kleinjan, D.E. Clay, F. Forcella, and W. Batchelor. 2005. Growth and fecundity of several weed species in corn and soybean. Agron. J. 97:294-305.

Clay, S.A., K.R. Banken, F. Forcella, M.M. Ellsbury, D.E. Clay, and A. E. Olness. 2006. Influence of yellow foxtail (*Setaria pumila*) on corn (*Zea mays*) growth and yield. Comm. Soil Sci. Plant Anal. (in press)

Conley, S.P. 2001. Interference among giant foxtail, common lambsquarters, and soybean. PhD dissertation. University of Wisconsin- Madison, Madison, WI

Conley, S.P., D.E. Stoltenberg, C.M. Boerboom, and L.K. Binning. 2003. Predicting soybean yield loss in giant foxtail (*Setaria faberi*) and common lambsquarters (*Chenopodium album*) communities. Weed Sci. 51:402-407.

Cousens, R. 1985. An empirical model relating crop yield to weed and crop density and a statistical comparison with other models. J. Agric. Sci. 105:513-521.

Cowan, P., S.E. Weaver, and C.J. Swanton. 1998. Interference between pigweed (*Amaranthus* spp.), barnyardgrass (*Echinochloa crus-galli*), and soybean (*Glycine max*). Weed Sci. 46:533-539.

Cowbrough, M.J., R.B. Brown, and F.J. Tardif. 2003. Impact of common ragweed (*Ambrosia artemisiifolia*) aggregation on economic thresholds in soybean. Weed Sci. 51:947-954.

Evans, S.P., S.Z. Knezevic, J.L. Lindquist, C.A. Shapiro, and E.E. Blankenship. 2003. Nitrogen application influences the critical period for weed control in corn. Weed Sci. 51:408-417.

Forcella, F., M.E. Westgate, and D.D. Warnes. 1992. Effect of row width on herbicide and cultivation requirements in row crop. Am. J. Altern. Agric. 7:161-167.

Lindquist, J.L., D.A. Mortesen, P. Westra, W.J. Lambert, T.T. Bauman, J.C. Fausey, J.J. Kells, S.J. Langton, R.G. Harvey, B.H. Bussler, K. Banken, S. Clay, and F. Forcella. 1999. Stability of corn (*Zea mays*) – foxtail (*Setaria* spp.) interference relationships. Weed Sci. 47:195-200.

Moechnig, M.J., C.M. Boerboom, D.E. Stoltenberg, and L.K. Binning. 2003. Growth interactions in communities of common lambsquarters (*Chenopodium album*), giant foxtail (*Setaria faberi*) and corn. Weed Sci. 51:363-370.

Ontario Ministry Agriculture and Food (OMAF) Staff. 2003. Publication 811, Agronomy guide for field crops. Online at: www.gov.on.ca/OMAFRA/english/crops/pub811. Accessed: June 2005.

Weaver, S.E. 2001. Impact of lambs quarters, common ragweed and green foxtail on yield of corn and soybean in Ontario. Can. J. Plant Sci. 81:821-828.

# USING CALCULUS TO CONDUCT AN ECONOMIC ANALYSIS OF A PLANT RESPONSE EXPERIMENT

**Chapter concepts:** Economic analysis, developing polynomial equations using Microsoft Excel Add-Trend Line and Linest functions, determining slope of polynomial equations using calculus, determining the point of maximum economic return.

**Chapter purpose:** To optimize management, an economic analysis of yield response data must be conducted. For experiments that include a single factor (N rates, herbicide rates, or plant populations) calculus can be used to solve the problem. This chapter shows how to conduct this analysis. This analysis can be separated into four steps. First, fit a polynomial equation, $y = b + mx + nx^2$, to the yield response data. Second, derive the economic optimum rate function, Third, determine the derivative of the polynomial equation. Fourth, conduct an economic analysis of the derivative.

## Fitting a Polynomial Equation Using Trend Line Analysis

Converting information into decisions requires that the data be converted into an equation. There are a number of different techniques to accomplish this objective. Previous chapters have showed how to determine polynomial equations using regression analysis and iteration. This discussion will use Add-Trend Line that is available in Microsoft Excel. This process is demonstrated below.

Type into the Excel spread sheet the hypothetical data shown in **Box 22-1.** When completed, the data should have the following format.

**Box 22-1.** Yield response data from a hypothetical experiment showing the relationship between yield and plant population.

| | A | B | C |
|---|---|---|---|
| 3 | 1 | plant pop | yield |
| 4 | 2 | 18 | 128 |
| 5 | 3 | 24 | 150 |
| 6 | 4 | 30 | 162 |
| 7 | 5 | 36 | 167 |

Trend-line analysis is conducted by the following steps.

1. Highlight the data in the spreadsheet and select Insert.
2. Select Chart and X-Y scatter, next, next, and finish.
3. Now, left click on the chart displayed on the page and right click on the first data point.
4. Now, left click on the first point and select Add-trend line.
5. Under type, select polynomial 2[nd] order.
6. Under options, mark the boxes for display equation on chart and display R-squared on chart.
7. Select OK.
8. When completed, the chart should look like **Box 22-2.**

Under step 6, the program was told to display the $R^2$ value on the graph. This value represents the amount of variation explained by the equation. In this example, the second order equation explained 99.95% of the variation.

**Box 22-2.** The completed chart showing the resulting equation and data set.

| | A | B | C | D | E | F | G | H | I |
|---|---|---|---|---|---|---|---|---|---|
| 1 | | | | | | | | | |
| 2 | | 18 | 128 | | | | | | |
| 3 | | 24 | 150 | | | | | | |
| 4 | | 30 | 162 | | | | | | |
| 5 | | 36 | 167 | | | | | | |
| 6 | | | | | | | | | |
| 7 | | | | | | | | | |
| 8 | | | | | | | | | |
| 9 | | | | | | | | | |
| 10 | | | | | | | | | |
| 11 | | | | | | | | | |
| 12 | | | | | | | | | |
| 13 | | | | | | | | | |
| 14 | | | | | | | | | |

**Exercise 22-1. Determine the second order polynomial equation for the following data using the Trend-Line tool.**

| 1,000 seeds X-values | Yield Y-values |
|---|---|
| 0 | 100 |
| 30 | 140 |
| 60 | 150 |
| 90 | 160 |
| 120 | 180 |
| 150 | 185 |

Solution:
Type the data into the Excel spread sheet. Follow steps 1-8 above. The resulting equation is y = $-0.0027x^2 + 0.93x + 104$, with a $R^2$ of 0.96. The sample Excel spread sheet is shown in Box 22-3.

**Box 22-3.** Microsoft Excel spread sheet for Exercise 22-1.

| | A | B | C | D | E | F | G | H | I | J |
|---|---|---|---|---|---|---|---|---|---|---|
| 1 | | X-values | Y-values | | | | | | | |
| 2 | | 0 | 100 | | | | | | | |
| 3 | | 30 | 140 | | | | | | | |
| 4 | | 60 | 150 | | | | | | | |
| 5 | | 90 | 160 | | | | | | | |
| 6 | | 120 | 180 | | | | | | | |
| 7 | | 150 | 185 | | | | | | | |
| 8 | | | | | | | | | | |
| 9 | | | | | | | | | | |
| 10 | | | | | | | | | | |



Equations can be inserted into the spreadsheet by using the Excel Linest Statistical function.

To use Linest, select cell B13 and with left mouse button down, drag across cells C13 and D13. When you are done highlighting cells B13 through D13, type =LINEST(C2:C7,B2:B7^{1,2}) (**Box 22-4; 22-5**). In this equation, C:2:C7 refers to the known y's, B2:B7 refers to known x's, and ^(1,2) tells the program that the resulting equation will contain two x terms (one for x and other for $x^2$). If the desire is to have two terms that are x and $x^3$, then ^(1,2) should be ^(1,3). The coefficients for the equation will show up in cell B13.

**Box 22-4.** The code for Table 2

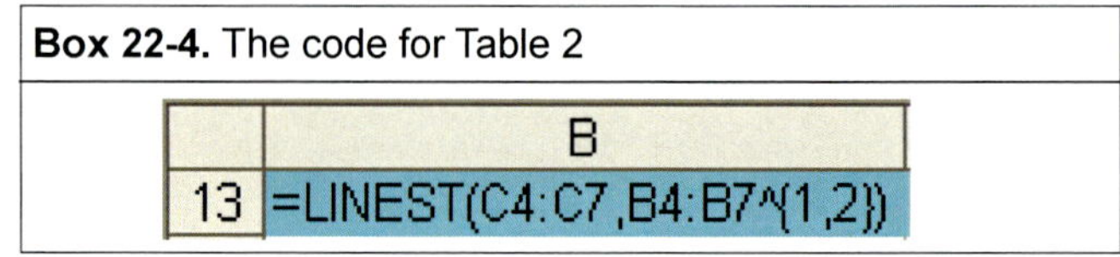

| | B |
|---|---|
| 13 | =LINEST(C4:C7,B4:B7^{1,2}) |

When the equation is entered, simultaneously push the "Shift", "Ctrl" , and "Enter" keys. Values for cells B13, C13, and D13 will fill automatically as shown in **Box 22-5**.

**Box 22-5.** The equation coefficients.

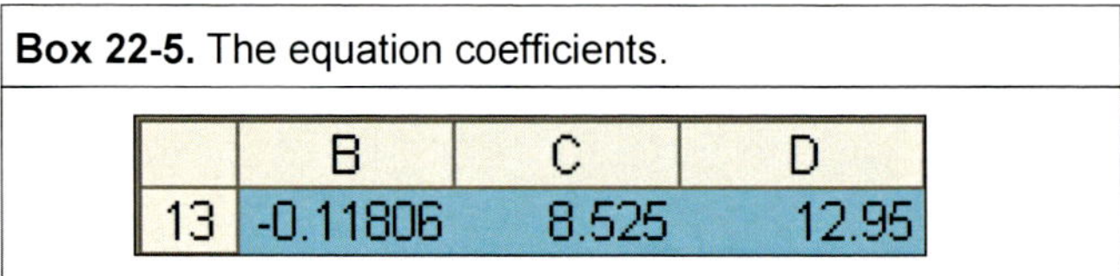

| | B | C | D |
|---|---|---|---|
| 13 | -0.11806 | 8.525 | 12.95 |

If only enter is used, then only the value in cell B13 will appear. The data in **Box 22-5** represents the equation,
yield = 12.95 + 8.525 • (seeds)-0.1181 • $(seeds)^2$.

## Developing the Economic Optimum Equation Function

To determine the point of maximum economic profit, the yield response equations above must be converted to a monetary basis. If yield is a function of population level, then yield can be defined by the equations,

yield = f(population level) where f(population level) = 12.95 + 8.525 • (seeds)-0.1181 • $(seeds)^2$.

The point of the maximum economic optimum population is the point on this curve where the change in the dollar cost of the inputs (seed) equals the change in the dollar value of the corn (yield). This point is defined by the equation,

**d**Yield$ = **d**Seeds$

which is to say that the change (d) in the dollars of Yield equals the change in dollars of seed, where

$$\text{Yield\$} = \left(\frac{\text{bu corn}}{\text{acre}}\right)\cdot\left(\frac{\$}{\text{bu corn}}\right) = \frac{\$}{\text{acre}}$$

$$\text{Seeds\$} = \left(\frac{\text{seeds}}{\text{acre}}\right)\cdot\left(\frac{\$}{\text{seeds}}\right) = \frac{\$}{\text{acre}}$$

**d**Yield$ = **d**Seeds$ can be rearranged to the form

$$\frac{\text{dYield\$}}{\text{dSeeds\$}} = 1$$

After substituting $\text{Yield\$} = \left(\frac{\text{bu corn}}{\text{acre}}\right)\cdot\left(\frac{\$}{\text{bu corn}}\right)$ and $\text{Seeds\$} = \left(\frac{\text{seeds}}{\text{acre}}\right)\cdot\left(\frac{\$}{\text{seeds}}\right)$ into $\frac{\text{dYield\$}}{\text{dSeeds\$}} = 1$

the equation, $$\frac{\text{dYield\$}}{\text{dInputs\$}} = \frac{d\left(\frac{\text{bu corn}}{\text{acre}}\cdot\frac{\$}{\text{bu corn}}\right)}{d\left(\frac{\text{seeds}}{\text{acre}}\cdot\frac{\$}{\text{seeds}}\right)}$$

was derived. If the value of the corn (\$/bu corn) and cost of the seed (\$/seeds) are viewed as constants, then this equation can be rearranged into the form,

$$\frac{\frac{\$}{\text{bu corn}}\cdot d\left(\frac{\text{bu corn}}{\text{acre}}\right)}{\frac{\$}{\text{seeds}}\cdot d\left(\frac{\text{seeds}}{\text{acre}}\right)} = \frac{\frac{\$}{\text{bu corn}}\cdot d(\text{Yield})}{\frac{\$}{\text{seeds}}\cdot d(\text{Seeds})} = 1,$$ which is

rearranged into $\frac{\text{dYield}}{\text{dSeeds}} = \frac{\frac{\$}{\text{seeds}}}{\frac{\$}{\text{bu corn}}}$.

If yield is defined by the function, $\text{yield} = a + b\cdot(\text{seeds}) + c\cdot(\text{seeds})^2 = 12.95 + 8.53\cdot(\text{seeds}) - 0.118\cdot(\text{seeds})^2$, then the derivative of this function, $\frac{\text{dYield}}{\text{dSeeds}}$, is equal to $\frac{\$/\text{seeds}}{\$/\text{bu corn}}$.

This solution defines economic optimum point to a function of the seed costs and corn value.

## Solving for the Derivative

Calculus is used to determine $\frac{\text{dYield}}{\text{dSeeds}}$. This value represents the slope of the line that relates yield on the y-axis to seeds on the x-axis. In calculus, the slope is the first derivative. For polynomial equations, determining the derivative is straight-forward. Let's first consider a linear function: $y = 3 + 2x$. In this function, y changes twice as fast as x, and therefore the slope is 2. However, we can also use a derivative to find this out. Let's use the notation of "d" to denote "change in" or "derivative". We want to find out how y changes as we change x. Another way of writing that is: dy/dx, or "how does y change for a change in x so small that it approaches 0?" Or, as mathematicians would say, "the derivative of y with respect to x." In functions that have a variable raised to a power, the derivative is found by:

1. Deleting any constants from the derivative.
2. Taking the power and multiplying it by the coefficient of the variable.
3. Reducing the power by 1, and raising the variable to this new power.

In our example, $y = 3 + 2x$, 3 is a constant and therefore not included in the derivative. The coefficient of the variable $x^1$ is 2. The power of x is 1 (remember that $x = x^1$). Now multiply the coefficient 2 by the power 1 x 2 = 2.

Now reduce the power of x by $1 - 1 = 0$.

We now have: $dy/dx = (1 \times 2)x^{1-1} = 2x^0$.

Because $x^0 = 1$, $dy/dx = 2x^0 = 2(1) = 2$. This means that y changes twice as fast as x, a result we found when we first looked at the equation.

As we move to higher order polynomial equations, the rate of change is not constant. We can determine the rate of change for these equations using the rules above. Let's consider the following example:

$y = 50 + 4n - 3n^2$

Using the rules discussed above,

dy/dn of 50 = 0
dy/dn of 4n = 4
dy/dn of $3n^2 = 3 \cdot 2 \cdot n = 6n$
so, dy/dn of 4 - 6n

These steps are summarized below.

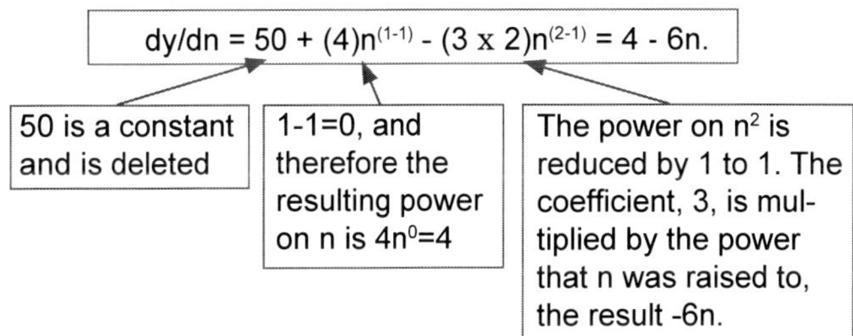


Now we are ready to determine the $\frac{\text{dYield}}{\text{dSeeds}}$ function for the equation,

$\text{Yield} = 12.95 + 8.53\cdot(\text{Seeds}) - 0.118\cdot(\text{Seeds})^2$.

Following the rules described above the derivative is,
$$\frac{\text{dyield}}{\text{dseed}} = 8.53\cdot(\text{seeds})^{(1-1)} - 2\cdot 0.1181\cdot(\text{seeds})^{(2-1)} = 8.525 - 0.236\cdot(\text{seeds})$$

The point for determining maximum profit is derived in the three steps described below.

1. Because $\frac{\text{dYield}}{\text{dSeeds}} = \frac{\frac{\$}{\text{seeds}}}{\frac{\$}{\text{bu corn}}}$, the $\frac{\text{dyield}}{\text{dseed}}$ expression can be replaced with $\frac{\frac{\$}{\text{seeds}}}{\frac{\$}{\text{bu corn}}}$, resulting in the equation, $\frac{\$/\text{seeds}}{\$/\text{bu corn}} = 8.525 - 0.236\cdot(\text{seeds})$

2. Solve for seeds in this equation.

$$\frac{\frac{\$}{\text{seeds}}}{\frac{\$}{\text{bu corn}}} = 8.525 - 0.236 \cdot (\text{seeds})$$

$$\text{seeds} = \left( \frac{\frac{\$}{\text{seeds}}}{\frac{\$}{\text{bu corn}}} - 8.525 \right) \Big/ -0.236$$

3. Because 8.525 is equal to the b coefficient and 0.236 is equal to -2 times the c coeffienct in the polynomial equation, a generalized solution for the point of maximum profitability is determined using a second order polynomial equation.

$$\text{seeds} = \left( \frac{\frac{\$}{\text{seeds}}}{\frac{\$}{\text{bu corn}}} - b \right) \Big/ (2 \cdot c)$$

## Determining the Economic Optimum Seeding Rate

When the slope of the yield response curve is known, selecting the appropriate population level will depend on two additional pieces of information, seed costs and value of the product being produced. Typically, the value of the product being produced is unknown at planting time. To account for this uncertainty a range of values can be used. In this example, four corn values ranging from $2.00/bushel to $3.00/bushel will be used, and 5 seed values ranging from $80.00 to $160/bag will be used. Each bag contains approximately 80,000 seeds and therefore for an $80.00/bag a 1,000 seeds cost about $1.00 and for $160/bag seed 1,000 seeds costs about $2.00 (**Box 22-6**). To determine the economic optimum seeding rates enter the following data and equations into the Excel spreadsheet.

**Box 22-6.** The data and equations that must be entered in the spreadsheet to calculate economic optimum seeding rates.

| | B | C | D |
|---|---|---|---|
| 13 | -0.11806 | 8.525 | 12.95 |

Yield = 12.95 + 8.53 • (seeds) - 0.118 • $(\text{seeds})^2$
This equation was inserted using linest fuction.

| | B | C | D |
|---|---|---|---|
| 19 | | | Population of maximum profitabil |
| 20 | | | |
| 21 | cost of seed | | |
| 22 | $/bag | $/1000 sd | 2 |
| 23 | 80 | =B23/80 | =($C23/D$22-$C$13)/(2*$B$13) |
| 24 | 100 | =B24/80 | =($C24/D$22-$C$13)/(2*$B$13) |
| 25 | 120 | =B25/80 | =($C25/D$22-$C$13)/(2*$B$13) |
| 26 | 140 | =B26/80 | =($C26/D$22-$C$13)/(2*$B$13) |
| 27 | 160 | =B27/80 | =($C27/D$22-$C$13)/(2*$B$13) |

$C23 is cost of 1,000 seeds when a bag cost $80. D$22 is the value of the corn when sold ($2/bu)

The equation in cells D23 through D27 represents the Excel version of $\text{seeds} = \left( \frac{\frac{\$}{\text{seeds}}}{\frac{\$}{\text{bu corn}}} - b \right) \Big/ (2 \cdot c)$.

In the cell D23, the value $C23 represents the cost in $ per 1,000 seeds, D$22 is the value of corn in $/bu, $C$13 is the coefficient 8.525 in the equation described above, and 2*$B$13 is 2 times the value -0.11906 described above. Optimum seeding rates for higher value corn is determined by placing different corn values in E22 through G22, and copying D23 through D26 to the right.

**Box 22-7.** The results matrix showing point of maximum profitability.

| | B | C | D | E | F | G |
|---|---|---|---|---|---|---|
| 19 | | | Population of maximum profitability | | | |
| 20 | | | | value of corn | | |
| 21 | cost of seed | | | $/bu corn comodity price | | |
| 22 | $/bag | $/1000 sd | 2 | 2.25 | 2.5 | 3 |
| 23 | 80 | 1 | 34.0 | 34.2 | 34.4 | 34.7 |
| 24 | 100 | 1.25 | 33.5 | 33.8 | 34.0 | 34.3 |
| 25 | 120 | 1.5 | 32.9 | 33.3 | 33.6 | 34.0 |
| 26 | 140 | 1.75 | 32.4 | 32.8 | 33.1 | 33.6 |
| 27 | 160 | 2 | 31.9 | 32.3 | 32.7 | 33.3 |

**Exercise 22-2. Use the approach described above to determine the economic optimum N rate for the following data for a variety of N cost ($0.1/lb N, $0.20/lb N, $0.30/lb, and $0.40/lb) and corn values ($2.00/bu, $2.25/bu, $2.50/bu, and $3.00/bu. The relationship between N application rate and yield is,**

| | A | B | C |
|---|---|---|---|
| 1 | 1 | Lb N | yield |
| 2 | 2 | 60 | 128 |
| 3 | 3 | 100 | 156 |
| 4 | 4 | 140 | 169 |
| 5 | 5 | 180 | 172 |

## Summary

In summary, an approach for conducting an economic analysis of a common expert is shown. The approach uses a 2nd order polynomial equation and calculus to solve the problem.

# USING PARTIAL DERIVATIVES TO CONDUCT AN ECONOMIC ANALYSIS ON THE IMPACT OF SEEDING DENSITY AND N RATES ON YIELD

**CHAPTER CONCEPTS:** Independent variable, Dependent variable, Partial derivatives, Factorial experiment Multiple regression.

**CHAPTER PURPOSE:** The purpose of the example in this chapter is use the regression package available under the Tools menu to develop a multivariate equation containing two independent variables. This example builds on the exercise shown in Chapter 19, where an equation for a single independent variable was developed. Economic analysis of this problem will be conducted by deriving an equation relating costs and value at the economic optimum point. Two partial derivatives for these equations will be determined and economic optimum values for the two independent variables determined.

## Solving Problems with Two Independent Variables

Solving problems with two variables is much more complex than solving problems with one variable. The previous chapter showed that a derivative (dy/dx) of a function relating a dependent and an independent variable can be used to determine the point of maximum economic viability (changes in input costs equal changes in yield value). **Independent variables** are those that represent the different treatments, or are manipulated, whereas **dependent variables** are only measured.

In this problem, two partial derivatives, one for each independent variable (N and population level) will be obtained. A **partial derivative** is calculated with respect to one of the variables when the other variable is held constant. The first partial derivative will be used to define the relationship of changes in yield relative to changes in the amount of N fertilizer applied under conditions when plant population is considered a constant. The second partial derivative will be used to define changes in yield relative to changes in plant population when the amount of N fertilizer applied is treated as a constant. Each partial derivative will be used to derive a line that defines costs of inputs and value of outputs. The intersection of the lines is the economically optimum value for both variables. This exercise will be separated into five steps.

1. The problem will be explained.
2. The function containing the two variables will be derived.
3. The lines based on the partial derivatives will be developed.
4. The equations relating costs, inputs, and outputs will be derived.
5. The point where the two partial derivatives intersect will be determined.

## Defining the Problem

Prior to initiating a new study or conducting extensive data analysis, it is important to review previous research findings. Research has shown that: 1) water and N fertilizer are the most limiting resources for crop growth in semi-arid landscapes; 2) the application of N fertilizer increases yields under some conditions and reduces yields under other conditions. 3) under high soil water conditions, N fertilizer is lost through leaching or denitrification. These research findings can be summarized into a relational diagram **(Box 23-1)** and mathematical equations.

**Box 23-1.** Relational diagram showing how N fertilizer rate, grain yields, climatic conditions (rainfall), and N fertilizer loss interact to influence yields.

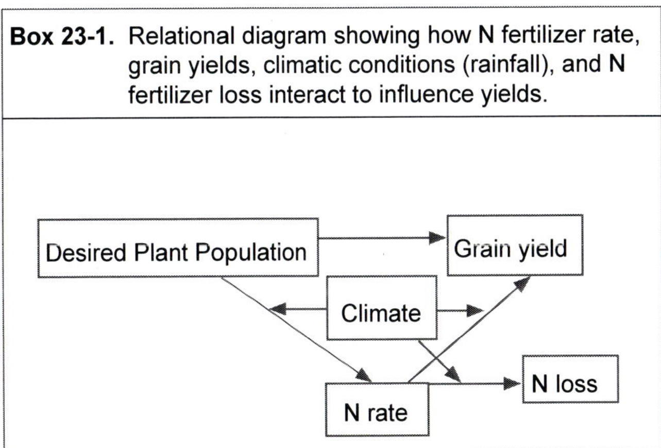

This relational diagram can be summarized by the expression, Corn yield = f (N, population, population by N interaction). This notation simply states that the yield of corn is a function (f) of N and population (the first two terms listed). But since we also know that these two factors interact, the third, generalized term is added. Relational diagrams and mathematical notation are very useful in designing experiments.

Based on the understanding of the problem, a **factorial experiment** that contains three replicates, three populations, and three N rates was conducted. Factorial means that all possible combinations between two treatments were included in the experiment. The total number of N and population level treatments was 3 N rates times 3 population levels or 9 treatments. If the experiment contained only one replicate (repeated measure), it would contain nine plots where the different treatments where applied. Replicating the experiment 3 times resulted in 27 plots. A plot is an area of land where a specific treatment is applied. Yields from these treated areas were measured.

In field experiments, the efficiency of the experiment can often be improved by using a blocked design. Each block represents a specific area of land that contains plots where each treatment was applied. In this case, the three 'blocks' are located next to each other (**Box 23-2**). If the treatment locations within a block are randomized, the experimental design would be classified as a randomized complete block. Note: there are many different types of experimental designs, each with their own strengths and weaknesses. To minimize problems, we recommend that students should consult with a statistician when setting up experiments. The size of individual plots can vary. Plot sizes should match the problem and the equipment. In this example, the farmer selected a plot size of 1 acre. This size was selected because it fits his/her equipment. The producer plans to harvest the plots with a combine equipped with a yield monitor. Yields from each of the plots are shown in **Box 23-3**.

**Box 23-2.** Example of a replicate or randomized test plot block. Each column is one block and each number within a replicate is one of the nine treatments.

| Block 1 | Block 2 | Block 3 |
|---|---|---|
| 1 | 9 | 2 |
| 2 | 8 | 1 |
| 3 | 7 | 8 |
| 4 | 6 | 9 |
| 5 | 4 | 7 |
| 6 | 5 | 3 |
| 7 | 3 | 5 |
| 8 | 2 | 4 |
| 9 | 1 | 6 |

**Box 23-3.** Results from an on-farm study where N and plant population (p) rates were varied. In this table, yield at 15.5% moisture is shown in column B, the amount of N in lb/acre is shown in column C, and the plant population in *1,000s is the planted population in plants/acre. This will be the data used later in the chapter using MS Excel.

| | A | B | C | D | E | F | G |
|---|---|---|---|---|---|---|---|
| 1 | | Yield | | | | | |
| 2 | | bu/ acre | N rate | Population *1,000/acre | N^2 | p^2 | N*p |
| 3 | Rep 1 | 99 | 40 | 16 | 1600 | 256 | 640 |
| 4 | | 140 | 40 | 24 | 1600 | 576 | 960 |
| 5 | | 150 | 40 | 32 | 1600 | 1024 | 1280 |
| 6 | | 115 | 120 | 16 | 14400 | 256 | 1920 |
| 7 | | 150 | 120 | 24 | 14400 | 576 | 2880 |
| 8 | | 155 | 120 | 32 | 14400 | 1024 | 3840 |
| 9 | | 120 | 200 | 16 | 40000 | 256 | 3200 |
| 10 | | 152 | 200 | 24 | 40000 | 576 | 4800 |
| 11 | | 160 | 200 | 32 | 40000 | 1024 | 6400 |
| 12 | Rep 2 | 89 | 40 | 16 | 1600 | 256 | 640 |
| 13 | | 130 | 40 | 24 | 1600 | 576 | 960 |
| 14 | | 140 | 40 | 32 | 1600 | 1024 | 1280 |
| 15 | | 105 | 120 | 16 | 14400 | 256 | 1920 |
| 16 | | 140 | 120 | 24 | 14400 | 576 | 2880 |
| 17 | | 145 | 120 | 32 | 14400 | 1024 | 3840 |
| 18 | | 110 | 200 | 16 | 40000 | 256 | 3200 |
| 19 | | 142 | 200 | 24 | 40000 | 576 | 4800 |
| 20 | | 150 | 200 | 32 | 40000 | 1024 | 6400 |
| 21 | Rep 3 | 109 | 40 | 16 | 1600 | 256 | 640 |
| 22 | | 150 | 40 | 24 | 1600 | 576 | 960 |
| 23 | | 160 | 40 | 32 | 1600 | 1024 | 1280 |
| 24 | | 125 | 120 | 16 | 14400 | 256 | 1920 |
| 25 | | 160 | 120 | 24 | 14400 | 576 | 2880 |
| 26 | | 165 | 120 | 32 | 14400 | 1024 | 3840 |
| 27 | | 130 | 200 | 16 | 40000 | 256 | 3200 |
| 28 | | 162 | 200 | 24 | 40000 | 576 | 4800 |
| 29 | | 170 | 200 | 32 | 40000 | 1024 | 6400 |

## Developing the Predictive Equations

Plotting the field results is the first step in analyzing the data. This step is important because it is used to identify problems and improve your understanding of the data set. A graphical representation of the data is shown in **Box 23-4**. This 3-D plot was developed using the graphing program "Sigma Plot". Two dimensional graphs can be developed using the Chart Wizard contained within Excel. Chart Wizard is accessed on the tool bar and then following directions. Using 2-D graphing programs, a three dimensional representation can be obtained by separating the data set into three population groups and then graphing each population group separately. Developing 2-D graphs in Chart Wizard is accomplished by: 1) highlighting two columns of data; 2) selecting x-y scatter under chart wizard; 3) selecting x-y scatter; 4) selecting appropriate chart sub-type; and 5) clicking on "next" and "finished".

**Box 23-4.** Relationship between population, N rate, and corn yields in the hypothetical experiment. This plot was made with the Sigma Plot graphing program.

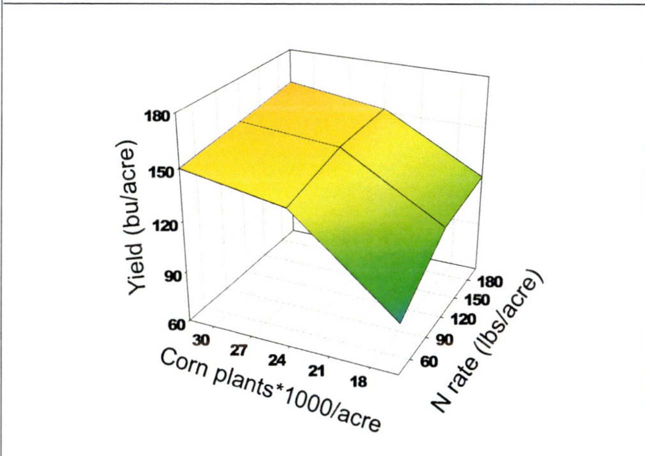


**Box 23-4** shows that for a given plant population level, yields increase with increasing N. Yields also increase with population level. A polynomial equation ($y = b + cp + dp^2 + eN + fN^2 + gNP$, where p is population level and N is nitrogen fertilizer rate) can be used to describe this relationship. The squared terms in the polynomial equations are included in columns E and F in **Box 23-3**. To capture a potential interaction between plant population and N rate, these values multiplied by each other are saved in column G.

To develop a regression model that predicts yields based on the N rate and population **multiple regression** will be conducted. Multiple regression is a statistical approach to define a mathematical model between a dependent variable and several independent variables. A number of different programs are available for developing this model. This example used a regression package available in Microsoft Excel. *Note: This program was selected because most students and agricultural professionals have this software on their computers.* If this program is not available, the Analysis Tool Pac can be downloaded from the program disk. This is accomplished by: 1) selecting Add-Ins under Tools; 2) clicking on the Analysis Tool Pak; and 3) following directions. Once the program is available, follow the steps listed below.

1. Open a blank worksheet in Microsoft Excel.
2. Beginning with cell A1, enter the data, exactly as shown, in columns A through D in **Box 23-3**. Although these data are hypothetical, they are based on our understanding of the system.
3. Calculate $N^2$, $p^2$, and N *p values for columns E, F, and G.

a. In cell E3, type the formula "=C3^2" and press Enter. This calculates the squared N value. Highlight cell E3, click on the fill handle, and drag to cell E29. This fills in the N value for the remaining yield data

b. In cell F3, type the formula "=D3^2". This calculates the squared P amount applied value. As in Step 3a, copy this formula into the remaining rows in column E.

c. In cell G3, type the formula "=C3 *D3". This calculates N rate times seed population level value. Fill in the remainder of column F with this formula.

4. Compare your spreadsheet with **Box 23-3**. They should be identical.
5. On the standard toolbar, click Tools | Data analysis | Regression, then click OK. Select $B$3:$B$29 for the Input Y Range and $C$3:$H$29 for the Input X Range. Under Output options, click New Worksheet Ply, then click OK. The regression coefficients we need for our multivariate equation are located in the bottom table of the newly created Excel sheet. They are located under the column entitled "Coefficients".

Note that there are 6 coefficients listed in cells B17: B22 in **Box 23-5**. Excel has calculated the coefficients of the variables C3:G29 in **Table 23-3**.

The statistical analysis provides a lot of information about the multiple regression equation. The R squared value ($R^2$=0.854) (cell B5, **Table 23-5**) represents the amount of yield variability explained by the multiple regression equation. This value ranges from 0 to 1 and provides an index on the models relative predictive power. The closer the R squared term is to one the higher the models predictive ability. The **Table 23-5** also has a statistic called adjusted R square (0.820) in cell B6. This statistic adjusts the $R^2$ value for the degrees of freedom in the model. The ANOVA analysis provides details about the significance level of the multiple regression model. The F value of 24.73 indicates that the model is very significant. The value (significant F = 3.88E-08) represents the $\alpha$ value. The $\alpha$ value represent the probability of accepting a false positive. Based on these analysis results the mathematical model for this data set is,

$$\text{Yield} = -73.805 + 0.31145 \cdot N + 13.8697 \cdot p - 0.0004947 \cdot N^2 - 0.221354 \cdot p^2 - 0.0042968 \cdot N \cdot p$$

In this equation, yield is bu/acre, N is the amount of the nitrogen fertilizer in lb/acre, and p is the seeding population in thousands of seeds/acre. The coefficients for this model are located in cells B17 through B22. This equation will be used in the discussion below.

## Determining the Partial Derivatives

In this analysis, two partial derivatives will be determined. The first equation is based on the change in corn yield (denoted $\delta y$) with respect to the change in N rate (denoted $\delta N$) when population is considered a constant and the other partial derivative is the change in corn yield with respect to the change in population (denoted $\delta p$) when N is considered a constant.

1. When taking a partial derivative any constants or variables that do not contain the variable of interest are eliminated from the equation;
2. The exponent power of the variable is reduced by 1 and the constant that the variable is multiplied by is the exponential power.

**Box 23-5.** A summary of the multiple regression statistical analysis of the N by population data as shown in **Box 23-3.**

| | A | B | C | D | E | F | G | H | I |
|---|---|---|---|---|---|---|---|---|---|
| 1 | SUMMARY | | | | | | | | |
| 2 | | | | | | | | | |
| 3 | Regression Stat. | | | | | | | | |
| 4 | Multiple R | 0.92456 | | | | | | | |
| 5 | R Square | 0.854811 | | | | | | | |
| 6 | Adjusted R Square | 0.820242 | | | | | | | |
| 7 | Standard Error | 9.372156 | | | | | | | |
| 8 | Observations | 27 | | | | | | | |
| 9 | | | | | | | | | |
| 10 | ANOVA | | | | | | | | |
| 11 | | df | SS | MS | F | Signif F | | | |
| 12 | Regression | 5 | 10860.08 | 2172.017 | 24.72773 | 3.88E-08 | | | |
| 13 | Residual | 21 | 1844.583 | 87.8373 | | | | | |
| 14 | Total | 26 | 12704.67 | | | | | | |
| 15 | | | | | | | | | |
| 16 | | Coefficients | Stand Error | t Stat | P-value | Lower 95% | Upper 95% | Lower 95.0% | Upper 95.0% |
| 17 | Intercept | -73.8056 | 35.49244 | -2.07947 | 0.050014 | -147.616 | 0.0050 | -147.61 | 0.0050 |
| 18 | X Variable 1 | 0.311458 | 0.177884 | 1.750905 | 0.094558 | -0.05847 | 0.6814 | -0.0585 | 0.6814 |
| 19 | X Variable 2 | 13.86979 | 2.927171 | 4.738292 | 0.000111 | 7.782406 | 19.957 | 7.7824 | 19.957 |
| 20 | X Variable 3 | -0.00049 | 0.000598 | -0.82763 | 0.417184 | -0.00174 | 0.00075 | -0.0017 | 0.00075 |
| 21 | X Variable 4 | -0.22135 | 0.059784 | -3.70257 | 0.00132 | -0.34568 | -0.0970 | -0.3457 | -0.097 |
| 22 | X Variable 5 | -0.0043 | 0.004227 | -1.01644 | 0.320986 | -0.01309 | 0.0045 | -0.0131 | 0.0045 |
| 23 | | | | | | | | | |

Not significant as 5%

Developing partial derivatives is best explained through an example. For the equation,

$y = 5 + 4x^2 - 2z^2$,

2 partial derivatives (δy/δx, and δy/δz) can be determined.

For the δy/δx partial derivative, z is treated as a constant, and therefore the 2 $z^2$ term is deleted from the resulting partial derivative. The 5 value is also a constant and therefore it also is deleted from the resulting partial derivative. The x term is squared and therefore the 4 coefficient is multiplied by 2 and exponent is reduced by 1. The end result is $2 \cdot 4x^{(2-1)}$ or 8x. These series of steps are summarized below.

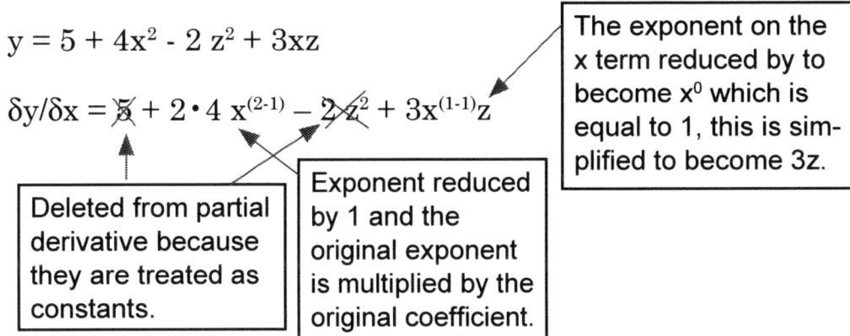


For the δy/δx the same logic is followed. For this partial derivative, 5 and $4x^2$ are deleted because they are considered constants, z is squared and therefore, the coefficient is multiplied by 2 and exponential power 2 is reduced by 1. The resulting partial derivative is $2 \cdot 2z^{(2-1)}$ or 4z.

**Exercise 23-1. Determine the partial derivatives for the equation, $y = 3 + 2x + 5x^2 + 4z + 3xz$.**

Solution:
δy/δx = 2 + 10x + 3z
δy/δz = 4 + 3x

**Exercise 23-2. Determine the partial derivative for the equation, $y = 6 + 3x^3 - 2x^4 + 2zx^2 + 5z^3$**

**Exercise 23-3. Determine the partial derivative for the equation, $y = 1 + 2x + 3z$.**

## Deriving Economic Equations Relating Yields, Inputs, and Costs

To determine the economically optimum values the predictive equation must be converted from amounts to costs. The following discussion provides guidance on how to accomplish this task. At the economic optimum point, input costs ($\delta N_\$$ and $\delta p_\$$) and corn value ($\delta y_\$$) can be defined by the equations, $\frac{\partial y_\$}{\partial p_\$} = 1$ and $\frac{\partial y_\$}{\partial N_\$} = 1$.

Since the total change in value of corn ($\delta y_\$$) is equal to the change in amount of corn produced (δy) times its selling price (val) the change in yield value can be defined by the equation, $\partial y_\$ = \partial y * val$. This equation can be rearranged into the equation, $\partial y = \frac{\partial y_\$}{val}$. This equation is important because it describes change in yield as a function of value and change in total corn worth ($\delta y_\$$).

This same process can be used to define the relationship between N rate and N costs at the economic optimum value. The change in cost of the fertilizer ($\delta N_\$$) is equal to the change in amount of N applied (δN) times the cost of the N (costN), which is expressed mathematically by the equation, $\partial N_\$ = \partial N *$ cost N. This equation can be rearranged into the form, $\partial N = \frac{\partial N_\$}{costN}$.

The change in population costs ($\delta p_\$$) is equal to change in population (δp) times seed costs (costp).

Mathematically this is expressed by the equation, $\partial p_{\$} = \partial p \cdot \text{cost p}$. Again this equation is rearranged into the form $\partial p = \frac{\partial p_{\$}}{\text{costp}}$.

Now, because, $\partial y = \frac{\partial y_{\$}}{\text{val}}$. and $\partial p = \frac{\partial p_{\$}}{\text{costp}}$. the value for δy/δp can be defined by the equation, $\frac{\partial y}{\partial p} = \frac{\frac{\partial y_{\$}}{\text{val}}}{\frac{\partial p_{\$}}{\text{cost p}}}$.

This expression can be reorganized into the expression, $\frac{\partial y}{\partial p} = \frac{\partial y_{\$}}{\partial p_{\$}} \cdot \frac{\text{cost p}}{\text{val}}$.

Since at the economic optimum point $\frac{\partial y_{\$}}{\partial p_{\$}} = 1$ the value for δy/δp at the economic optimum rate can be defined by the equation, $\frac{\partial y}{\partial p} = \frac{\text{cost p}}{\text{val}}$, and by using the same logic $\frac{\partial y}{\partial N} = \frac{\text{cost N}}{\text{val}}$.

These derivations are very important because they provide the tools to define changes in yield, N rate, and population in terms of changes in corn value, changes in N fertilizer costs, and population (seed) costs. These equations will be used in the example below to determine the economically optimum values for N and seed population rate.

**Exercise 23-4. Derive** $\frac{\partial y}{\partial N} = \frac{\text{cost N}}{\text{val}}$.

Solution:
The δy/δN value is can be expressed by the equation

$$\frac{\partial y}{\partial N} = \frac{\frac{\partial y_{\$}}{\text{val}}}{\frac{\partial N_{\$}}{\text{cost N}}}$$

which can be reorganized into the equation

$$\frac{\partial y}{\partial N} = \frac{\partial y_{\$}}{\partial N_{\$}} \cdot \frac{\text{cost N}}{\text{val}}$$

At the economic optimum N rate $\frac{\partial y_{\$}}{\partial N_{\$}} = 1$, and therefore at the economic optimum N rate $\frac{\partial y}{\partial N} = \frac{\text{cost N}}{\text{val}}$.

## Determining the δy$/δN$ and δy$/δp$ Values

To determine the δy$/δN$ and δy$/δp$ function, the δy/δN and δy/δp partial derivative must be determined. These steps are summarized below.

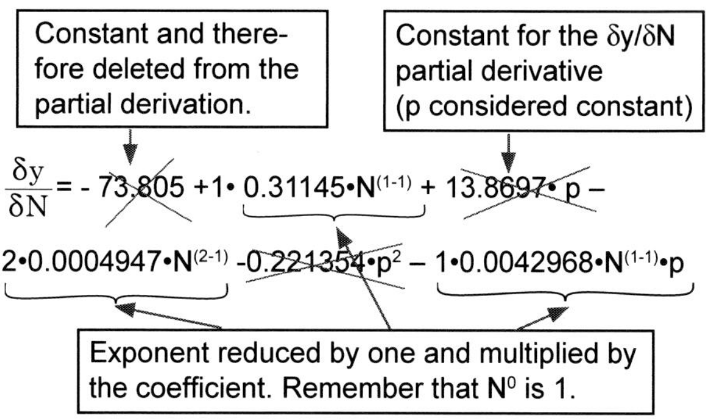


After these operations, the resulting partial derivative is,

δy/δN = 0.31145 – 2 • 0.0004947 • $N^{(1)}$ – 1 • 0.0042968 • p

which is simplified to

δy/δN= 0.31145 – 0.0009894 N – 0.0042968 p

The same process should be completed to determine the δy/δp function. This process is shown below

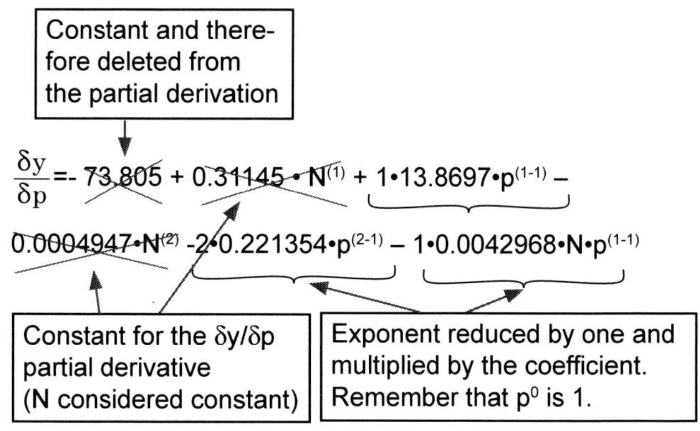


The resulting partial derivative is

δy/δp = 13.8697 $p^0$ – 2 • 0.221354$p^1$-0.0042968N $p^0$

which is simplified to (note: $p^0$=1)

δy/δp = 13.8697 – 0.442687 p – 0.0042968 N

$= \frac{\text{Cost p}}{\text{Value Corn}}$

The δy/δN and δy/δp partial derivatives can be redefined as the costN/corn value and costp/corn value ratios, respectively. These substitutions are shown on the next page.

$$\partial y/\partial N = 0.31145833 - (0.00048479)\,N - 0.00429687\,p = \frac{\text{cost N}}{\text{val}}$$

for val = \$2.50/bu (which represents the value of the corn) and costN= \$0.25/lb (which represents the cost of the N fertilizer), this equation becomes

$$\frac{\text{cost N}}{\text{val}} = 0.31145833 - (0.00049479 \cdot 2)\,N^{(2-1)} - 0.00429687\,p = \frac{0.25}{2.50} = 0.10$$

which is rearranged into the form, 0.00098958 N + 0.00429687 p = 0.21145833, which is reorganized into

$$N = \frac{0.2114 - 0.00429\,p}{0.000989}$$

This equation is important because it defines the δy/δN at the economic optimum N rate when population is considered a constant. A graphical representation of this line is shown in **Box 23-6.**

The line based on the δy/δp partial derivative is described below. The equation relating seed (costp) and corn selling price (val) was

$$\frac{\partial y}{\partial p} = 13.8697917 - 0.00429687\,N - (2 \cdot 0.22135417)\,p^{(2-1)} = \frac{\text{cost p}}{\text{val}}$$

for cost p = \$1.00/1,000 seeds and val = \$2.5/bu equation becomes

$$13.8697917 - 0.00429687\,N - (0.44270833)\,p = \frac{1.00}{2.50}$$

which can be reorganized into the form,

$$p = \frac{13.47 - 0.00429\,N}{0.4427}$$

This equation is important because it defines the δy/δp at the economic optimum p rate when N is considered a constant. This line is also shown in **Box 23-6**.

**Box 23-6.** A graphical representation of the two partial derivative. The optimum N and population level is the point where the two lines intersect.

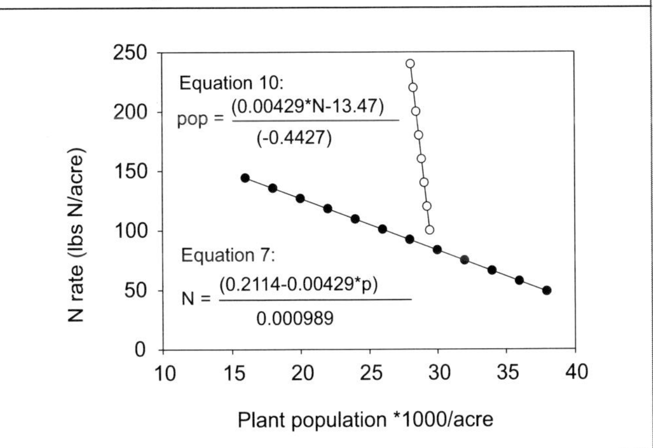


## Determining the Intersection of the Partial Derivative

To determine the intersection of the partial derivatives they must be solved simultaneously.

First, solve for N in $N = \frac{0.2114 - 0.00429\,p}{0.000989}$. This value is determined by adding -0.00429687•p to both sides of the equation and then dividing both sides of the equation by 0.00098958. The resulting equation is, N=(0.21145833-0.00429687p)/0.00098958.

Second, this value for N is then substuted into the N value in $p = \frac{13.47 - 0.00429\,N}{0.4427}$. The resulting equation is,0.00429687((0.21145833-0.00429687p)/0.00098958)+0.44270833p=13.46979167

Third, the optimum P value is determined by solving this expresion for P which is 29.6 or 29,600 seeds/acre.

Fourth, this value for P (29.6) is substituted into, N = (0.21145833-0.00429687p)/0.00098958 which is solved to determine the the optimum N rate. These steps are shown below,

N = (0.21145833-0.00429687p)/0.00098958

p = 29.6

N = (0.211458-0.0042969•29.6)/0.00098958

N = 85 lb N/acre

Using these original data, the equations (when solved for N and P when fertilizer cost is \$0.25/lb and seed cost \$1.00/1,000 and corn sells for \$2.50/bushel), the optimum seeding rate is 29,600 seeds/acre and the optimum N rate is 85 lb N/acre. By inspection, you can see that changing any or all of these values will result in slightly different optimum points. Other questions that can be asked are: 1) How fast will these values change? 2) Which value (seed, N, or corn value) is most important? Once you understand the concept behind this problem-solving exercise you can experiment with different values to see which changes fastest. Remember, the original data may or may not be appropriate for your area. Setting up experiments, collecting data, and analyzing the results will help you fine tune results for your area.

## Exercise 23-5. What are independent and dependent variables?

Solution:
Independent variables are those that represent the different treatments or are manipulated, whereas dependent variables are only measured.

## Exercise 23-6. What is a multiple regression equation? Why is it used?

Solution:
Multiple regression is a statistical approach where the coefficients in the polynomial equation are defined. Multiple regression equations are used to mathematically define relationships among variables.

## Exercise 23-7. Using the data provided in the chapter, calculate the optimum N and seeding rate if N costs were $0.30/lb N, seed costs $1.00/1,000 seeds, and corn value was $2.50/bu?

Solution:

Equation $N = \frac{0.2114 - 0.00429\,p}{0.000989}$ becomes,

$$\frac{\text{cost N}}{\text{val}} = 0.31145866 - (0.00099)\,N^{(2-1)} - 0.0043\,p = \frac{0.30}{2.50}$$

which simplifies to $0.000990\,N + 0.00430\,p = 0.19145$

Equation $p = \frac{13.47 - 0.00429\,N}{0.4427}$ remains

$$\frac{\text{cost p}}{\text{val}} = 13.47 - 0.00430\,N - (0.443)\,p^{(2-1)} = \frac{1.00}{2.50}$$

which simplifies to 0.004297 N + 0.443 p = 13.47

the equation, 0.000990•N + 0.00430•p = 0.19145 is solved for N resulting in the equation,

N = (0.19145 - 0.00429687•p)/.00098958

This value for N is substituted into the equation

0.00430 N + 0.443 p = 13.47

Resulting in the equation

0.00430 ((0.1915 -0 .00430 p) /0.000990) + 0.443 p = 13.47. Based on this equation, an optimum population level is determined. In this example, the optimum population is 29,800 plants/acre (29.8).

This value of the population is substituted back into the equation, N = (0.191 - 0.00430•29.8) / 0.000990,which when solved, results in an optimum N rate of 64.1 lb N/acre.

## Exercise 23-8. What would be the optimum N and seeding rate if N costs were $0.25/lb N, seed costs where $1.25/1,000 seeds, and corn value was $2.50/bu?

Solution:

$$\frac{\text{cost p}}{\text{val}} = 0.311 - (0.000990)\,N^{(2-1)} - 0.00430\,p = \frac{0.25}{2.50}$$

This equation simplifies to,

0.000990 N + 0.00430 p = 0.211

By solving for N, the equation,

N= (0.211 - 0.00430•p) /0.000990

was derived. This value is then substituted into the equation derived from the δy/δp equation.

The δy/δp equation was

$$\frac{\text{cost p}}{\text{val}} = 13.87 - 0.00430\,N - (0.443)\,p^{(2-1)} = \frac{1.25}{2.50}$$

which simplifies to

0.00430 N + 0.443 p = 13.37

Substituting the value for N derived above into this equation results in the equation,

0.00430 ( (0.2115 - 0.00430 p) / 0.000990) + 0.443 p = 13.37

The optimum seed population level is determined by solving this equation for p. In this example, p is 29.36 • 1,000 seeds/acre. This value is then substituted back into the equation derived from the δy/δN equation. When solved for N, the optimum N rate is 86.2 lb N/acre.

**Exercise 23-9. What would be the optimum N and seeding rates if N costs were \$0.35/lb N, seed costs \$1.25/1,000 seeds, and corn value was \$3.00/bu?**

**Exercise 23-10. Determine the multiple regression equation for model, yield = A + B • N + C • p + D • N² + E • p² + F • N • p, if the data set is:**

| Yield | N rate | Seeding |
|---|---|---|
| bu/acre | lb/acre | *1,000/acre |
| 99 | 40 | 16 |
| 140 | 40 | 24 |
| 142 | 40 | 32 |
| 115 | 100 | 16 |
| 145 | 100 | 24 |
| 150 | 100 | 32 |
| 120 | 140 | 16 |
| 150 | 140 | 24 |
| 151 | 140 | 32 |

Solution:
First, set up the data set. To do this problem using the multiple regression program in Excel, columns of numbers for the $N^2$, $p^2$, and N times p values must be developed.

Second, use the regression program under data analysis/tools to solve the problem. The resulting equation is:

Yield = -82.836 + 0.4399 N + 14.602 p – 0.0006944 $N^2$ - 0.24219 $p^2$ – 0.007566 Np

**Exercise 23-11. What are the two partial derivatives for the equation developed in 23-10 earlier?**

**Exercise 23-12. What are the resulting equations that must be solved simultaneously if N costs \$0.25/lb, seed costs \$1.00/1,000 seeds, and corn value was \$2.50/bu?**

Solution:

$$\frac{.25}{2.50} = 0.43991228 - (0.000694444 \cdot 2) \cdot N^{(2-1)} - 0.007565789 \cdot p$$

$$\frac{1.00}{2.50} = 14.60197368 - 0.007565789 \cdot N - (2 \cdot 0.2421875) \cdot p^{(2-1)}$$

The final answer after some substitution and rearrangement is the economically optimum plant population = 25.75 x 1,000 plants/acre and economically optimum N rate of 104.3 lb N/acre.

## Summary

In summary, an approach for solving problems where yield is influenced by two factors is shown. This solution involves the use of partial derivatives.

# USING CONCEPTUAL MODELS AND RELATIONAL DIAGRAMS TO DEVELOP MECHANISTIC MODELS FOR DETERMINING SOIL ORGANIC CARBON TURNOVER RATES

**Chapter concepts:** Carbon turnover, conceptual models, relational diagrams, mechanistic models.

**Mathematical skill:** Developing mathematical model from relational diagram.

**Chapter purpose:** To ensure the long-term sustainability of agricultural systems producing biomass for food, fiber, and energy, soil quality must be maintained. To determine accurate soil organic C maintenance requirements accurate estimates of soil organic carbon (SOC) turnover and mineralization rate are needed. This chapter shows how a conceptual model can be expanded to produce relational diagram, equilibrium relationships, and simple models for assessing the impact of management on soil carbon sustainability.

In a carbon constrained economy, the ability to accurately predict the impact of agricultural management, climate, and landscape variability on soil carbon turnover is needed. The impact of management on C turnover can be estimated if conversion factors (rate constants) from one pool to another pool are known. A number of different approaches have been used to calculate rate constants. Once the rate constants are defined, the impact of a management change can be calculated directly from the data set. Defining soil organic carbon (SOC) and non-harvested carbon (NHC) mineralization rate constants requires accurate measures of organic carbon inputs, outputs, and a clear mechanistic understanding of the C turnover processes. Obtaining good measures of above ground biomass is relatively easy and is typically accomplished by weighing the amount of biomass returned or estimating the value from the harvest index. However, obtaining accurate measures of below ground biomass is very difficult (Kuzyakov and Domanski, 2000; Amos and Walters, 2006). In the past, nearly all efforts have underestimated below ground biomass because they do a poor job at measuring small roots, root exudates, and below ground biomass derived $CO_2$. Quantifying non-structural root biomass is important because the amount of root derived respired carbon can be substantial (Kuzyakov, 2001), and root exudation may reduce the mineralization of other carbon sources. Efforts to measure root respiration and the impact of root exudates on soil respiration have relied on the measurement of $CO_2$ release in areas with and without plants. Kuzyakov and Domanski (2000) suggested that approximately ½ of the below-ground C is incorporated into root tissue, $^1/_3$ is respired by roots and rhizosphere microorganisms, with the remaining $^1/_6$ incorporated into soil and microorganisms.

Below ground biomass is typically estimated from the root to shoot ratio (Johnson et al., 2006; Bolinder et al., 2007). Extreme care must be used when using root to shoot ratios because: 1) scientists define root-to-shoot ratios differently (Johnson et al., 2006; Amos and Walters, 2006); 2) genotype and stress impact root production; (Amos and Walters, 2006; Johnson et al., 2006); and 3) the accuracy and precision of the reported values are influenced by the measurement method. Once the non-harvested C (NHC) inputs and temporal changes in SOC are known, this information can be used to calculate relic SOC and NHC mineralization rate constants (Clay et al., 2005). The mineralization rate constants can be used to compare results from different studies.

## Calculating Soil Organic Carbon Maintenance Requirements

Over the past 100 years, thousands of soil carbon studies have been conducted. Over this time period, a mechanistic approach for calculating the mineralization rate constants for relic soil organic carbon and non-harvested fresh biomass was not available, and therefore most studies did not calculate rate constants. Rate constants are needed for comparison and prediction.

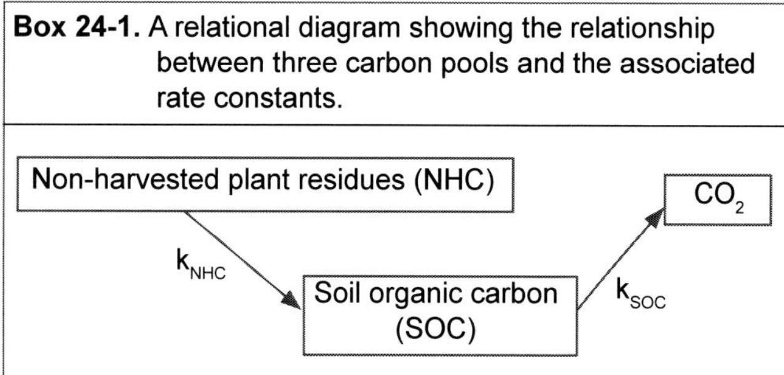


**Box 24-1.** A relational diagram showing the relationship between three carbon pools and the associated rate constants.

Clay et al. (2006) developed an approach to calculate these values. The approach is based on measured changes in soil organic carbon over time and measured amounts of non-harvested carbon returned to soil. This approach has been used to compare carbon turnover kinetics at sites using different cultural practices. Based on this relational diagram above, two equations,

$$\frac{dSOC}{dt} = k_{NHC}[NHC_a - NHC_m] \text{ and}$$

$$k_{SOC} \cdot SOC_C = k_{NHC} \cdot NHC_m$$

were defined. These equations say that at the equilibrium point, the change in carbon over time (dSOC/dt) is equal to the difference between the amount of carbon applied ($NHC_a$) and the maintenance requirement ($NHC_m$) multiplied times the proportion of non-harvested carbon converted to SOC ($k_{NHC}$), and that the carbon inputs must equal the carbon outputs. In these equations, dSOC/dt was the change in SOC with time, $SOC_e$ was the amount of SOC at equilibrium, and $NHC_m$ was the non-harvested C maintenance requirement. These equations are combined to produce the equation $\frac{NHC_a}{SOC_e} = \frac{k_{SOC}}{k_{NHC}} + \frac{dSOC}{dt}\left[\frac{1}{k_{NHC} \cdot SOC_e}\right]$.

This equation is solved by defining $\frac{NHC_a}{SOC_e}$ as y and $\left[\frac{dSOC}{dt}\right]$ as x. Based on this solution the,

SOC maintenance requirement ($NHC_m = b \cdot SOC_e$),

Non-harvested carbon conversion factor to SOC ($k_{NHC}$) = $k_{NHC} = 1/(m \cdot SOC_e)$, and

Relic carbon mineralization rate ($k_{SOC}$) $k_{SOC} = b/(m \cdot SOC_e)$

constants are calculated.

These rate constants can be used to define two equations.

$$\text{SOC equilibrium} = \frac{k_{NHC} NHC}{k_{SOC}}$$

$$SOC_{+} = SOC_{+-1} + k_{NHC}\, NHC - k_{SOC}\, SOC_{+-1}$$

where $SOC_{+}$ was SOC at year +.

These equations can be used to define the expected SOC content for a given combination of rate constants and NHC values, as well as calculate SOC at sometime in the future. The SOC equilibrium equation can be easily integrated into an Excel spread sheet.

**Exercise 24-1. Determine the non-harvested carbon rate constant for the conversion of NHC to SOC and relic carbon mineralization rate for the following data (modified from Larson et al., 1972). For these calculations the amount of soil organic carbon in the surface 15 cm at the beginning of the experiment was 26,750 kg C/ha.**

| | A | B | C |
|---|---|---|---|
| 1 | dSOC/dt | | NHC/SOC |
| 2 | -389.04 | | 0 |
| 3 | -277.35 | | 0.07306 |
| 4 | -175.35 | | 0.106315 |
| 5 | -106.63 | | 0.136492 |
| 6 | 120.56 | | 0.195823 |
| 7 | 599.46 | | 0.314508 |
| 8 | -242.51 | | 0.105714 |
| 9 | -66.536 | | 0.13439 |
| 10 | 53.737 | | 0.19475 |
| 11 | 598.86 | | 0.314093 |

Solution:

Determine the linear equation between NHC/SOC (y) and dsoc/dt (x)

1. Type in data
2. Select tools
3. Select data analysis
4. Select regression
5. Highlight C2 to C11 under y range
6. Highlight A2 to A11 under x range
7. Select OK

The following is an output of the resulting regression.

| SUMMARY OUTPUT | | | | | | | | |
|---|---|---|---|---|---|---|---|---|
| | | | | | | | | |
| Regression Statistics | | | | | | | | |
| Multiple R | 0.98104 | | | | | | | |
| R Square | 0.962439 | | | | | | | |
| Adjusted R Square | 0.957744 | | | | | | | |
| Standard Error | 0.020584 | | | | | | | |
| Observations | 10 | | | | | | | |
| | | | | | | | | |
| ANOVA | | | | | | | | |
| | *df* | *SS* | *MS* | *F* | *Significance F* | | | |
| Regression | 1 | 0.086853 | 0.086853 | 204.9862 | 5.53E-07 | | | |
| Residual | 8 | 0.00339 | 0.000424 | | | | | |
| Total | 9 | 0.090243 | | | | | | |
| | | | | | | | | |
| | Coefficients | Standard Error | t Stat | P-value | Lower 95% | Upper 95% | Lower 95.0% | Upper 95.0% |
| Intercept | 0.154229 | 0.006513 | 23.67912 | 1.08E-08 | 0.139209 | 0.169249 | 0.139209 | 0.169249 |
| X Variable 1 | 0.000285 | 1.99E-05 | 14.31734 | 5.53E-07 | 0.000239 | 0.000331 | 0.000239 | 0.000331 |
| | | | | | | | | |

Based on this equation, the y-intercept is 0.154 and the slope is 0.000285.

The rate constants and maintenance requirement are determined by substituting these values into the appropriate equation.

SOC maintenance requirement:

$(NHC_m = b \times SOC_e) = 26{,}750 \times 0.154 = 4{,}120$ kg C/(ha x year)

Non-harvested carbon conversion factor:

$\{k_{NHC}$ [= 1/ (m x $SOC_e$)]}= 1/(0.000285 x 26,750) = 0.131 g K/(kg NHC)

Relic carbon mineralization rate:

$\{k_{SOC}$ [= b/(m x $SOC_e$)]= 0.154/(0.000285 x 26,750) = 0.0202 kg C/kg C

These values mean that annually 2% of the relic carbon is mineralized, 13% of the non-harvested carbon is converted into SOC, and 4,120 kg C/(ha x year) must be returned annually.

**Exercise 24-2. Determine $\delta SOC/\delta t$ for the following data collected from an 11-year experiment.**

| MgC/ha (final) | MgC/ha(initial) |
|---|---|
| 22.4715 | 26.750895 |
| 23.7 | 26.750895 |
| 24.822 | 26.750895 |
| 25.578 | 26.750895 |
| 28.077 | 26.750895 |
| 33.345 | 26.750895 |
| 24.0833 | 26.750895 |
| 26.019 | 26.750895 |
| 27.342 | 26.750895 |
| 33.3384 | 26.750895 |

Solution:
Subtract SOCfinal from SOCinitial and divide the difference by 11 (the number of years).

| MgC/ha (f) | MgC/ha(i) | dSOC/dt |
|---|---|---|
| 22.4715 | 26.750895 | -389.04 |
| 23.7 | 26.750895 | -277.35 |
| 24.822 | 26.750895 | -175.35 |
| 25.578 | 26.750895 | -106.63 |
| 28.077 | 26.750895 | 120.56 |
| 33.345 | 26.750895 | 599.46 |
| 24.0833 | 26.750895 | -242.51 |
| 26.019 | 26.750895 | -66.536 |
| 27.342 | 26.750895 | 53.737 |
| 33.3384 | 26.750895 | 598.86 |

**Exercise 24-3. Calculate the amount of SOC contained in the surface 15 cm of soil from the following information.**

1. **The percentage of SOC at the end of the experiment is 1.42%.**
2. **The bulk density is 1.055 g/cm³.**

Solution:

$$15\text{ cm}\cdot\frac{1.42\text{ g C}}{100\text{ soil}}\cdot\frac{1.055\text{ g}}{\text{cm}^3}\cdot\frac{10,000\text{ cm}^2}{\text{m}^2}\cdot\frac{10,000\text{ m}^2}{\text{ha}}\cdot\frac{\text{kg}}{1,000\text{ g}}=\frac{22,471\text{ kg C}}{\text{ha}}$$

**Exercise 24-4. Calculate the amount of non-harvested biomass if the average yield is 4,442 kg grain/ha, harvest index is 49% (HI = grain/grain + stover), and the root to shoot ratio is 0.55.**

Solution:

$$\left(\frac{4,442}{0-0.49}\right)-4,442=4,623\text{ kg Non-harvested Biomass}$$

Total Above Ground Biomass = 4,442 + 4,623 = 9,065 kg

$$0.55=\frac{\text{roots}}{9,065\text{ kg Above Ground Biomass}}$$

$$\text{roots}=\frac{4,986\text{ kg Biomass}}{\text{ha}}$$

$$\text{Non-harvested Biomass}=4,986+4,623=\frac{9,609\text{ kg}}{\text{ha}}$$

**Exercise 24-5. Derive the equation used by Larson et al. (1972) to calculate non-harvested biomass maintenance requirements.**

Solution:
In Larson et al. (1972), (dSOC/dt) is plotted on the y-axis and non-harvested biomass is plotted on the x-axis. The maintenance requirement is defined as the non-harvested biomass at the point where dSOC/dt is zero. Soil organic turnover can be conceptualized by using a relational diagram (**Box 24-1**). In **Box 24-1** the non-harvested plant residues (NHC) is the annual additions of organic carbon to the system. A portion of this plant material is converted into soil organic carbon. The rate constants ($k_{NHC}$ and $k_{SOC}$) represent the rate that carbon is transferred from one pool to the next.

Based on this relational diagram, the equation $\frac{dSOC}{dt}=k_{NHC}[NHC_a-NHC_m]$ was defined.

This equation can be rearranged into the form,

$$\frac{dSOC}{dt}=k_{NHC}[NHC_a-NHC_m]$$

which was rearranged into the form,

$$\frac{dSOC}{dt}=k_{NHC}[NHC_a-NHC_m]=k_{NHC}NHC_a-k_{NHC}NHC_m.$$

This equation was converted to the linear form (y=mX + b),

*...continued*

by defining $\frac{dSOC}{dt}$ as y, $NHC_a$ as x, and $k_{NHC}$ as m. The m and b values are calculated by determining the zero order rate equation of the line where $\frac{dSOC}{dt}$ is plotted on the y axis and NHCa is plotted on the x axis. The y-intercept using this derivative is $k_{NHC} \cdot NHC_m$ and the slope is $k_{NHC}$. Using this derivative the relic carbon mineralization rate ($k_{SOC}$) constant can not be determined.

**Exercise 24-6. What is the dSOC/dt if soil organic carbon is 1,000 and 900 kg C/ha at years 0 and 5 during a study?**

**Exercise 24-7. An experiment is conducted to determine the SOC maintenance requirements. Determine the nonharvested carbon maintenance requirement for this 5 year study using the Larson et al. (1972) approach (asoc/dt = $k_{NHC}$ NHC a + b). The data from this experiment is as follows.**

| SOC | Initial | dSOC/dt | NHCa |
|---|---|---|---|
| 5 years | C (kg/ha) | y-term (kg/ha • yr) | x-term (kg/ha) |
| 9,500 | 10,000 | -100 | 1,000 |
| 10,000 | 10,000 | 0 | 2,000 |
| 10,500 | 10,000 | 100 | 3,000 |
| 11,000 | 10,000 | 200 | 4,000 |

Solution:

To determine the maintenance requirement, determine the zero order equation where dSOC/dt is plotted against annual carbon inputs (NHC). In this example, regression analysis available in Excel was used to determine this equation. For this analysis, select tools, data analysis, regression, highlight x and y ranges, and OK. The resulting analysis is

SUMMARY OUTPUT

| Regression Statistics | |
|---|---|
| Multiple R | 1 |
| R Square | 1 |
| Adjusted R Square | 1 |
| Standard Error | 7.07E-15 |
| Observations | 4 |

ANOVA

| | Df | SS | MS | F | Significance F |
|---|---|---|---|---|---|
| Regression | 1 | 50,000 | 50,000 | 9.99E+32 | 1E-33 |
| Residual | 2 | 1E-28 | 5E-29 | | |
| Total | 3 | 50,000 | | | |

| | Coefficients | Standard Error | t Stat | P-value | Lower 95% | Upper 95% | Lower 95.0% | Upper 95.0% |
|---|---|---|---|---|---|---|---|---|
| Intercept | -200 | 8.66E-15 | -2.3E+16 | 1.88E-33 | -200 | -200 | -200 | -200 |
| X Variable 1 | 0.1 | 3.16E-18 | 3.16E+16 | 1E-33 | 0.1 | 0.1 | 0.1 | 0.1 |

Based on this regression analysis the equation is dSOC/dt = -200 + 0.1 • $NHC_a$

$k_{NHC}$ = 0.10

$SOC_m$ = 200/0.10 = 2,000 kg C

**Exercise 24-8. What is the maintenance requirement for soils containing 30,000, 40,000, and 50,000 kg C/ha if NHC/SOC = 0.000284(dSOC/dt) + 0.13?**

**Exercise 24-9. What is the maintenance requirement and $k_{NHC}$, and $k_{SOC}$ values for a soil containing 35,000 kg C/ha if NHC/SOC = 0.000284(dSOC/dt) + 0.13?**

Solution:
Maintenance requirement = 0.13 • 35,000 = 4,550
$k_{NHC}$ = (1/SOC • 0.000284) = 0.10 g NHC-C/yr • g - SOC
$k_{SOC}$ = [0.13/(SOC • 0.000284)] = 0.013 g SOC-C/yr • g - SOC

**Exercise 24-10. What is the maintenance requirement, $k_{SOC}$, and $k_{NHC}$ values if NHC/SOC = 0.09 + 6.67 • $10^{-5}$ (dSOC/dt) when SOC is 50,000 lb C/a?**

**Exercise 24-11. If $k_{NHC}$ = 0.2998, and $k_{SOC}$ = 0.0269, what does this mean?**

**Exercise 24-12. Using the approach described in Exercise 24-1, what is the maintenance requirement for data from Barber et al. (1976)? Data for the 0 to 15 cm soil depth are summarized below. These numbers are based on a 50% harvest index and a 0.55 root to shoot ratio (total root C/total above ground biomass). This soil contains 34,900 kg C/ha.**

| | years 1-6 | | years 6-12 | |
|---|---|---|---|---|
| Treatment | dSOC/dt | NHC/SOC | dSOC/dt | NHC/SOC |
| Removed | -388.081 | 0.0960 | -116.424 | 0.2051 |
| Returned | 0 | 0.1866 | 116.4244 | 0.1965 |
| 2x | 776.1628 | 0.2685 | -93.1395 | 0.1695 |
| Fallow | -776.163 | 0.0000 | 279.4186 | 0.2353 |

**Exercise 24-13. If $k_{SOC}$ and $k_{NHC}$ are 0.016 and 0.15, and NHC is 3,000 kg C/ha, what is the SOC level at the equilibrium level?**

Solution:
As discussed above,

$$\frac{dSOC}{dt} = k_{NHC}[NHC_a - NHC_m]$$

$$k_{SOC} \cdot SOC_e = k_{NHC} \cdot NHC_m$$

If $NHC_m$ is defined as 3,000 kg C/ha then SOC can is equal to

$$SOC_e = \frac{3,000\ kg\ C}{C} \cdot \frac{0.15\ g\ C}{1\ g\ C} \cdot \frac{1\ g\ C}{0.016\ g\ C} =$$

$$\frac{28,125\ kg}{ha} SOC_e = 3,000\ kg\ C/ha$$

**Exercise 24-14. If $k_{NHC}$ is 0.20, NHC is 4,000 kg C/ha, and $SOC_e$ is 50,000 kg SOC/ha, what is $k_{SOC}$?**

**Exercise 24-15. Over the past 20 years, what are the $k_{NHC}$, and $k_{SOC}$ values for a field in continuous corn if SOC in the surface 30 cm is 40,000 kg C/ha, grain removed is 6,000 kg/ha, stover + cobs returned to the surface 30 cm are 6,000 kg/ha, and roots returned to the surface 30 cm are 6,600 kg C/ha? Soybeans are planted in the field for 3 years. During this period of time, SOC in the surface 30 cm decreases to 39,400 and NHC is 4,500 kg/ha.**

Solution:

We solve this problem by making the assumption that after 20 years SOC is approaching equilibrium and therefore $\frac{dSOC}{dt}$ and $NHC_m = NHC_a$. Based on these assumptions,

$$k_{SOC} \cdot SOC_e = k_{NHC} \cdot NHC_m$$

$$k_{SOC} \cdot 40,000 = k_{NHC} \cdot 6000$$

$$k_{SOC} = \frac{k_{NHC} \cdot 6,000 \text{ kg C/ha}}{40,000 \text{ kg C/ha}}$$

This equation can not be solved because it contains two unknowns.

In years 21-24

$$\frac{dSOC}{dt} = k_{NHC}[NHC_a - NHC_m]$$

$$\frac{(40,000 - 39,400 \text{ kg C/ha})}{3 \text{ years}} = k_{NHC}(6,000 - 4,500 \text{ kg C/ha})$$

$$\frac{200 \text{ kg C/yr}}{1,500 \text{ kg C/ha}} = \frac{0.133 \text{ g C}}{\text{g C}}$$

Insert into above equation

$$k_{SOC} = \frac{k_{NHC} \cdot 6,000 \text{ kg C/ha}}{40,000 \text{ kg C/ha}}$$

$$k_{SOC} = \frac{0.133 \text{ gC/g} \cdot 6,000 \text{ kg C/ha}}{40,000 \text{ kg C/ha}} = 0.01995 \text{ g C/g C}$$

This discussion shows how the rate constants can be estimated relatively quickly following a management change. Once the rate constants are known, many other factors can be calculated. See above.

**Exercise 24-16. What are the assumptions associated with Exercise 24-15?**

Solution:

1. That the conversion rates are constant
2. Above and below ground biomass have similar mineralization kinetics
3. The amount of below ground biomass is known

**Exercise 24-17. The long-term average for the amount of above ground biomass returned to soil at two sites located in a field are 9,000 and 6,000 kg C (ha/year). What is the potential to sequester additional C in the soils if they contain 60,000 and 50,000 kg C/ha? If it is assumed that 50% of the roots are contained in the sampling zone, the harvest index is 50%, the root-to-shoot ratio is 0.55, and biomass contains 40% carbon, then NHC for site one is 7,560 kg C/ha (NHC = 9 • 0.4 + 18 • 0.55 • 0.4) and NHC for site 2 is 5,040 kg C/ha (NHC = 6 • 0.4 + 12 • 0.55 • 0.4).**

Solution:

If the rate constants are known ($k_{SOC}$ =0.02 and $k_{NHC}$ = 0.20) then the equation,

$$k_{SOC} \cdot SOC_e = k_{NHC} \cdot NHC_m$$

can be used to solve for the SOC content at the equilibrium point.

In this example,

Site1: $SOC_e$ is 75,600 kg C/ha [(7,560 • (0.2/0.02)] and

Site 2: $SOC_e$ is 50,400 kg C/ha [5,040 • (0.2/0.020] kg C/ha.

If these sites contain 60,000 and 50,000 kg C/ha, then the potential to sequester additional C is 15,600 kg C/ha (75,600 – 60,000) at site 1 and 400 kg C/ha (50,400 – 50,000) at Site 2.

**Exercise 24-18. How long will it take to reach the equilibrium points if $SOC_i$ is 20,000 kg/ha and $k_{SOC}$ is 0.015 and $k_{NHC}$ is 0.2, and NHC is 7,000 kg C/ha?**

Solution:

SOC =($k_{NHC}$/$k_{SOC}$) •NHC = 93,000 kg SOC-C/ha

Using the Excel file below, it was determined that it would take 267 years to reach 93,000 kg SOC-C/ha. This program is based on the model $SOC_{+} = SOC_{+-1} + k_{NHC}NHC - k_{SOC} SOC_{+-1}$.This was accomplished by running the simulation through 500 years. The simulation below only is conducted through year 5. To make the calculation work, type code in D3, D4, and D5 into cells B3, B4, and B5. To make the calculation work, copy A6 and B6 down.

| | A | B | C | D |
|---|---|---|---|---|
| 1 | year | SOC | NHC | $SOC_{+}$ |
| 2 | 1 | $SOC_{+-1}$ | 7000 | =B2-B2•0.015 + C2 • 0.20 |
| 3 | 2 | $SOC_{+}$ | 7000 | =B3-B3•0.015 + C2 • 0.20 |
| 4 | 3 | 22183.5 | 7000 | =B4-B4•0.015 + C2 • 0.20 |
| 5 | 4 | 23250.75 | 7000 | =B5-B5•0.015 + C2 • 0.20 |
| 6 | 5 | 24301.99 | 7000 | =B6-B6•0.015 + C2 • 0.20 |

$k_{SOC}$ = 0.015
$k_{NHC}$ = 0.2

**Exercise 24-19. What is the potential for a soil to sequester carbon if $k_{soc}$ is 0.015g/(g SOC •year), $k_{NHC}$ is 0.16 g/(g NHC• year), and NHC is 6,000 lb C/(acre• year)?**

Solution:

SOC sequestering potential = $\left(\frac{k_{NHC}}{k_{SOC}}\right)(NHC) = \frac{64{,}000 \text{ lb SOC}}{\text{acre}}$

**Exercise 24-20. What is the SOC half life if $k_{SOC}$ is 0.015g/(g SOC • year)?**

Solution:

Half life = ln 2/k = 0.693/0.015 = 46.2 years

## Additional Information


Amos B. and D.T. Walters. 2006. Maize root biomass and net rhizodeposited carbon: An analysis of the literature. Soil Sci. Soc. Am. J. 70, 1489-1503.

Barber, S.A. and J.K. Martin. 1976. The release of organic substance by cereal roots in the soil. New Phytol 76:69-80.

Bolinder, M.A., H.H. Janzen, E.G. Gregorich, D.A. Angers, and A.J. Vanden Bygaart. 2007. An approach for estimating net primary productivity and annual carbon inputs to soil for common agricultural crops in Canada. Agri. Ecosystem. Environ. 118, 29-42.

Clay D.E., C.G. Carlson, S.A. Clay, J. Chang, and D.D. Malo, 2005. Soil organic C maintenance in a corn (*Zea mays* L.) and soybean (*Glycine max* L.) as influenced by elevation zone. Journal Soil Water Conser. 60, 342-348.

Clay, D.E., C.G. Carlson, S.A. Clay, C. Reese, Z. Liu, and M.M. Ellsbury. 2006. Theoretical derivation of new stable and non-isotopic approaches for assessing soil organic C turnover. Agron. J. 98:443-450.

Larson, W.E., C.E. Clapp, W.H. Pierre, and Y.B. Morachan. 1972. Effect of increasing amounts of organic residues on continuous corn: Organic carbon, nitrogen, phosphorous, and sulfur. Agron. J. 64:204-208.

Johnson, J.M.F., R.R. Allmaras, and D.C. Reicosky. 2006. Estimating source carbon from crop residues, roots, and rhizodeposits using the national grain-yield data-base. Agron. J. 98:622-636.

Kuzyakov, Y.V. 2001. Tracer studies of carbon translocation by plants from the atmosphere into the soil. Eurasian Soil Sci. 34, 28-42.

Kuzyakov Y. and G. Domanski. 2000. Carbon inputs by plants into the soil. Review. J. Plant Nutr. Soil Sci. 163, 421-431.

# CALCULATING PARTITION COEFFICIENTS AND SORPTION ISOTHERMS

**Key problems:** The distribution of a chemical at equilibrium in a liquid vs solid phase of a slurry by calculating the partition coefficient ($K_d$) or Freudlich isotherm ($K_f$).

**Chapter concepts:** Retention and availability characteristics of nutrients and pesticides influence plant uptake and degradation kinetics, which significantly influences their environmental fate. Calculating $K_d$ and $K_f$ values and interpreting data from simple experiments will help managers understand fertilizer and pesticide retention and give insight into availability and leaching potential.

Pesticides, nutrients, and growth regulating chemicals are applied to crops to control pests and increase yield potential. The environmental fate of agrichemicals is influenced by the reaction and retention with the soil particles and organic matter, which in turn can influence their ability to leach or decompose. Agrichemical molecules can be attracted to and attach or adhere to the surface of the soil particle. Adsorption is defined as the enrichment of a chemical species in the region of the solid interface. In addition, some of the molecules may move into the matrix (like water into a sponge) and this process is called absorption. Often, experiments cannot distinguish between adsorption and absorption, so the general term sorption is used and refers to the simultaneous processes of adsorption and absorption. The result of either type of process is the removal of ions and molecules from solution.

The distribution of chemical between the water and soil gives an indication of chemical availability. As sorption increases, the concentration of the sorbed species is greater at the interface than in the bulk fluid (water that contains soluble soil materials) surrounding the particle. The chemical that is retained by the soil is less available to plants so there may be less uptake and less available to soil organisms, increasing the residence time in the soil and slowing degradation, and the chemical has lower potential to leach from the application area.

Sorption is controlled by properties of the chemical of interest, including the water solubility, pH, $pK_a$, octanol/water partition coefficient, and other factors (Weber, 1995). The sorption of the chemical also is affected by soil properties, including water, organic matter, clay, sand, and oxide contents, and soil pH. Soils high in sand generally sorb much less chemicals than loamy or clay type soils. Sorption generally increases as organic matter and clay contents increase, with organic matter generally sorbing 10 times greater chemical amounts than clay particles. Agricultural practices that involve modifying soil organic matter content will increase chemical retention by the soil.

Measuring the amount of chemical retained, or sorbed, by the soil helps to understand its potential fate in the environment and its availability to growing organisms. The distribution of chemicals between the solution and solid phase at equilibrium is expressed in terms of the partition coefficient or $K_d$. The $K_d$ is equal to the ratio between the amount of compound held in the soil per unit of soil and the concentration of the chemical in the solution after equilibrium.

Numerous studies have used the partition coefficient to assess the potential for a variety of compounds (N, P, K, radioactive nucleotides, pesticides, and pharmaceuticals) to leach through soil. Partition coefficients are generally measured using a 'batch or slurry' technique. In these experiments, solution containing a known amount of chemical is added to a fixed amount of soil. After an equilibration period, some of the solution is removed from the soil and the chemical content is determined. The amount sorbed is calculated as the difference between the amount added in solution and the amount in solution after equilibration. Based on the measured solution values, the $K_d$ value is determined by dividing concentration in the solid phase by the concentration in the solution phase.

The $K_d$ values can range from 0 (compound not sorbed to soil) to very large positive values and is dependent on chemical and soil characteristics. Chemicals with values <5 are considered mobile and have a high potential for leaching, whereas compounds with very high (>100) values are strongly retained by soil and typically have low leaching potential.

**Exercise 25-1. Based on the following information, determine the $K_d$ value for atrazine. Ten mL of solution that contains a total of 10 mg of atrazine (1 mg of atrazine/mL of solution) is added to 15 g of sandy soil (62% sand, 12% clay content, and the organic carbon content is 1%). After shaking for 24 hr to reach equilibrium, the soil slurry is centrifuged and 3 mL of solution is analyzed for atrazine. The measured amount of atrazine from the solution is 0.7 mg of atrazine/1 mL of solution. Determine the $K_d$ of atrazine for this soil.**

Solution:

First, determine the amount of atrazine sorbed to soil. Starting with 1 mg/mL atrazine in solution and ending with 0.7 mg/mL, the amount adsorbed from each mL of solution is

1mg/mL-0.7mg/mL = 0.3 mg/mL.

Next, calculate the total amount of atrazine sorbed:

10 mL of solution was added to the soil and from each 1 mL 0.3 mg was sorbed, therefore:

10 mL • 0.3 mg/mL = 3 mg total sorbed. This is the total amount of atrazine sorbed from solution onto the 15 g of soil. These steps are summarized in the equation,

$$\text{amount of sorbed atrazine} = (1.0\text{ mg} - 0.7\text{ mg})(10\text{ mL}) = 3\text{ mg}$$

To calculate the $K_d$:

$$K_d = \frac{\text{amount of sorbed chemical/unit of soil}}{\text{Concentration after equilibrium}} = \frac{(3\text{ mg atrazine/15 g soil})}{(0.7\text{ mg atrizine/1 mL solution})}$$

$$K_d = \frac{0.2\text{ mg atrazine/g soil}}{0.7\text{ mg atrizine/mL solution}} = 0.285$$

A $K_d$ value of 0.285 suggests that for every 1 mg of atrazine measured in solution, there is 0.28 mg atrazine sorbed to the soil.

**Exercise 25-2. Determine the $K_d$ value for a clay loam soil (35% sand and 35% clay content, and the organic C content is 3%). Ten mL of solution that contains a total of 10 mg of atrazine (1 mg of atrazine/mL of solution) is added to 15 g of soil. After 24 hr of equilibration, the soil slurry is centrifuged and 3 mL of solution is analyzed for atrazine. The measured amount of atrazine from the solution is 0.2 mg of atrazine/1 mL of solution.**

**a) Calculate the $K_d$ value for the clay loam soil**

**b) Compare the $K_d$ value for this clay loam soil with the $K_d$ value for the sandy soil in Exercise 25-1.**

**c) What components of the soil tend to increase sorption?**

## Organic Matter Impact on the Partition Coefficient

For many compounds, the partition coefficient is assumed to be most impacted by the soil organic matter content. The partition coefficient can be modified to compare the amount sorbed based on the organic carbon using the equation,

$$K_{oc} = \frac{K_d}{\text{Fraction organic carbon}} \text{ or } \frac{K_d \cdot 100}{\text{\% organic carbon}}$$

Calculated $K_{oc}$ values are used to normalize the sorption of chemical on soils with different organic matter contents.

**Exercise 25-3. Determine the $K_{oc}$ value for the soil in Exercise 25-1.**

Solution:

$$K_{oc} = \frac{0.28}{0.01} = 28 \text{ or } \frac{0.28 \cdot 100}{1} = 28$$

**Exercise 25-4. Determine the $K_{oc}$ value for the soil in Exercise 25-2.**

**Exercise 25-5. Determine the $K_d$ and $K_{oc}$ values if 10 mL of solution containing a total of 20 mg of chemical is added to 10 g of soil containing 2% organic carbon. After equilibration, total chemical left in solution is calculated to be 15 mg.**

Solution:

Total amount added = 20 mg

Total amount remaining in solution = 15 mg

Amount remaining/mL of solution = 15 mg/10 mL = 1.5 mg/mL

Total retained on soil = 5 mg/10 g soil = 0.5 mg/g

$$K_d = \frac{0.5 \text{ mg/g}}{1.5 \text{ mg/mL}} = 0.33 \text{ mL/g}$$

$$K_{oc} = \frac{0.33}{0.02} = 16.7 \text{ mL/g}$$

The $K_d$ indicates that for each mg of chemical measured in solution, 1.33 mg is retained by soil. This value is fairly low and the chemical would most likely be plant available and has high leaching potential.

**Exercise 25-6. Determine the $K_d$ and $K_{oc}$ value if 10 mL of solution containing 100 mg of chemical/mL of solution is added to 5 g of soil containing 10% C. After equilibration, 1 mg/mL of chemical remains in solution.**

**Exercise 25-7. Determine the $K_d$ and $K_{oc}$ value if 15 mL of solution containing 100 ppm of chemical are added to 7 g of soil containing 5% organic C. After equilibration, measure 80 ppm of chemical in solution.**

## Chemical Dependency on Partition Coefficients Using the Freundlich Equation

Chemical sorption is often dependent on the amount of chemical added to the soil. Therefore, instead of calculating the partition coefficient from just a single concentration of chemical added to soil, a range of concentrations are tested. In many situations, the relationship between the amount of chemical sorbed vs the amount of chemical applied is not linear. Non-linearity is attributed to decreasing sorption of the substance with increasing concentration. This may be due to decreased numbers of free sorption sites with increasing concentrations of chemical.

Several different models can be used to discribe sorption. The simplest is the linear relationship between sorption and the chemical concentration. The equation is

$$x/m = a+b(c),$$

Many sorbates exhibit linear isotherms at low concentrations.

A Langmuir isotherm can be used if sorption increases to a maximum value. The equation,

$$\frac{x}{m} = \frac{b \cdot K_L C}{1 + K_L C}$$

is used for this isotherm. In soils, sorption often increases with concentration, with decreasing slope. A maximum value may not be reached.

The Freundlich equation is used to describe this sorption characteristic. The equation is

$$x/m = K_f C^{1/n},$$

where x/m is the weight of the adsorbate per unit weight of the absorbent, C is the absorbate concentration in solution, and two empirical constants are used to fit the equation, $K_f$ (Freundlich value), and n. This equation is linearized by taking the log of both sides of the equation. The resulting equation is,

$$\log (x/m) = (1/n)(\log C) + \log K_f$$

The Freundlich value ($K_f$) provides an indication of the average sorption capacity of the soil. The constant '1/n' is the slope of the line. A slope = 1 indicates that sorption over the range of chemical concentrations used in the experiment is not affected by the concentration of the chemical in solution. A slope < 1 indicates that there is proportionally more chemical sorbed when the concentration is low and less chemical sorbed when the concentration is high.

## Exercise 25-8. Determine the $K_f$ value for the following data.

Solution:

A series of concentrations of a chemical is used for soil batch equilibration studies. The amount of soil used in each is 3 g, the amount of solution is 5 mL, and there are 4 initial chemical concentrations. After equilibration, the results are:

| Initial solution concentration | Final solution concentration |
|---|---|
| 1 μg/mL | 0.3 μg/mL |
| 5 μg/mL | 3.2 μg/mL |
| 10 μg/mL | 5.8 μg/mL |
| 50 μg/mL | 26 μg/mL |

First, calculate the amount sorbed from solution.

Ccn 1 = (1 μg/mL – 0.3 μg/mL ) • 5 mL = 3.5 μg

Ccn 2 = (5 μg/mL – 3.2 μg/mL) • 5 mL = 9 μg

Ccn 3 = (10 μg/mL – 5.8 μg/mL) • 5 mL = 21 μg

Ccn 4 = (50 μg/mL – 26 μg/mL) • 5 mL = 120 μg

Next, calculate the amount of chemical sorbed per g of soil. In this case, 3 g of soil were used, therefore,

Ccn 1 = 3.5 μg/3g = 1.17 μg/g

Ccn 2 = 9 μg/3 g = 3 μg/g

Ccn 3 = 21 μg/3 g = 7 μg/g

Ccn 4 = 120 μg/ 3 g = 40 μg/g

Next, take the log of the solution and soil concentrations.

| | Solution | log solution | soil | log soil |
|---|---|---|---|---|
| Ccn 1 | 0.3 | -0.52288 | 1.17 | 0.068186 |
| Ccn 2 | 3.2 | 0.50515 | 3.0 | 0.477121 |
| Ccn 3 | 5.8 | 0.763428 | 7.0 | 0.845098 |
| Ccn 4 | 26 | 1.414973 | 40.0 | 1.60206 |

Using the data analysis function in Excel, select regression, the y-data is the log soil data and the x-data is the log solution data. Place the results in an appropriate data cell. For the above exercise, the results are:

SUMMARY OUTPUT

| Regression Statistics | |
|---|---|
| Multiple R | 0.953031 |
| R Square | 0.908267 |
| Adjusted R Square | 0.862401 |
| Standard Error | 0.241766 |
| Observations | 4 |

ANOVA

| | df | SS | MS | F | Significance F |
|---|---|---|---|---|---|
| Regression | 1 | 1.157468 | 1.157468 | 19.80246 | 0.046969 |
| Residual | 2 | 0.116901 | 0.058451 | | |
| Total | 3 | 1.274369 | | | |

| | Coefficients | Standard Error | t Stat | P-value | Lower 95% | Upper 95% |
|---|---|---|---|---|---|---|
| Intercept | 0.331569 | 0.152888 | 2.168704 | 0.162361 | -0.32626 | 0.989394 |
| X Variable 1 | 0.771144 | 0.173291 | 4.449996 | 0.046969 | 0.025533 | 1.516755 |

The log $K_f$ is the 'Intercept' and the 'X variable 1' is the slope (1/n) of the line. The 1/n value is 0.77, which is the slope of the line. The value is not very close to 1, so the interpretation is that at lower concentrations there is proportionally more chemical sorbed by soil than at higher concentrations. The adjusted $R^2$ value is 0.86, which is a reasonable but not excellent fit of the data to the line and indicates that 86% of the amount sorbed to soil is described by the amount found in solution. The regression F and significance of F is 19.8 and 0.047, respectively. These values indicate that the model is significant at $p = 0.05$.

The plot of the data (solid line with symbols for data points) and the regression line (dotted line) are shown in **Box 25.1**. Based on these data, the most variation in the data set occurs at the second concentration, where more chemical than expected is in solution and not sorbed to soil. In addition, the regression equation slightly underestimates the amounts sorbed at both the upper and lower limits of the equation. When using these data to interpret other data from the same soil, it is important to remember that only solution concentrations that are similar to those that fall within the bounds of the experiment can be extrapolated with some confidence. If solution concentrations are either much lower than the lowest concentration or much higher than the upper concentration, another set of experimental concentrations that fall within the ranges to be extrapolated to should be examined for greater confidence with the experimental data.

**Box 25-1. Graphical representation of the sorption study (Exercise 25-8) employing the linearized Freundlich isotherm.**

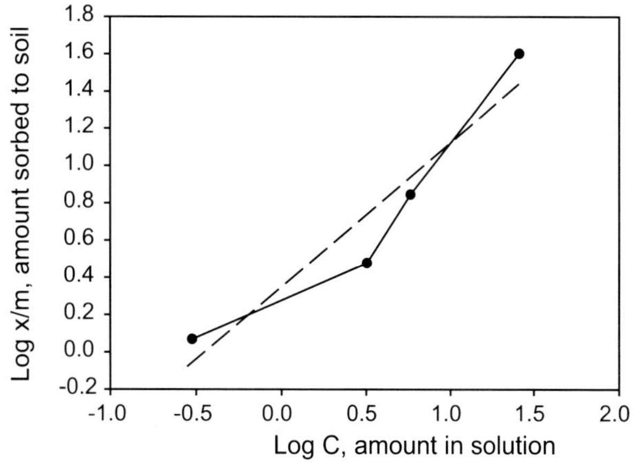

The antilog of the intercept is the $K_f$ value for the chemical. In Excel, the antilog can be calculated by typing in '=10^0.33156' which is 2.14, the $K_f$ value. The upper and lower 95% confidence limits for the log $K_f$ value are on the Excel printout and in this example are -0.326 and 0.989. The antilogs of these values are calculated to give the lower and upper confidence limits for the sorption coefficient and are 0.47 (= 10^-0.326) and 9.75 (=10^0.989). The interpretation is that for each 1 µg/ml measured in solution, 2.14 µg/g on average are sorbed to soil with a confidence range of 0.47 to 9.75 µg/g. Several replications of each concentration and/or several more concentrations may be used to decrease the range of the confidence interval.

A variety of units have been used in the determination of concentrations in sorption studies. The international scientific (SI) units used are mols of chemicals instead of g, i.e. µmol/L for solution C values and µmol/kg for sorption amounts. If these units are used the molecular weight of the compound is needed for conversion.

### Exercise 25-9. Determine the $K_f$ and 1/n values for the following batch experiment.

A batch sorption experiment is conducted with three initial concentrations of chemical that has a molecular weight of 560 µg/µmol. The amount of soil was 15 g, the amount of solution was 15 ml, and the initial chemical concentrations in solution were 50, 75, and 100 µg/ml of solution. The final amounts of chemical in solution were measured at 28, 40, and 54 µg/ml. Convert the chemical units to µmol/L and µmol/kg.

## Additional Information


Weber, J. B. 1995. Physiochemical and mobility studies with pesticides. Pg. 99-116. *In* Leng, M.L., E.M.K. Loevey, and P.L. Zubkoff. Agrochemical environmental fate. State of the art. CRC Press. Boca Raton, FL.

# APPENDIX 1
# BASIC AGRONOMIC INFORMATION



Key problems: Quantifying field values is the first step in solving problems. Commonly used techniques for staging crop growth, estimating field pest population, and calibrating sprayers are discussed.

Chapter purpose: This section provides guidance on how to quantify field numbers. Included here are discussions of plant growth stages, quantifying insect defoliation, and soil sampling.

## Growth Stages

Natural resource management requires the proper identification of growth stage. Different plants use different approaches to define the growth. The ability to correctly determine the stage of the crop is important for communicating crop status to other agronomic professionals and proper weed, insect, disease, and fertility management. Recordkeeping that includes specific dates and growth stages in relation to planting can be used from year to year for many comparisons. Each crop has a different approach for determining growth stage.

## Corn Growth Stages

Historically and in the literature, many authors have developed methods for determining the stage of a corn plant's development. An early method is/was the "corn leaf method", commonly used by hail adjusters. This method counts the number of leaves, starting from the bottom leaf and counting up to the last leaf that is more than 50% emerged. Determining 50% emerged is subjective and is usually taken to be the leaf that has emerged enough so that its tip is starting to point down. Using this method, we do not count leaves younger than this one, even though we can easily see them in the whorl.

The "corn leaf collar method" is a method that is used more often today. The leaf collar method relates well to the physiological stages of vegetative, "V", plant growth. In this method, count the number of leaves with visible collars. The collar is the part of the leaf that joins the leaf blade and leaf sheath. Collars are not visible until the leaves are developed enough to emerge from the whorl. In this publication, we will use the collar method for staging vegetative growth. Reproductive stages begin when silks begin to emerge from the ear, where these stages are noted with the letter "R". Short descriptions of each growth stage and are provided in **Box A1.1** along with a diagram of growth stage progress during normal crop growth and development (Grey, 1999). Further information is available from Ritchie et al. (1992).

| **Box A1-1.** Corn growth stages. | |
|---|---|
| Diagnostic feature | Growth stage |
| Seed planted | V0 |
| Emergence, coleoptile above the soil | VE |
| Collars for coleoptile and one additional leaf exposed | V2 |
| Number of collars exposed | V2-V14 |
| The tassel emerges | VT |
| The silks on the ear emerge | R1 |
| Brown silk, cob full sized, blister | R2 |
| Kernels are at milk stage | R3 |
| Kernels are at soft dough | R4 |
| Few kernels are dented | R5 |
| Black layer | R6 |

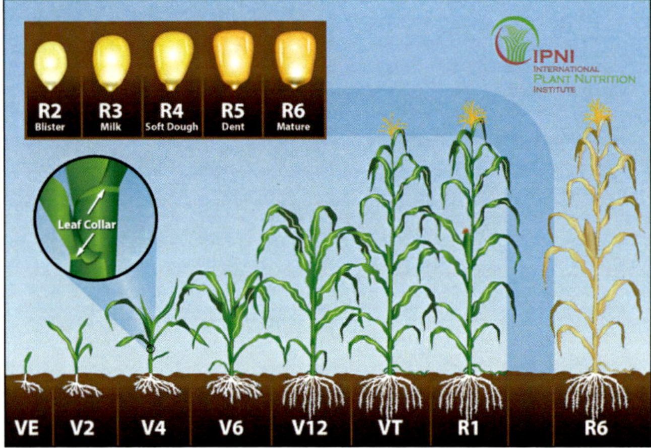


| **Box A1-2.** Soybean growth stages. | |
|---|---|
| Diagnostic feature | Growth stage |
| Cotyledons above the soil surface | VE |
| Unifoliolate leaves unrolled | VC |
| Fully developed trifoliate leaf | V1 |
| Number of fully developed trifoliates (n) | V(n) |
| One flower open at any node | R1 |
| Open flower at one of the two upper nodes (full bloom) | R2 |
| Pod at one of the four uppermost nodes | R3 |
| Pod ($\frac{3}{4}$ in.) at one of four uppermost nodes (full) | R4 |
| Seed in a pod at one of four uppermost nodes | R5 |
| Pod containing a green seed that fills the cavity | R6 |
| One normal pod on main stem has the color of a mature pod | R7 |
| 95% of the pods have reached mature pod color | R8 |
| From http://www.planthealth.info/diag_soygrowth.htm | |

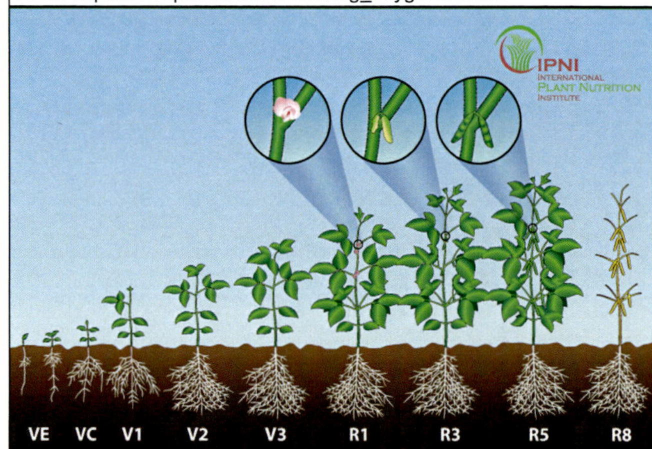


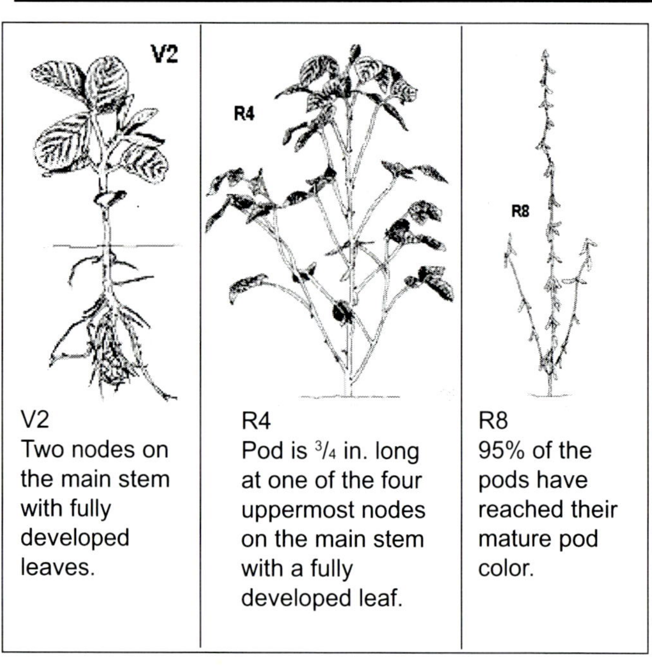


V2
Two nodes on the main stem with fully developed leaves.

R4
Pod is $\frac{3}{4}$ in. long at one of the four uppermost nodes on the main stem with a fully developed leaf.

R8
95% of the pods have reached their mature pod color.

## Soybean Growth Stages

Soybean varieties can be separated into two (2) general groups: determinate and indeterminate. Determinant varieties will not begin to flower until the plant has fully developed vegetatively, while indeterminant varieties begin to flower as the plant continues to develop vegetative.

There are thirteen (13) maturity group ratings ranging from “000”, “00”, “0”, and I (one) through X (ten). Varieties with lower numbers 000 to IV are considered indeterminate varieties adapted to northern climate regions while group V through X are determinant varieties adapted to southern climate regions. Vegetative growth is described by the number of nodes on the plant at which petioles and leaves develop from the stem. Diagrams for identifying soybean growth stages are available at Grey (1999). Detailed information for growth and development of soybean plants is available from Petersen (2004).

## Wheat Growth Stages

Several small grain growth and development scales are used in research and production agriculture. The most common are the Feekes and Zadoks scales. The Zadoks scale is more definitive and begins with dry seed, where the Feekes scale is more general and begins with emergence. The Feekes scale is the most common system used in the upper Great Plains. Diagrams for wheat growth stages are available at this website: >http://nue.okstate.edu/GSchart.htm<.

**Box A1-3.** Wheat growth stages; Feekes compared to Zadoks.

| Diagnostic feature | | Feekes Stage | Zadoks Stage |
|---|---|---|---|
| One shoot | Tillering | 1 | 1 |
| Tillering begins | | 2 | 21 |
| Tillers reformed | | 3 | 26 |
| Leaf sheaths lengthen | | 4 | 30 |
| Sheath erect | | 5 | 30 |
| First stem node visible | stem | 6 | 31 |
| 2nd stem node visible | elongation | 7 | 32 |
| Last leaf just visible | | 8 | 37 |
| Ligule of last leaf visible | | 9 | 39 |
| Ear swollen but not visible (boot) | | 10 | 45 |
| First ear just visible | heading | 10.1 | 50 |
| 1/4 heading process complete | | 10.2 | 53 |
| 1/2 heading process complete | | 10.3 | 55 |
| 3/4 heading process complete | | 10.4 | 57 |
| Heading complete | | 10.5 | 59 |
| Begin flowering | | 10.5.1 | 60 |
| Flowering complete | | 10.5.4 | 71 |
| Milky ripe | ripening | 11.1 | 75 |
| Mealy ripe-kernel soft | | 11.2 | 85 |
| Kernel hard | | 11.3 | 91 |
| Straw dead | | 11.4 | 92 |

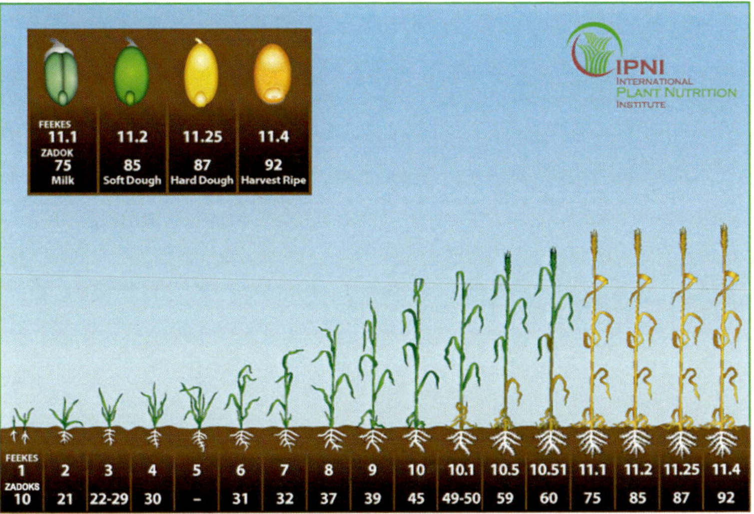


## Alfalfa Growth Stages

Detailed information about alfalfa growth and development is available in Mueller and Teuber (2007). Alfalfa growth is controlled by an interaction between the genetic potential of the plant and the environment. Many management decisions should be based on stage of growth, not age (Mueller and Teuber, 2007). Forage quality of alfalfa changes over time. Generally, the amount increases and the quality declines as the plant matures.

**Box A1-4.** Alfalfa growth and development. Taken from Mueller and Teuber, 2007.

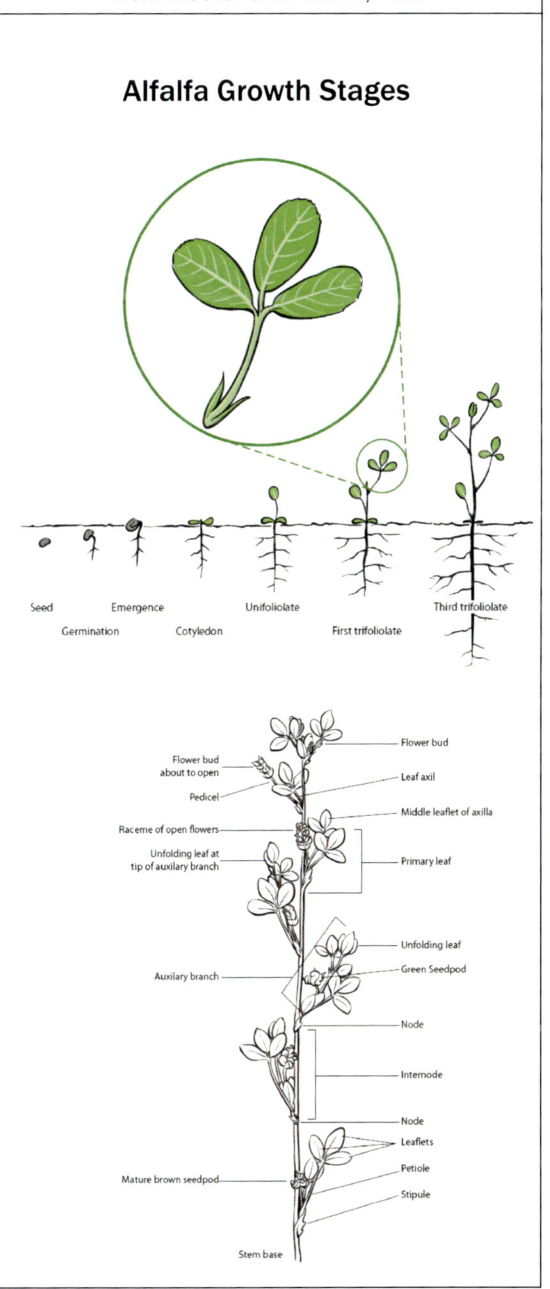

| Box A1-5. Alfalfa growth stages. | | |
|---|---|---|
| Stage | Growth stage and characteristics | |
| Stage 0 | Early vegetative | Stem length < 6 in., no visible buds, flower, or seed pods |
| Stage 1 | Mid-vegetative | Stem length 6 to 12 in., no visible buds, flowers, or seed pods |
| Stage 2 | Late-vegetative | Stem length >12 in., no visible buds, flowers, or seed pods |
| Stage 3 | Early bud | 1 to 2 nodes with visible buds, no-flowers |
| Stage 4 | Late bud | 3 or more nodes with visible buds, no-flowers |
| Stage 5 | Early flower | One node with open flower, no seed pods |
| Stage 6 | Late flower | 2 or more nodes with open flowers, no seed pods |
| Stage 7 | Early seed pod | 1 to 3 nodes with green seed pods |
| Stage 8 | Late seed pod | 4 or more nodes with green seed pods |
| Stage 9 | Ripe seed pod | Nodes mostly brown, mature seed pods |

## Crop Defoliation

Defoliation can occur from insects and hail. The impact of defoliation-causing events on yield are influenced by timing and crop. Hail in the spring can kill many newly emerged plants. To assess early year hail damage, it is necessary to wait 4 to 10 days before assessing the number of killed plants. It is important to wait because it will be easier to identify dead tissue. Assessments should be conducted at several locations in the field. The larger the area assessed, the more accurate the information. For soybean, hail damage can be determined by counting the number of live, dead, and questionable plants along a specific length of row (> 3 ft). Repeat several times over the field. The number of live plants is estimated by adding half of the questionable plant count to the live plant total. If the plant is cut off below the cotyledons (thick bottom seed leaves), there is no-remaining tissue and it shows no growth, or the plant is severely bruised and is bent over, it most likely will not re-grow. Charts are available for assessing the impact of defoliation in corn (Vost, 1993) and soybean (Fawcett and Schmitt, 2001).

Mathematical relationships between defoliation and yield losses have also been developed. For example, in corn, Adee (2005) developed an empirical relationship between % yield loss and the percentage of severity integrated over the grain filling period remaining. The Adee (2005) equation was,

$$\% \text{ expected corn yield} = -0.0005858X + 98.7,\ r^2 = 0.68,$$

where X = CDPC (a unitless number describing defoliation severity over potential grain fill period remaining). For example, if a hybrid that requires 2,700 growing degree days (50 to 86 °F) to reach maturity has a 25% defoliation at 1,440 GDD, the expected yield is calculated as shown,

$$77.6\% \text{ yield} = [(1{,}440 \times 25) \times -0.0005858] + 98.7$$

The expected yield loss is calculated by subtracting the expected yield from 100%. In this example, the expected yield loss was 22.4% (100-77.6) (Adee, 2005). Growing degree days are defined in **Chapter 3**.

## Defoliation Caused by Pests

A different approach is needed to assess defoliation damage from pests. For soybean, defoliation is estimated by clipping 10 trifoliate leaves from the upper and lower portion of the plant at five locations. Pick the leaves at random and estimate foliage loss on each trifoliate by comparing the leaves with known examples (Grey, 1999). Examples and directions for estimating soybean defoliation are available in Rice (2002). Estimated soybean yield losses relative to the percent defoliation are below.

In corn, different pests feed on different parts of the plant and therefore require different protocols. For example, for the first generation of corn borers, examine at least 25 corn plants at several locations (minimum 4) in a field. Note the percent of total plant whorls with feeding damage; unroll several damaged whorls, record the number of live worms/plant. Note size of worms (Wright and Witkowski, 1999). Control depends on price of corn, yield potential, cost of application, and number of larvae. About 3 to 5% loss per borer that reaches maturity per plant is expected. For corn rootworms, a different protocol is used. For these pests dig up 2 plants at each of five locations with the soil from 6 to 8 in. around the plant. Sift soil over a sheet of black plastic looking for larvae 1/32 to ½ in. long. Larvae can also be counted using the flotation method (1 lb salt/1 gal $H_2O$). Control is often recommended when 2 to 3 larvae per plant are found.

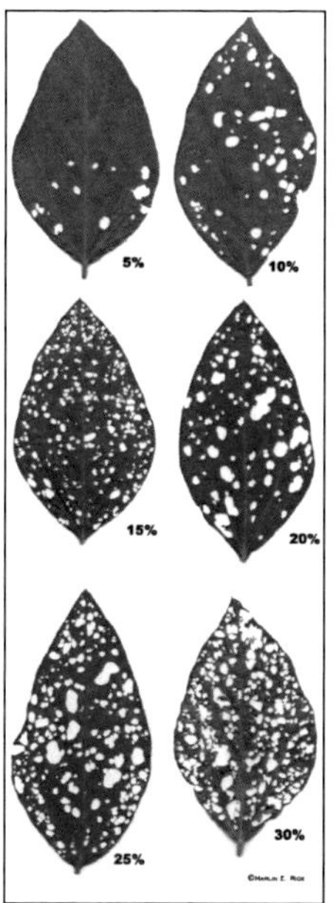


**Box A1-6.** Estimating foliage damage in soybean (taken from Rice, 2002) and the expected yield loss when % leaf area is destroyed.

| Plant growth stage | % leaf area destroyed | | | | | | | | | |
|---|---|---|---|---|---|---|---|---|---|---|
| | 10 | 20 | 30 | 40 | 50 | 60 | 70 | 80 | 90 | 100 |
| VC-V# | 0 | 0 | 0 | 0 | 0 | 0 | 0 | 0 | 0 | 0 |
| R1 | 0 | 1 | 2 | 3 | 3 | 4 | 5 | 6 | 8 | 12 |
| R2 | 0 | 2 | 3 | 5 | 6 | 7 | 9 | 12 | 16 | 23 |
| R3 | 2 | 3 | 4 | 6 | 8 | 11 | 14 | 18 | 24 | 33 |
| R4 | 3 | 5 | 7 | 9 | 12 | 16 | 22 | 30 | 39 | 56 |
| R5 | 4 | 7 | 10 | 13 | 17 | 23 | 31 | 43 | 58 | 75 |
| R6 | 1 | 6 | 9 | 11 | 14 | 18 | 23 | 31 | 41 | 53 |

From Crop Insurance Service, "Soybean Loss Instructions",pub. 6302, Rev 99.

## Quantifying Pest Populations: Basic Insect Scouting Procedures

Several approaches are used when scouting for insects. The first approach is to look for the insects or their damage on the leaf or root tissues. This can be accomplished by digging up the plant and inspecting the roots or inspecting the above-ground tissue. Damage can be recorded by determining the percentage affected. For some situations, it may be necessary to document the degree of damage. Treatment decisions can be based on both types of information.

Insect populations in fields can be measured by using a beat cloth or net. A beat cloth can be different sizes. The most common is 2-ft long and wide enough to stretch between the rows. The crop residue is then pulled over the cloth and shaken/beaten to dislodge any insects (Grey, 1999). Insects on the cloth are counted. For example, if 21 cloverworms are counted on a cloth that is 3 ft long, then the field contains 7 Cloverworms (21/3) per foot of row.

Sweep nets are used if there is a solid stand of plants. A sweep net is a standardized hoop (usually 15 in.) with a net on it. When the net is swept through the foliage, it collects insects. Standardized protocols are used with sweep nets (Grey, 1999).

## Quantifying Pest Populations: Weed Scouting Procedures

A weed is a plant that is not intentionally being grown. Weeds reduce crop yields by competing with the plants for resources. For many plants, there are weed-free periods where yield can be reduced even though the weeds are not directly competing with the plant. To plan an effective weed management strategy, a manager should have some understanding of the amount, species, and relative competiveness of the weeds in the field.

Within the first couple of weeks following crop emergence, a weed map should be developed. For this map to be useful, it should be accurate. This map can be created by separating the field into management zones or grid zones for which weed assessments have been made.

**Box A1-7.** Weed grid zones for mapping.

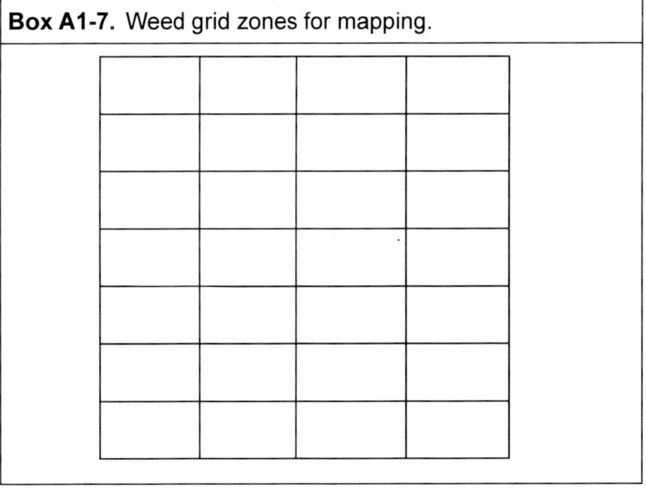

Within each zone, weed populations can be determined by counting the number of weeds contained within a specified area or the number of weeds per 100 ft of row. These populations can be compared to the economic threshold level for that weed. Based on these values, management or no-management can be recommended.

| **Box A1.8.** Corn and soybean yield losses resulting from selected weeds. Modified from Gray (1999). | | | | | | | | | | | | |
|---|---|---|---|---|---|---|---|---|---|---|---|---|
| | % corn yield loss | | | | | | % soybean yield loss | | | | | |
| | 1% | 2% | 4% | 6% | 8% | 10% | 1% | 2% | 4% | 6% | 8% | 10% |
| | weeds/ft$^2$ | | | | | | | | | | | |
| Giant ragweed | 4 | 8 | 16 | 28 | 34 | 40 | 1 | 2 | 4 | 6 | 8 | 10 |
| Redroot pigweed | 12 | 25 | 50 | 100 | 125 | 150 | 2 | 4 | 6 | 10 | 15 | 20 |
| Velvet leaf | | | | | | | 8 | 16 | 24 | 32 | 40 | 50 |
| Volunteer corn | | | | | | | 1 | 2 | 3 | 4 | 5 | 6 |

## Quantifying Soil Nutrient Amounts: Collecting Soil Samples

Fertilizer recommendations based on soil samples are only as good as the soil sample that was collected. When collecting soil samples for fertilizer recommendations, the precision of the recommendation must be balanced with the cost of obtaining the sample. For fertilizer recommendations, the soil nutrient concentrations must be calibrated with plant responses. Many fertilizer recommendations start with the soil sample. General guidance for collecting soil samples are:

- Determine the soil sampling protocol for your region and problem.
- Attempt to collect representative samples. For example, avoid sampling old feedlots and edges of fields.
- Maximize the number of individual cores composited into a single sample. The precision of the recommendation increases with the number of cores taken to mix into a composite sample. A general rule of thumb is to composite 15 to 20 individual cores.
- Preserve the sample by storing in a cooler prior to analysis.

Additional information for obtaining soil samples is available in Clay et al. (2002).

Once the sample has been collected, it should be sent to an appropriate laboratory for analysis.

## Quantifying Plant Populations: Row Crops

The plant population can be estimated by counting the number of plants contained in a predetermined area, followed by conversion to the number of plants per acre. One of the easiest approaches is to count the number of plants contained in 1/1,000 of an acre. For this 1/1,000 of an acre, the length of row that should be counted depends on the row width. The relationship between length of row and 1/1,000 of an acre is shown below (Box A1-9). The plant population is determined by multiplying the plant population in the specified row length times 1,000. For example, if 15 plants are counted within the distance that is equal to 1/1,000 of an acre, then the plant population is calculated to be 15,000 plants per acre (see **Chapter 10** for examples).

| **Box A1-9.** The length of row relative to the row width required for 1/1,000 of an acre. | |
|---|---|
| Row spacing, in. | length of row required for 1/1000 acre, ft |
| 7 | 74 ft and 8.4 in. |
| 10 | 52 ft and 3.6 in. |
| 15 | 34 ft and 10 in. |
| 20 | 26 ft and 2 in. |
| 22 | 23 ft and 8 in. |
| 28 | 18 ft and 8 in. |
| 30 | 17 ft and 5 in. |
| 32 | 16 ft and 4 in. |
| 34 | 15 ft and 5 in. |
| 36 | 14 ft and 6 in. |
| 38 | 13ft and 9 in. |

## Estimating Corn Yields

Corn yield can be estimated with a 4 step process.

1. Count the number of ears for 1/1,000 of an acre. The length of row is influenced by row spacing.
2. On several ears, count the number kernels per row and the number of rows per ear.
3. Multiply the number of kernels per row by the number of rows multiply this value by the number of plants in 1/1,000 of an acre.
4. The estimated yield in units of bu/acre is determined by dividing the number of kernels per 1/1,000 of an acre by 80. This assumes that there are 80,000 kernels/acre. (See **Chapter 10** for examples).

## Quantifying Plant Populations: Solid-Seeded Crops

The "hula hoop" method can be used in fields where crops are solid-seeded. This method can be used in solid-seeded soybean or pastures. This method requires a perfectly circular object (hula hoop, barrel hoop). The hoop should be placed (randomly tossed) at multiple locations (at least five) in a field or management zone. The number of plants within the hoop should be counted. Multiply the average number of plants in the enclosure (the number of plants in the hoop/ the number of in.$^2$ in the hoop) by a factor to determine the number of plants per acre. The factor is equal to,

Factor = 43,560/[(radius of hoop in inches)$^2$ • 3.14]/144

A table showing the relationship between this factor and diameter of the hoop are shown below.

**Box A1-10.** The relationship between the diameter of the hoop and the factor used to calculate the number of plants/acre. This approach is suitable for solid seeded plants.

| Hoop diameter, in. | Factor when multiplied by the number of plants in the hoop equals plants/acre |
|---|---|
| 18 | 24,662 |
| 21 | 18,119 |
| 24 | 13,872 |
| 27 | 10,961 |
| 30 | 8,878 |
| 33 | 7,337 |
| 36 | 6,165 |

An alternative approach for determining populations of solid-seeded plants is provided below.

| Number of plants in hoop | Inside diameter of hula hoop (in.) | | | | |
|---|---|---|---|---|---|
| | 30 | 32 | 34 | 36 | 38 |
| | number (1,000) plants/acre | | | | |
| 10 | 89 | 78 | 69 | 62 | 55 |
| 12 | 107 | 94 | 83 | 74 | 66 |
| 14 | 124 | 109 | 97 | 86 | 77 |
| 16 | 142 | 125 | 110 | 99 | 89 |
| 18 | 160 | 140 | 124 | 111 | 100 |
| 20 | 178 | 156 | 138 | 123 | 111 |
| 22 | 196 | 172 | 152 | 136 | 122 |
| 24 | 213 | 187 | 166 | 148 | 133 |
| 26 | 231 | 203 | 179 | 160 | 144 |
| 28 | 249 | 218 | 193 | 173 | 155 |
| 30 | 266 | 234 | 207 | 185 | 166 |
| 32 | 284 | 250 | 221 | 197 | 177 |
| 34 | 302 | 265 | 235 | 209 | 188 |
| 36 | | 281 | 249 | 222 | 199 |
| 38 | | 297 | 362 | 234 | 210 |
| 40 | | | 277 | 247 | 221 |
| 42 | | | 290 | 259 | 232 |
| 44 | | | 304 | 271 | 243 |
| 46 | | | | 284 | 255 |

## Simple Technique to Calibrate a Sprayer

To insure that the proper amount of chemical is applied, sprayers should be calibrated and proper nozzles selected. If the nozzles are selected, the 1/128$^{th}$ acre approach can be used for calibration. This method works because there are 128 oz in a gallon. This means that the application of 1 oz/128$^{th}$ acre = 1 gal/acre. This method contains four steps.

1. Measure the width covered by one nozzle (called the spray nozzle swath width). This is the distance from the center of one nozzle to the center of the next nozzle.
2. Measure the length of time required to drive 1/128$^{th}$ of an acre. These distances are provided in **Box A1-11** below.
3. Using an ounce-delineated measuring container, with your sprayer loaded only with water (no chemical mixed), collect spray from one nozzle for the required time to drive the calibration course. As an example, if your nozzles are 18 in. apart and if you will drive your sprayer at 5 miles/hour, then you will collect all the discharge from one nozzle for 30.8 seconds.
4. Determine the ounces collected. Each ounce equals 1 gal/acre. From the example above, if you collected 10 oz of water, you would be applying 10 gal of water/acre.

**Box A1-11.** The relationship between the swath width of a spray nozzle, distance required to cover 1/128 acre, and length of time required to collect samples for an applicator traveling at different speeds.

| Spray nozzle swath width | Distance for 1/128 acre | mph | | | |
|---|---|---|---|---|---|
| | | 5 | 10 | 15 | 20 |
| in. | ft | seconds | | | |
| 6 | 681 | 92.9 | 46.4 | 31.0 | 23.2 |
| 8 | 507 | 69.1 | 34.6 | 23.0 | 17.3 |
| 10 | 408 | 55.6 | 27.8 | 18.5 | 13.9 |
| 12 | 340 | 46.4 | 23.2 | 15.5 | 11.6 |
| 14 | 292 | 39.8 | 19.9 | 13.3 | 10.0 |
| 16 | 255 | 34.8 | 17.4 | 11.6 | 8.7 |
| 18 | 226 | 30.8 | 15.4 | 10.3 | 7.7 |
| 24 | 170 | 23.2 | 11.6 | 7.7 | 5.8 |

A similar approach is used to calibrate drop fertilizer applicators. For these applicators, a pan that fits under the applicator and collects all of the material that is discharged from the spreader for the given length of pan is needed. The amount of chemical spread per unit area is determined.

| **Box A1-12.** Relationship between travel distance and the pan width below the drop spreader required for a range of areas. | | | | |
|---|---|---|---|---|
| Pan width, | 1/100 acre | 1/50 acre | 1/25 acre | 1/10 acre |
| ft | ft | | | |
| 2.5 | 174.24 | 348.48 | 696.96 | 1742.4 |
| 5 | 87.12 | 174.24 | 348.48 | 871.2 |
| 10 | 43.56 | 87.12 | 174.24 | 435.6 |
| 20 | 21.78 | 43.56 | 87.12 | 217.8 |
| 30 | 14.52 | 29.04 | 58.08 | 145.2 |

Calibration is a five step process.

1. Measure the width of the pan which will be used to collect fertilizer.
2. Attach a catch pan under the applicator so that it captures all of the fertilizer flowing from the width of the pan.
3. Drive the distance needed for a given area and applicator width.
4. Weigh the material collected
5. Calculate the application rate by multiplying the weight times the amount of area collected. For example, if 5 lb were collected under a 30 ft wide applicator that traveled 145.2 ft, then the application rate is 50 lb/acre (5 • 10).

| **Box A1-13.** Ranges of seeds per pound for selected plants. | |
|---|---|
| Plant | seeds/lb |
| Alfalfa | 200,000-250,000 |
| Barley | 12,000-15,000 |
| Corn grain | 1,000-1,500 |
| Flax | 70,000-90,000 |
| Kentucky bluegrass | 2,000,000-2,500,000 |
| Pea | 1,500-3,600 |
| Reed canarygrass | 500,000-5,500,000 |
| Soybean | 2,500-3,000 |
| Wheat (hard red) | 12,000-15,000 |

| **Box A1-14.** Pounds of seed per acre as influenced by population and seeds per pound. | | | | | |
|---|---|---|---|---|---|
| | Desired population | | | | |
| Seeds/lb | 120,000 | 150,000 | 175,000 | 200,000 | 225,000 |
| | lb/acre | | | | |
| 3,400 | 37 | 44 | 51 | 59 | 66 |
| 3,600 | 35 | 42 | 49 | 56 | 62 |
| 3,800 | 33 | 39 | 46 | 53 | 59 |
| 4,000 | 31 | 38 | 44 | 50 | 56 |
| 4,200 | 30 | 36 | 42 | 48 | 54 |
| 4,400 | 28 | 34 | 40 | 45 | 51 |
| 4,600 | 27 | 33 | 38 | 43 | 49 |

| **Box A1-15.** Typical seeding rates for various plants. | | | | |
|---|---|---|---|---|
| Plant | Climate | Seeding rate lb/acre | Weight/ bushel | Growth type |
| Alfalfa | humid | 10-20 | 60 | P |
| | irrigation | 10-15 | | |
| | semi-arid | 8-10 | | |
| Barley | | 72-96 | 48 | A |
| Field bean | | 40-75 | 60 | A |
| Kentucky bluegrass | | 15-25 | 14 | P |
| Big bluestem | | 15-20 | 28 | P |
| Little bluestem | | 12-20 | 28 | P |
| Buckwheat | | 36-60 | 48 | A |
| Buffalograss | | 15-20 | | P |
| Corn-field grain | | 6-18 | 56 | A |
| Corn-field silage | | 8-18 | 56 | A |
| Corn- sweet | | 12-18 | 50 | A |
| Crested wheat grass | | 15-25 | | P |
| Fescue-tall | | 10-25 | | P |
| Field pea | | 120-180 | 60 | A |
| Flax-seed | | 28-42 | 56 | A |
| Gramagrass-black | | 7-9 | | P |
| Lentil | | 5-8 | 60 | A |
| Oat-common | | 48-128 | 32 | A |
| Orchardgrass | | 20-25 | 14 | P |
| Peanut | | 20-40 | 20-30 | A |
| Potato | | 1000 | 60 | P(A) |
| Rape | | 3-4 | | A |
| Reed canarygrass | | 10-12 | 14 | P |
| Rice | | 67-160 | 45 | A |
| Rye | | 28-122 | 56 | A |
| Soybean-large seeded | | 30-45 | 60 | A |
| Sudangrass | | 20-35 | 40 | A |
| Sugarbeet | | 2-8 | 15 | B |
| Sunflower | | 3-10 | 24 | A |
| Timothy | | 8-12 | 45 | P |
| Wheat-common | | 30-120 | 60 | A |
| Wheatgrass-crested | | 12-20 | | P |
| Wheatgrass-western | | 12-20 | 35 | P |

**A is annual crops**

**P is perenial crop**

| **Box A1-16.** Plants/acre as influenced by distance between plants and row spacing. | | | | | | | | | |
|---|---|---|---|---|---|---|---|---|---|
| No. plants/in. of row | Plants per acre (width between rows, in.) | | | | | | | | |
| | 8 | 12 | 26 | 30 | 36 | 38 | 40 | 42 | 60 |
| 1 | 784,080 | 522,720 | 241,255 | 209,088 | 174,240 | 165,069 | 156,816 | 149,349 | 104,544 |
| 2 | 392,040 | 261,360 | 120,628 | 104,544 | 87,120 | 82,535 | 78,408 | 74,674 | 52,272 |
| 3 | 261,360 | 174,240 | 80,418 | 69,696 | 58,080 | 55,023 | 52,272 | 49,783 | 34,848 |
| 4 | 196,020 | 130,680 | 60,314 | 52,272 | 43,560 | 41,267 | 39,204 | 37,337 | 26,136 |
| 5 | 156,816 | 104,544 | 48,251 | 41,818 | 34,848 | 33,014 | 31,363 | 29,870 | 20,909 |
| 6 | 130,680 | 87,120 | 40,209 | 34,848 | 29,040 | 27,512 | 26,136 | 24,891 | 17,424 |
| 8 | 98,010 | 65,340 | 30,157 | 26,136 | 21,780 | 20,634 | 19,602 | 18,669 | 13,068 |
| 10 | 78,408 | 52,272 | 24,126 | 20,909 | 17,424 | 16,507 | 15,682 | 14,935 | 10,454 |
| 12 | 65,340 | 43,560 | 20,105 | 17,424 | 14,520 | 13,756 | 13,068 | 12,446 | 8712 |
| 14 | 56,006 | 37,337 | 17,233 | 14,935 | 12,446 | 11,791 | 11,201 | 10,668 | 7467 |
| 16 | 49,005 | 32,670 | 15,078 | 13,068 | 10,890 | 10,317 | 9801 | 9334 | 6534 |
| 18 | 43,560 | 29,040 | 13,403 | 11,616 | 9680 | 9171 | 8712 | 8297 | 5808 |
| 20 | 39,204 | 26,136 | 12,063 | 10,454 | 8712 | 8253 | 7841 | 7467 | 5227 |
| 22 | 35,640 | 23,760 | 10,966 | 9504 | 7920 | 7503 | 7128 | 6789 | 4752 |
| 24 | 32,670 | 21,780 | 10,052 | 8712 | 7260 | 6878 | 6534 | 6223 | 4356 |
| 26 | 30,157 | 20,105 | 9279 | 8042 | 6702 | 6349 | 6031 | 5744 | 4021 |
| 28 | 28,003 | 18,669 | 8616 | 7467 | 6223 | 5895 | 5601 | 5334 | 3734 |
| 30 | 26,136 | 17,424 | 8042 | 6970 | 5808 | 5502 | 5227 | 4978 | 3485 |
| 40 | 19,602 | 13,068 | 6031 | 5227 | 4356 | 4127 | 3920 | 3734 | 2614 |
| 50 | 15,682 | 10,454 | 4825 | 4182 | 3485 | 3301 | 3136 | 2987 | 2091 |
| 60 | 13,068 | 8712 | 4020 | 3485 | 2904 | 2751 | 2614 | 2489 | 1742 |
| 70 | 11,201 | 7467 | 3447 | 2987 | 2489 | 2358 | 2240 | 2134 | 1493 |

Populations are rounded off to whole numbers. Percent germination is not considered in this table.

| **Box A1-17.** The weight per bushel as impacted by crop and moisture percent. | | | | | | |
|---|---|---|---|---|---|---|
| | Grain moisture, % | | | | | |
| Crop | 20% | 18% | 15.50% | 13% | 10% | 0% |
| | Weight, lb/bu | | | | | |
| Corn | 59.15 | 57.71 | 56 | 54.39 | 52.58 | 47.32 |
| Soybeans | 65.25 | 63.65 | 61.78 | 60 | 58.0 | 52.2 |
| Wheat | 64.88 | 63.29 | 61.42 | 60 (13.5 % moist.) | 57.67 | 51.9 |

| **Box A1-18.** Commonly used base and maximum values when calculating heat units (growing degree days). | | | |
|---|---|---|---|
| | Commonly used Base values | | Commonly used Base values |
| **Crop** | °F | **Weeds** | °F |
| Corn | 50* | Barnyardgrass | 50 |
| Wheat | range from 32 to 40 | Yellow foxtail | 55 |
| Soybean | 50 | Common cocklebur | 40 |
| Sunflower | 44 | Sowthistle | 47 |
| **Insects** | | Wild buckwheat | 40 |
| European corn borer | 54 | | |
| Alfalfa weevil | 48 | | |
| Western corn rootworm | 53 | | |
| Gypsy moth | 50 | | |
| Stalk borer | 41 | | |

* $T_{max}$ value for corn is 86 °F.

**Box A1-19.** Fertilizer grades of selected fertilizer materials.

| | N, % | $P_2O_5$, % | $K_2O$, % | Density*, lb/gal |
|---|---|---|---|---|
| **Solid fertilizers** | | | | |
| Ammonium sulfate | 21 | 0 | 0 | |
| Potassium sulfate | 0 | 0 | 50-52 | |
| Potassium magnesium sulfate | 0 | 0 | 22 | |
| Ammonium nitrate | 33-34 | 0 | 0 | |
| Diammonium phosphate (DAP) | 18-21 | 46-54 | 0 | |
| Monoammonium phosphate (MAP) | 10-11 | 48-55 | 0 | |
| Potassium chloride | 0 | 0 | 60-62 | |
| Potassium nitrate | 13 | 0 | 44 | |
| Urea | 45-46 | 0 | 0 | |
| **Liquid fertilizers** | | | | |
| 7-21-7 multigrade | 7 | 21 | 7 | 11.3 |
| 9-18-9 multigrade | 9 | 18 | 9 | 11.1 |
| Urea-ammonium-nitrate (UAN) | 28-32 | 0 | 0 | 10.6-11.1 |
| Ammonium polyphosphate (APP) | 10-12 | 34-41 | 0 | 12 |
| Phosphoric acid | 0 | 48-53 | 0 | 14.1 |
| **Gas fertilizers** | | | | |
| Anhydrous ammonia | 82 | 0 | 0 | |

* Densities are reported at the standard temperature of 20 °C (68 °F). Higher temperatures have lower densities than those reported, while lower temperatures have higher densities.

**Box A1-20.** The amount of nutrients removed by several commonly grown plants. (http://www.ipni.net/article/IPNI-3296)

| Crop | Unit | Removal, lb/unit | | | |
|---|---|---|---|---|---|
| | | N | $P_2O_5$ | $K_2O$ | S |
| Alfalfa (DM) | ton | 51 | 12 | 49 | 5.4 |
| Alsike Clover (DM) | ton | 41 | 11 | 54 | 3.0 |
| Bahiagrass | ton | 43 | 12 | 35 | |
| Barley grain | bu | 0.99 | 0.40 | 0.32 | 0.09 |
| Barley straw | bu | 0.40 | 0.16 | 1.2 | 0.10 |
| Barley straw | ton | 13 | 5.1 | 39 | 3.0 |
| Beans (dry) | bu | 3.0 | 0.79 | 0.92 | 0.52 |
| Bermuda grass | ton | 46 | 12 | 50 | -- |
| Birdsfoot trefoil (DM) | ton | 45 | 11 | 42 | -- |
| Bluegrass (DM) | ton | 30 | 12 | 46 | 5.0 |
| Bromegrass (DM) | ton | 32 | 10 | 46 | 5.0 |
| Buckwheat | bu | 0.83 | 0.25 | 0.22 | -- |
| Canola grain | bu | 1.6 | 0.80 | 0.40 | 0.25 |
| Corn grain | bu | 0.67 | 0.35 | 0.25 | 0.08 |
| Corn silage (67% water) | bu | 1.6 | 0.51 | 1.2 | 0.18 |
| Corn silage (67% water) | ton | 9.7 | 3.1 | 7.3 | 1.1 |
| Corn stover | bu | 0.45 | 0.16 | 1.1 | 0.07 |
| Corn stover | ton | 16 | 5.8 | 40 | 2.6 |
| Cotton (lint) | bale | 32 | 14 | 19 | -- |
| Cotton stover | ton | 19 | 6.7 | 22 | -- |
| Fescue (DM) | ton | 37 | 12 | 54 | 5.7 |
| Flax grain | bu | 2.5 | 0.70 | 0.60 | 0.19 |
| Flax straw | bu | 0.7 | 0.16 | 2.2 | 0.15 |
| Millet grain | bu | 1.4 | 0.40 | 0.40 | 0.08 |
| Millet straw | ton | 15 | 4.3 | 39 | -- |
| Mint oil | lb | 1.9 | 1.1 | 4.5 | -- |
| Oat grain | bu | 0.77 | 0.28 | 0.19 | 0.07 |
| Oat straw | bu | 0.31 | 0.16 | 0.94 | 0.11 |
| Oat straw | ton | 12 | 6.3 | 37 | 4.5 |
| Orchardgrass (DM) | ton | 36 | 13 | 54 | 5.8 |
| Peanut nuts | ton | 70 | 11 | 17 | -- |
| Peanut stover | ton | 33 | 6.8 | 24 | -- |
| Potato tuber | cwt | 0.30 | 0.15 | 0.65 | 0.03 |
| Potato above-ground stems & leaves | cwt | 0.2 | 0.05 | 0.3 | 0.02 |
| Red clover (DM) | ton | 45 | 12 | 42 | 3.0 |
| Reed canarygrass (DM) | ton | 31 | 13 | 25 | -- |
| Rice grain | bu | 0.57 | 0.30 | 0.16 | -- |
| Rice straw | ton | 17 | 5.5 | 41 | -- |
| Rye grain | bu | 1.4 | 0.46 | 0.31 | 0.10 |
| Rye straw | bu | 0.80 | 0.21 | 1.5 | 0.14 |
| Rye straw | ton | 12 | 3.0 | 22 | 2.0 |
| Ryegrass (DM) | ton | 43 | 12 | 43 | -- |
| Sorghum grain | bu | 0.66 | 0.39 | 0.27 | 0.06 |
| Sorghum stover | bu | 0.56 | 0.16 | 0.83 | 0.12 |
| Sorghum stover | ton | 28 | 8.3 | 42 | 5.9 |
| Sorghum-sudan (DM) | ton | 30 | 9.5 | 34 | 5.8 |
| Soybean grain | bu | 3.3 | 0.73 | 1.2 | 0.18 |
| Soybean hay (DM) | ton | 45 | 11 | 25 | 5.0 |
| Soybean stover | bu | 1.1 | 0.24 | 1.0 | 0.17 |
| Soybean stover | ton | 40 | 8.8 | 37 | 6.2 |

*...continued on page 198*

**Box A1-20.** *continued from page 197*

| Crop | Unit | Removal, lb/unit N | $P_2O_5$ | $K_2O$ | S |
|---|---|---|---|---|---|
| Sugarbeet root | ton | 3.7 | 2.2 | 7.3 | 0.45 |
| Sugarbeet top | ton | 7.4 | 4.0 | 20 | 0.40 |
| Sugarcane | ton | 2.0 | 1.2 | 3.5 | -- |
| Sunflower grain | cwt | 2.7 | 0.97 | 0.9 | 0.25 |
| Sunflower stover | cwt | 2.8 | 0.24 | 4.1 | 0.6 |
| Sunflower stover | ton | 23 | 2.0 | 34 | 5.0 |
| Switchgrass (DM) | ton | 22 | 12 | 58 | -- |
| Timothy (DM) | ton | 25 | 11 | 42 | 2.0 |
| Tomatoes | ton | 2.5 | 0.92 | 5.7 | -- |
| Tobacco leaves | cwt | 3.6 | 0.90 | 5.7 | 0.6 |
| Vetch (DM) | ton | 57 | 15 | 49 | -- |
| Wheat straw | bu | 0.7 | 0.16 | 1.2 | 0.14 |
| Wheat straw | ton | 15 | 3.7 | 29 | 5.4 |
| Wheat (spring) grain | bu | 1.5 | 0.57 | 0.33 | 0.10 |
| Wheat (winter) grain | bu | 1.2 | 0.48 | 0.29 | 0.10 |

**Notes:** Nutrient removal refers to the quantity of nutrient removed from the field at crop harvest.
Reported nutrient removal coefficients may vary regionally depending on growing conditions. Use locally available data whenever possible.
DM = dry matter basis; otherwise moisture content is standard marketing convention or at the stated moisture content.

**Box A1-21.** Plant available water in inches per foot of soil.

| Soil texture | Plant available water inch / ft soil |
|---|---|
| Fine sands | 0.7-1.0 |
| Loamy sands | 0.9-1.5 |
| Sandy loams | 1.3-1.8 |
| Loam | 1.8-2.5 |
| Silt loam | 1.8-2.6 |
| Clay loam | 1.8-2.5 |

**Box A1-22.** The approximate cation exchange capacity (CEC) of several different soil separates.

| Soil phase | Approximate CEC |
|---|---|
| | $cmol_c$ /kg |
| Organic matter | 200 |
| Clays | |
| Vermiculite | 150 |
| Smectite | 100 |
| Mica | 30 |
| Kaolinite | 9 |
| Gibbsite | 4 |

**Box A1-23.** Salt classification for several methods to extract salts.

Salt classification for silt-loam to clay loam soil using different methods.

| | Saturated paste | 1:1 soil to water ratio |
|---|---|---|
| | dS/m | |
| non-saline | 0-2 | 0-1.3 |
| slightly | 2.1-4.0 | 1.4-2.5 |
| moderate | 4.1-8.0 | 2.6-5.0 |
| strongly | 8.1-16.0 | 5.1-10.0 |
| very | >16.1 | >10.1 |

Sodic soils SAR > 13 (see **Chapter 10** for examples)

**Box A1-24.** $pK_a$ for important agricultural acids.

| Weak acid | $pK_a$ |
|---|---|
| $H_3PO_4 \rightarrow H^+ + H_2PO_4^-$ | 2.23 |
| $H_2PO_4^- \rightarrow H^+ + HPO_4^{2-}$ | 7.2 |
| $HPO_4^{2-} \rightarrow H^+ + PO_4^{3-}$ | 12.3 |
| $H_2CO_3 \rightarrow H^+ + HCO_3^-$ | 6.4 |
| $HCO_3^- \rightarrow H^- + CO_3^{2-}$ | 10.33 |
| $NH_4 \rightarrow H^+ + NH_3$ | 9.2 |
| $H_2BO_3 \rightarrow HBO_3^- + H^+$ | 9.3 |
| $H_2O \rightarrow H^+ + OH^-$ | 14 |

**Box A1-25.** Estimated cost of production for corn grown in Iowa, South Dakota, and Nebraska.

Iowa estimated cost of production (Duffy, 2011)

| | Corn following corn | | | | Corn following soybean | | | | Soybean following corn | | | |
|---|---|---|---|---|---|---|---|---|---|---|---|---|
| | 2006 | 2007 | 2008 | 2009 | 2006 | 2007 | 2008 | 2009 | 2006 | 2007 | 2008 | 2009 |
| Machinery | 100.07 | 102.94 | 110.98 | 115.99 | 97.39 | 100.12 | 107.88 | 113.98 | 45.9 | 46.76 | 48.50 | 55.80 |
| Seed, chemical, etc | 201.62 | 222.22 | 271.97 | 387.4 | 169.26 | 189.33 | 230.35 | 344.03 | 106.79 | 107.58 | 126.06 | 202.85 |
| Labor | 29.93 | 31.35 | 31.35 | 31.35 | 27.3 | 28.6 | 28.60 | 28.60 | 25.73 | 26.95 | 26.95 | 26.95 |
| Land | 145 | 155 | 190.00 | 205.00 | 145 | 155 | 190.00 | 205.00 | 145 | 155 | 190.00 | 205.00 |
| Total cost/ acre $/acre | 476.61 | 511.51 | 604.20 | 739.77 | 438.95 | 473.05 | 556.83 | 691.61 | 323.41 | 336.29 | 391.51 | 490.60 |
| Assumed yield, (bu/acre) | 140 | 145 | 145 | 145 | 155 | 160 | 160 | 160 | 45 | 50 | 50 | 50.00 |
| Cost/bu, $/bu | 3.4 | 3.53 | 4.17 | 5.10 | 2.83 | 2.96 | 3.48 | 4.32 | 7.19 | 6.73 | 7.83 | 9.81 |

| | Iowa (estimated) | | |
|---|---|---|---|
| Alfalfa hay, annual costs | 2006 | 2007 | 2008 |
| One-third planting costs, $/acre | 36.83 | 37.27 | 38.97 |
| Annual fertilizer | 103.36 | 103.46 | 294.60 |
| Harvest machinery | 107.1 | 90.4 | 102.90 |
| Labor | 56 | 58.67 | 58.67 |
| Land | 95 | 100 | 125.00 |
| Total cost/acre | 398.29 | 389.79 | 632.27 |
| Assumed yield, tons | 6 tons | 6 tons | 6 tons |
| Cost/ton, $/ton | 66.38 | 64.97 | 105.38 |
| From Duffy, 2011 | | | |

Nebraska cost of production 2006 (Seely and Klein, 2006)

| Crop | Previous crop | | | | Goal/acre | Cost/acre |
|---|---|---|---|---|---|---|
| Corn | Soybean | Dryland | No-tilled | Bt | 110 bu | 271.45 |
| Corn | Corn | Irrigated | No-tilled | Bt | 195 bu | 538.21 |
| Soybean | Corn | Dryland | No-tilled | | 40 bu | 166.96 |
| Wheat | Row crop | Dryland | No-tilled | | 40 bu | 176.37 |
| Grass hay | | Dryland | | | 2 tons | 72.72 |

## Additional Information

Adee, E.A. 2005. Yield loss for corn hybrids to incremental defoliation. Plant Science Network. Available at http://www.plantmanagementnetwork.org/pub/cm/research/2005/defoliate/.

Clay, D.E., N. Kitchen, C.G. Carlson, J.L. Kleinjan, and W.A. Tjentland, 2002. Collecting representative soil samples for N and P fertilizer recommendations. Online. Crop Management doi:10.1094/CM-2002-12XX-01-MA. Available at http://www.plantmanagementnetwork.org/pub/cm/management/np/

Duffy, M. 2011. Estimated cost of production in Iowa. Iowa State University. Available at: http://extension.instate.edu/agdm/crops/pdf/a1-20.pdf.

Fawcett, J. and V. Schmitt. 2001. Evaluating hail damage on soybean. Iowa State University Extension. Available at http://www.extension.iastate.edu/Pages/eccrops/hailsoy.html.

Gray, M.E. 1999. Field crop scouting manual. University of Illinois Extension (X880C). University of Illinois, Campaign Il.

Mueller, S.C. and L.R. Teuber. 2007. Alfalfa growth and development. University of California. #8289. available at http://alfalfa.ucdavis.edu/IrrigatedAlfalfa/pdfs/UCAlfalfa8289GrowthDev.pdf.

Rice. M.R. 2002. Estimating soybean defoliation. Integrated Crop Management IC-488 (19). Available at http://www.ipm.iastate.edu/ipm/icm/2002/7-29-2002/soydefoliation.html.

Seely, R.A. and R.N. Klein. 2006. Nebraska crop budgets - 2006 (EC872) University of Nebraska.

Vost, T.V. 1993. Assessing hail damage to corn. National Corn Handbook, NCH-1. Available at http://www.ces.purdue.edu/extmedia/NCH/NCH-1.html.

Wrogjt, R.J. and J.F. Witkowski. 1999. Corn insects-quick reference. University of Nebraska Cooperative Extension. EC 98-1562-B. Available at http://ianrpubs.unl.edu/insects/ec1562 htm.

# INDEX

## D



## E



## F

## G



## H



## I



## K



## L

**Q**



**R**



**S**

## T